The Leatherback Turtle

The Leatherback Turtle

Biology and Conservation

Edited by James R. Spotila
and Pilar Santidrián Tomillo

JOHNS HOPKINS UNIVERSITY PRESS | BALTIMORE

Printed in the United States of America on acid-free paper
9 8 7 6 5 4 3 2 1

Johns Hopkins University Press
2715 North Charles Street
Baltimore, Maryland 21218-4363
www.press.jhu.edu

Library of Congress Cataloging-in-Publication Data

The leatherback turtle : biology and conservation / edited by James R. Spotila and Pilar Santidrián Tomillo.
pages cm
Includes bibliographical references and index.
ISBN 978-1-4214-1708-0 (hardcover : alk. paper) — ISBN 1-4214-1708-1 (hardcover : alk. paper) — ISBN 978-1-4214-1709-7 (electronic) — ISBN 1-4214-1709-X (electronic) 1. Leatherback turtle. 2. Leatherback turtle—Conservation. I. Spotila, James R., 1944- editor. II. Santidrián Tomillo, Pilar, 1974- editor.
QL666.C546L43 2015
597.92'89—dc23 2014036887

A catalog record for this book is available from the British Library.

Special discounts are available for bulk purchases of this book. For more information, please contact Special Sales at 410-516-6936 or specialsales@press.jhu.edu.

Johns Hopkins University Press uses environmentally friendly book materials, including recycled text paper that is composed of at least 30 percent post-consumer waste, whenever possible.

To Laurie Dittrich Spotila

Contents

List of Contributors *ix*
Preface *xi*

Part I. BIOLOGY

1 **Introduction:** *Phylogeny and Evolutionary Biology of the Leatherback Turtle* 3
PETER C. H. PRITCHARD

2 Phylogeny, Phylogeography, and Populations of the Leatherback Turtle 8
PETER H. DUTTON AND KARTIK SHANKER

3 Diving Behavior and Physiology of the Leatherback Turtle 21
NATHAN J. ROBINSON AND FRANK V. PALADINO

4 Anatomy of the Leatherback Turtle 32
JEANETTE WYNEKEN

Part II. LIFE HISTORY AND REPRODUCTION

5 Reproductive Biology of the Leatherback Turtle 51
DAVID C. ROSTAL

6 Nesting Ecology and Reproductive Investment of the Leatherback Turtle 63
KAREN L. ECKERT, BRYAN P. WALLACE, JAMES R. SPOTILA, AND BARBARA A. BELL

7 Egg Development and Hatchling Output of the Leatherback Turtle 74
PILAR SANTIDRIÁN TOMILLO AND JENNIFER SWIGGS

8 Sex Determination and Hatchling Sex Ratios of the Leatherback Turtle 84
CHRISTOPHER A. BINCKLEY AND JAMES R. SPOTILA

Part III. POPULATION STATUS AND TRENDS

9 Leatherback Turtle Populations in the Atlantic Ocean 97
MARC GIRONDOT

10 Leatherback Turtle Populations in the Pacific Ocean 110
SCOTT R. BENSON, RICARDO F. TAPILATU, NICOLAS PILCHER, PILAR SANTIDRIÁN TOMILLO, AND LAURA SARTI MARTÍNEZ

11 Leatherback Turtle Populations in the Indian Ocean 123
RONEL NEL, KARTIK SHANKER, AND GEORGE HUGHES

Part IV. FROM EGG TO ADULTHOOD

12 Leatherback Turtle Eggs and Nests, and Their Effects on Embryonic Development 135
PAUL R. SOTHERLAND, BRYAN P. WALLACE, AND JAMES R. SPOTILA

13 Leatherback Turtle Physiological Ecology: *Implications for Bioenergetics and Population Dynamics* 149
BRYAN P. WALLACE AND T. TODD JONES

14 Movements and Behavior of Adult and Juvenile Leatherback Turtles 162
GEORGE L. SHILLINGER AND HELEN BAILEY

15 Relation of Marine Primary Productivity to Leatherback Turtle Biology and Behavior 173
VINCENT S. SABA, CHARLES A. STOCK, AND JOHN P. DUNNE

Part V. THE FUTURE OF THE LEATHERBACK TURTLE

16 Warming Climate: *A New Threat to the Leatherback Turtle* 185
JAMES R. SPOTILA, VINCENT S. SABA, SAMIR H. PATEL, AND PILAR SANTIDRIÁN TOMILLO

17 Impacts of Fisheries on the Leatherback Turtle 196
REBECCA L. LEWISON, BRYAN P. WALLACE, AND SARA M. MAXWELL

18 Conclusion: *Problems and Solutions* 209
JAMES R. SPOTILA AND PILAR SANTIDRIÁN TOMILLO

Index 213

Color plates follow pages 48 and 128.

Contributors

Helen Bailey
Chesapeake Biological Laboratory
University of Maryland Center for Environmental Science
Solomons, MD, USA

Barbara A. Bell
The Leatherback Trust
Monterey, CA, USA

Scott R. Benson
NOAA National Marine Fisheries Service
Southwest Fisheries Science Center
Moss Landing Marine Laboratory
Moss Landing, CA, USA

Christopher A. Binckley
Department of Biology
Arcadia University
Glenside, PA, USA

John P. Dunne
National Oceanic and Atmospheric Administration
Office of Oceanic and Atmospheric Research
Geophysical Fluid Dynamics Laboratory
Princeton University Forrestal Campus
Princeton, NJ, USA

Peter H. Dutton
NOAA National Marine Fisheries Service
Southwest Fisheries Science Center
La Jolla, CA, USA

Karen L. Eckert
Wider Caribbean Sea Turtle Conservation Network (WIDECAST)
Ballwin, MO, USA

Marc Girondot
Laboratoire Ecologie, Systematique and Evolution
Université Paris-Sud
Paris, France

George Hughes
183 Amber Valley
Private Bag X30
Howick, South Africa

T. Todd Jones
JIMAR-UH-NOAA
Kewalo Research Facility
Honolulu, HI, USA

Rebecca L. Lewison
Biology Department
San Diego State University
San Diego, CA, USA

Sara M. Maxwell
Department of Biological Sciences
Old Dominion University
Norfolk, VA, USA

Ronel Nel
Department of Zoology
Nelson Mandela Metropolitan University
Port Elizabeth, South Africa

Frank V. Paladino
Biology Department
Indiana-Purdue University at Fort Wayne
Fort Wayne, IN, USA

Samir H. Patel
Coonamessett Farm Foundation
East Falmouth, MA, USA

Nicolas Pilcher
Marine Research Foundation
Kota Kinabalu
Sabah, Malaysia

Peter C. H. Pritchard
Chelonian Research Institute
Oviedo, FL, USA

Nathan J. Robinson
Biology Department
Indiana-Purdue University at Fort Wayne
Fort Wayne, IN, USA

David C. Rostal
Department of Biology
Georgia Southern University
Statesboro, GA, USA

Vincent S. Saba
NOAA National Marine Fisheries Service
Northeast Fisheries Science Center
Geophysical Fluid Dynamics Laboratory
Princeton University Forrestal Campus
Princeton, NJ, USA

Pilar Santidrián Tomillo
Population Ecology Group
Institut Mediterrani d'Estudis Avançats
Esporles, Mallorca, Spain

Laura Sarti Martínez
Comisión Nacional de Áreas Naturales Protegidas
SEMARNAT
México DF, México

Kartik Shanker
Centre for Ecological Sciences
Indian Institute of Science
Bangalore, India

George L. Shillinger
The Leatherback Trust
Monterey, CA, USA

Paul R. Sotherland
Department of Biology
Kalamazoo College
Kalamazoo, MI, USA

James R. Spotila
Department of Biodiversity, Earth and Environmental Science
Drexel University
Philadelphia, PA, USA

Charles A. Stock
National Oceanic and Atmospheric Administration
Office of Oceanic and Atmospheric Research
Geophysical Fluid Dynamics Laboratory
Princeton University Forrestal Campus
Princeton, NJ, USA

Jennifer Swiggs
Biology Department
Indiana-Purdue University at Fort Wayne
Fort Wayne, IN, USA

Ricardo F. Tapilatu
Marine Laboratory
The State University of Papua (UNIPA)
Manokwari, Papua Barat Province, Indonesia

Bryan P. Wallace
Stratus Consulting
Boulder, CO, USA

Jeanette Wyneken
Department of Biological Sciences
Florida Atlantic University
Boca Raton, FL, USA

Preface

WITH THEIR MASSIVE SIZE AND STRANGE APPEARANCE, LEATHERBACK turtles resemble species that inhabited Earth in the distant past. They are the sole extant members of the family *Dermochelyidae*, which includes at least eight other extinct species. Only leatherback turtles survive. These ancient reptiles have in recent decades introduced us to fascinating biological aspects of one of the oceans' most amazing creatures. Over the last 40 years, growing interest in the species has also resulted in an exponential increase in the number of scientific publications and in the number of conservation projects established by governments, nongovernmental organizations, academic institutions, and sometimes by local individuals and groups.

The idea of producing a book came after several scientific discussions on a nesting beach in Costa Rica. At the time, there were already books available on the biology of sea turtles in general, sea turtles by region, as well as on particular species such as loggerhead and olive ridley turtles. We realized that there was a need to compile the latest and finest information on leatherback turtles. Considering the state of knowledge, high interest, and uniqueness of the species, the need for a book became evident and the time for its production also seemed right.

This book constitutes a thorough analysis of the biology and conservation of leatherback turtles. We have brought together world experts on the subject to build a book that can be used for academic purposes, but is also of general interest for herpetologists and people in other fields. Because leatherback turtles are both fascinating to scientists and an icon of ocean conservation, we especially wanted to make a strong connection between biology and conservation. As scientists, leatherback turtles have opened doors to the past for us. As conservationists, we need to open doors to the future for them.

Many things have happened over the course of editing the book since we first presented the plan to John Hopkins University Press in 2010. Laurie Spotila was an inspiration from the very beginning, providing kindness, patience, and tolerance. Unfortunately, she passed away while the book was in press. Several family members and friends have also been especially supportive during the process. Some of them are no longer with us. The book would not have been possible without the help and support of many people who kindly offered their services. We are especially indebted to Vincent Burke for believing in the project and making the book a reality. We would like to acknowledge Joanna Alfaro, Ana Barragán, Christopher Binckley, Gabriela Blanco, Brian Bowen, Paul Fiedler, Robert George, Shaya Honarvar, Susan Kilham, Laurie Mauger, Michael O'Connor, David Owens, Frank Paladino, Aliki Panagopoulou, Edwin Price, John Roe, Vince Saba, Mike Salmon, Amanda Southwood, John Spotila, Laurie Spotila, Anthony Steyermark, Jack Suss, Manjula Tiwari, Bryan Wallace, Kris Williams, and Blair Witherington for their contributions to improving the chapters.

Finally, no scientific information would exist today without the dedicated work of thousands of biologists, conservationists, and volunteers who collect data over countless nights and days at many nesting beaches around the world, and in the oceans. To all of them, our most sincere thanks.

Publication of this book was supported by the Leatherback Trust and the Betz Chair of Environmental Science of Drexel University.

The Leatherback Turtle

Part I BIOLOGY

Introduction

Phylogeny and Evolutionary Biology of the Leatherback Turtle

PETER C. H. PRITCHARD

In the eastern Yap State of Micronesia, in the tropical western Caroline Islands, hard-shelled marine turtles are widely distributed, and frequently caught for human consumption (Pritchard 1977). However, the leatherback turtle, *Dermochelys coriacea* (i.e., the "leathery, skinny turtle"), while rarely seen in Yap waters, has a name there: "*Wongera,*" which translates to "whale turtle." There is no other turtle whose vernacular name contrasts the species in question with a cetacean, but there is some justification in this case.

First, the size of the animal is remarkable. There is no other chelonian (turtle) as large as a leatherback, which is bigger than the smallest cetaceans. Leatherbacks seem always to be big, both males and females, though subadult or posthatchling specimens are so rare that few have ever seen them. I have seen two or three half-grown mounted specimens of *Dermochelys* in museums in China; these were probably caught locally but without data. Also, Scott Eckert (2002) was able to locate records of a dozen juvenile leatherbacks. But juveniles are rare everywhere, and incidental captures in fisheries off Peru and Chile provide some of the only clues to the "lost years" for this species.

Leatherbacks have anterior flippers that are uniquely long and powerful, reminiscent of those of the humpback whale. They are so long that hatchlings racing to the sea may turn complete somersaults on the damp sand. The corselet skin on the carapace (upper shell) and plastron (lower shell), and around the neck and shoulders of adults has a rubbery surface texture otherwise unknown among chelonians, but similar to that of a dolphin—or a truck tire.

Other special leatherback features are the rapid growth rate, which is not yet definitively quantified but is considerably greater than that of cheloniid turtles (the family Cheloniidae comprises marine turtles other than leatherbacks). It seems that this is facilitated by the presence of blood vessels running through the cartilaginous ends of the bones, a rare chelonian feature that makes rapid growth possible (Rhodin et al. 1996).

The "shell," hardly an accurate description of the corselet encircling the body, has a unique structure. In the cheloniids, this structure is not greatly different from that of turtles of the various freshwater families.

In most turtles, the ribs are present although they are somewhat embryonic in hatchlings. They expand progressively, eventually closing off the intercostal fontanelles (the spaces between the ribs) and forming a complete bony carapace as growth proceeds. In the leatherback, the intercostal fontanelles remain open throughout growth, and like all the other bones of the leatherback, the ribs remain embryonic in form throughout life.

Similarly, while the plastron may close completely in testudinids and many other chelonian families, to various degrees it does not close completely with growth in cheloniids. The leatherback plastron retains only a flimsy peripheral bony ring, with the remainder of the plastron encircling an extensive central vacuity filled in life with oily cartilage (chapter 4). Uniquely, except for mud turtles of the genus *Kinosternon*, the entoplastron (median bony plate of plastron) is absent.

Tolerance of cold conditions by leatherbacks, at least among adults, is remarkable indeed. Especially cold winters in Florida may cause Kemp's ridleys (*Lepidochelys kempii*), green turtles (*Chelonia mydas*), and loggerheads (*Caretta caretta*) to undergo "cold stunning," which can kill them if they are not rescued quickly, whereas subarctic temperatures that would immobilize and kill cheloniid turtles are tolerated indefinitely by adult leatherbacks. This resistance to cold is made possible by countercurrent venous and arterial structures located at the bases of the limbs, which allow cold venous blood returning to the body to exchange heat with the adjacent arterial blood as it enters the extremities (chapter 13). Nevertheless, adults still need tropical or in some cases subtropical beaches for oviposition (egg laying), because of temperature-dependent sex determination (chapters 6, 8).

Leatherbacks may nest on the same beaches as cheloniid turtles, although sometimes they start earlier or later in the season than green and hawksbill turtles (*Erethmochelys imbricata*), perhaps to reduce competition. The eggs are laid at sufficient depth such that they are unlikely to be found by predators or exposed by other nesting turtles (chapter 6). While cheloniid turtles are often seen copulating on the water's surface near nesting beaches, this does not seem to be the case with leatherbacks; they may mate at considerable depths well away from the nesting beaches, although this is speculative at best and there are few actual observations.

Leatherback females produce the largest eggs among extant reptiles, and the clutch size in the Atlantic is, on average, greater than in Pacific populations (chapters 5, 12). A unique feature of these clutches is the production by the nesting female of a variable number of undersized shelled albumin gobs (SAGs) that were once called yolkless eggs (although they contain no embryo or yolk); these vary greatly in size and are laid toward the later stages of oviposition. The selective advantage of these SAGs is unclear, but they occur in every leatherback clutch, whereas cheloniid turtles hardly ever produce them. Possibly, there is a tendency for fine sand of some beach substrates to filter down between the large eggs, displacing the oxygen needed for egg and hatchling development. In this scenario, the discharge of a dozen or two undersized, miscellaneous SAGs after most of the normal eggs have been laid may preserve the gaps between the viable eggs and thus prevent loss of oxygen. Despite numerous attempts to discern the role of SAGS by observation and experiments in the field, their mystery continues.

Other unique features of the leatherback include the diet, which consists almost entirely of gelatinous marine fauna of a wide range of species. The jaws of the leatherback are highly adapted for a diet of gelatinous material, such as jellyfish, Ctenophora, and Urochordata. The leatherback's esophagus contains large pointed papillae that point inward and retain prey items when water is expelled (chapter 4). In contrast, other marine turtle species either consume various species of plant material (the green turtle), heavy-shelled molluscs (loggerheads), sponges (hawksbills), or shrimp and other smaller invertebrates (ridleys). The leatherback jaw has a sharp-edged cuspid, a scissorlike structure that works well for slicing medusae, but is fundamentally useless for handling the organisms eaten by cheloniids.

Another totally unique feature of the leatherback is the extraordinary layer of mosaic, interdigitating bones; thousands of them are millimeters below the skin of the corselet, and continuous on the dorsal side but interrupted or reduced to just a few on the plastron. Individually, these bones are rigid, but there appears to be some flexibility in the subdermal structure as a whole, which may absorb rapid pressure changes when the turtle dives very deep and very fast (chapter 3).

In the face of this extraordinary range of innovative and specialized features, it seems to be clear that the leatherback is not closely related to any living chelonian genus or family (chapter 2). Perhaps one should search for profound variation within the species *Dermochelys coriacea*. Over the years, there have been many efforts to describe subspecies or other secondary taxa of leatherbacks, and Fritz and Havas (2007) listed no fewer than thirty-five synonyms of *Dermochelys coriacea*. Several of these address the subject of a possible divide between the leatherbacks of the Atlantic and Pacific Oceans, the latter being addressed as the subspecies and the former as *Dermochelys coriacea coriacea*. However, subspecies are not now recognized in this species.

But the truth is that the leatherback is one of the most wide-ranging of all reptiles, and is as capable of reaching oceanic corners of planet Earth as *Homo sapiens*. The fact that there is only one recognizable species of leatherback is not a suggestion of endangered status, but of universality. Some minor expressions of geographic variation do exist, specifically with the East Pacific populations that nest in western Mexico and Costa Rica (Eckert et al. 2012). These differences reflect the overall body size of nesting females (the East Pacific ones being 140 to 150 cm in carapace length and smaller than those elsewhere), a smaller clutch size, and a tendency in some Pacific specimens to carry substantial numbers of barnacles; these crustaceans apparently are rarely present, or only in token numbers, on Atlantic leatherbacks (chapters 6, 7). These trivial differences may reflect some component of the ecosystems in which the turtles live (chapters 14, 15) but do not have any taxonomic significance. It is noteworthy that in the 1970s I noted a statistically significant difference in size (carapace length) between female leatherbacks in the heavily exploited Guyana population (151.9 cm) and in the (then) protected French Guiana and Surinam population (158.5 cm). It is not clear if this reflects a difference in age of these turtles or if it is related to protection of the latter population in some other way.

Fossil Leatherbacks

With this background of a species that has achieved global success we need to examine the fossil record to see whether any phylogenetic insight may be deduced. This process, however, is far from easy. A pelagic animal like *Dermochelys* has, upon its demise, very little likelihood of becoming a fossil. It will more likely fall apart in deep water and be dismembered by predators and scavengers, or perhaps it will wash up on a beach frequented by vultures and again be torn apart. On a beach, the unique stench of a decomposing leatherback will attract more and more vultures from great distances. The decomposition of the leatherback will be much more rapid than it would be in the case of a cheloniid, in which the skull remains in one piece (except in juveniles). In *Dermochelys*, the skull rapidly becomes completely disarticulated after death, and the ribs, mosaic bones, and plastral bones are quickly destroyed or scattered.

Thus, the paleontologist must search for traces of turtles that fall apart after death but that have persisted many millions of years after their demise—this is a difficult task, but not entirely impossible. Entire leatherback fossil skeletons will surely remain elusive, but from time to time a section or a chunk of a mosaic carapace, barely identifiable as part of a turtle, comes to light and is not to be confused with the surface of a dried up mudhole, in which the mud has cracked in a polygonal pattern not unlike the subdermal mosaic bones of the turtle. The mud and the turtle patterns are not exactly the same, in that dermochelyid mosaic bones are not completely random in their arrangement. The mosaic is instead distorted here and there to form ridges running the entire length of the carapace; these ridges themselves consist of individual mosaic bones that are significantly larger than those in the flat areas of the carapace.

Making sense of such rare fragments is thus difficult, but Roger Wood and colleagues have brought together all available fossilized chunks of leatherback fossil and published a useful compendium (Wood et al. 1996). As it happens, quite a few fragments had already been declared the holotypes of new species during the preceding decades and centuries, especially when skull fragments or intact humerus bones were available. The humeri of dermochelyid turtles are massive structures, powerful enough to propel the leatherback quite rapidly from one continent to another and are significantly different from those of cheloniid turtles, and from other chelonians as a whole. Wood et al. (1996) reviewed the literature, examined new specimens and presented the following summary of known fossil leatherback taxa.

RECENT: *Dermochelys coriacea* (only). The unique features of the extant genus include the following components:

1. Seven longitudinal ridges along the carapace, with crests bearing elongate nodules scattered along the length of each keel. The median ridge is always the highest.
2. The crests of the ridges undulate up and down.
3. The carapace mosaic is thickest underneath the axes of the ridges, and becomes very thin in the intervening stretches.
4. The visceral surface of the carapace arches upward beneath each keel.
5. Individual carapace ossicles are very small.

PLIOCENE (12 million years ago): A single indeterminate ossicle has been described, referable to Dermochelyidae.

MIOCENE (25 million years ago), two known taxa:

1. *Psephophorus polygonus*
2. *"Psephophorus" calvertensis*

OLIGOCENE (34 million years ago):
1. *Natemys peruvianus*
2. *"Psephophorus" rupeliensis*
3. A South Carolina specimen, not yet described
4. Probably also present: a taxon with tectiform ridges on the carapace

EOCENE (58 million years ago):
1. *Cosmochelys dolloi*
2. *Egyptemys eocanus* and *E. oregonensis*
3. *Eosphargis gigas* and *E. breineri*
4. A *"Psephophorus"* from New Zealand
5. Shell with tectiform ridges
6. Shell from Alabama USNM 23699

PALEOCENE (63 million years ago):
No dermochelyid material known

The list suggests that leatherbacks were morphologically varied and distributed around the world during the Eocene. The variety progressively diminished over time, leaving only *Dermochelys* surviving to modern times.

The best-preserved leatherback fossils are probably those appertaining to the genera *Eosphargis* and *Cosmochelys*. Particularly interesting is the shell morphology of a specimen of *Eosphargis*, in which the size is greatly reduced, pleurals have disappeared, and the plastral bones are represented by rodlike elements; this is a combination of cheloniid and dermochelyid components, suggesting that the one might be ancestral to the other.

The Oligocene species *Natemys peruvianus* has a unique feature in that the holotype appears to be somewhat flattened and shows both carapacial and plastral surfaces, each with the characteristic mosaic of small interlocking bones. This presence of mosaic bones in both surfaces is unknown elsewhere among leatherbacks, both fossil and modern, although there may be small, isolated mosaic bones in the plastron of *Dermochelys*. Having examined and dissected many contemporary leatherback carcasses washed up on beaches in the Guianas, I find it not inconceivable that a relatively large section of the surface tissue of the carapace, with mosaic bones intact, might have been folded upon itself postmortem to give the "sandwich" effect shown by the *Natemys* specimen.

In general, the overall difficulty of identifying and defining fossil leatherback turtles is the extremely small sample size of the various taxa. Only the holotype is available in many cases, and even then, it is probably just a fragment of the whole animal.

Dermochelys coriacea in Today's World

Only a few decades ago, the leatherback was considered the ultimate in marine rarities. In his epochal book *Handbook of Turtles*, Archie Carr (Carr 1952) reported that the only man in the world to have seen them nesting regularly was P.E.P. Deraniyagala in Sri Lanka, author of *The Tetrapod Reptiles of Ceylon* (Deraniyagala 1939). In 1989 we did not even know the metabolic rate of the leatherback (Paladino et al. 1990) and did not have a complete idea of its nesting locales. Since then we have made considerable progress in understanding leatherback turtles.

For most species of sea turtle, it is striking that reproducing individuals do not occupy an entire ocean basin and are instead notably concentrated into just a corner of one. For example, reproducing flatback turtles (*Natator depressus*) are limited to coastal areas of the northern half of Australia. In the Atlantic, Kemp's ridley nests only in extreme southern Texas and the Mexican State of Tamaulipas. In the Indian Ocean, many loggerheads nest in a very small area of Western Australia near Shark Bay, but the only really concentrated nesting in very large numbers is restricted to the small Island of Masirah, Oman. In the Pacific Ocean, there are few olive ridley (*Lepidochelys olivacea*) nesting grounds of any note except for the arribada sites in Costa Rica at Playas Ostional and Nancite; Playa la Flor in Nicaragua; and on the Mexican beaches of Escobilla and Morro Ayuta in Oaxaca, and nearby Ixtapilla in Michoacan. In the Indian Ocean, olive ridley nesting occurs in extraordinary numbers in the State of Orissa, where three arribada sites have been identified, in modest numbers on the eastern coast of India, and in scattered locations over parts of West Africa.

The other two species, the green turtle and hawksbill turtle, have circumtropical distributions; the former is extensively migratory and the latter much less so. Nesting is both on continental shorelines and on oceanic and continental islands. The eastern Pacific green turtle, also called the black turtle (*Chelonia mydas agassizii*), nests in good numbers in parts of Baja California and Michoacan, Mexico; the Galápagos Islands; and Costa Rica.

Which of these patterns of distribution is typical of the leatherback? It is certainly migratory, although its migratory patterns usually follow concentrations of floating prey (primarily jellyfish) rather than connecting fixed points between nesting and feeding grounds (chapter 14). Several decades ago, there was only one known major nesting ground of leatherbacks anywhere—in Terengganu, Malaysia—and it is salutary to note that not only has this colony diminished, it has completely disappeared (chapter 10). Today, there are not more than three or four major colonies, and the largest are in French Guiana, Trinidad, and a lengthy stretch of beach in Gabon, West Africa (chapter 9). All of these sites were quite recently discovered. The beach on the north coast of Trinidad is a small one, and two or three decades ago attracted only small numbers of leatherbacks; now it has the highest nesting density of any beach in the world. There have been changes in Guyana as well. Thirty years ago, only a small fraction of the turtles nesting there were leatherbacks; today, well over 80% of the nesters there are leatherbacks. Nesting in the Indian Ocean is primarily restricted to South Africa and the Andaman and Nicobar Islands, although the beaches there were destroyed by the tsunami in 2004 (chapter 11).

To summarize this picture, leatherback nesting in recent years has been abundant in the Atlantic, with consistently expanding populations both on mainland shores and on many of the Caribbean islands. Meanwhile, the East Pacific colonies have essentially collapsed, even though there has been extensive beach management and turtle protection in the Mexican and Costa Rican nesting areas. The disappearance of the Terengganu turtles can be easily explained by the many decades of nearly 100% collection of eggs. On the other hand, there is no simple explanation for the surprising recrudescence of the Atlantic populations, in that the same conservation efforts that had scant good results in the Atlantic have failed in the Pacific. The difference between the fates of the Atlantic and East Pacific populations is apparently tied to both ocean productivity (chapter 15) and the effect of fisheries in the two oceans (chapter 17). There are still many mysteries to unravel in the biology of leatherback turtles and many problems to solve in the conservation of these magnificent animals. One of the biggest problems facing leatherbacks in the twenty-first century is a changing climate due to global warming (chapter 16). Solving this problem is a key to survival of leatherbacks and all sea turtles into the twenty-second century.

LITERATURE CITED

Carr, A. 1952. Handbook of turtles: The turtles of the United States, Canada, and Baja. California. Cornell University Press, Ithaca, NY, USA.

Deraniyagala, P.E.P. 1939. The tetrapod reptiles of Ceylon, Vol. I. Testudinates and crocodilians. Dulau, London, UK.

Eckert, K. L., B. P. Wallace, J. G. Frazier, S. A. Eckert, and P.C.H. Pritchard. 2012. Synopsis of the biological data on the leatherback sea turtle (*Dermochelys coriacea*). US Department of the Interior, Fish and Wildlife Service, Biological Technical Publication BTP-R4015-2012. Fish and Wildlife Service, Washington, DC, USA.

Eckert, S. A. 2002. Distribution of juvenile leatherback sea turtle *Dermochelys coreiacea* sightings. Marine Ecology Progress Series 230: 289–293.

Fritz, U., and P. Havas. 2007. Checklist of chelonians of the world. CITES nomenclature committee. Vertebrate Zoology 57: 149–368.

Paladino, F. V., M. P. O'Connor, and J. R. Spotila. 1990. Metabolism of leatherback turtles, gigantothermy, and thermoregulation of dinosaurs. Nature 344: 858–860.

Pritchard, P.C.H. 1977. Marine turtles of Micronesia. Chelonia, San Francisco, CA, USA.

Rhodin, J.A.G., A.G.J. Rhodin, and J. R. Spotila. 1996. Electron microscopic analysis of vascular cartilage canals in the humeral epiphysis of hatchling leatherback turtles, *Dermochelys coriacea*. Chelonian Conservation and Biology 2: 250–260.

Wood, R. C., J. Johnson-Gove, E. F. Gaffney, and K. F. Maley. 1996. Evolution and phylogeny of the leatherback turtles (Dermochelyidae), with descriptions of new fossil taxa. Chelonian Conservation and Biology 2: 266–286.

2

Phylogeny, Phylogeography, and Populations of the Leatherback Turtle

PETER H. DUTTON
AND KARTIK SHANKER

Leatherback turtles (*Dermochelys coriacea*) belong to the family Dermochelyidae, one of two surviving sea turtle families, which both originated in the Cretaceous (Gaffney and Meylan 1988; Pritchard 1997). Leatherbacks, like other sea turtles, have several remarkable features in their life cycle that have affected their evolutionary history, global distribution, and population genetic structure. They are able to migrate large distances, typically between nesting and feeding areas, even more so than other sea turtles because of their ability to withstand cold waters. This probably affected their distribution and phylogeography over evolutionary time.

Little is known, however, about the reproductive behavior of these animals, because they are hard to observe in the wild. Female sea turtles display natal philopatry, returning to their natal beaches to nest. Early genetic studies on sea turtles demonstrated the role of natal philopatry in structuring populations (see Bowen and Karl 1997 for a review). However, precision of natal homing and its consequences for population genetic structure varies among species. Genetic studies that provide clues about mating systems are, therefore, extremely valuable for understanding the biology of these species.

In this chapter, we provide an overview of sea turtle phylogeny and the position of leatherback turtles within this phylogeny. We then examine the global phylogeography of leatherback turtles. We also examine population level studies of leatherback turtles, particularly in the context of identification of the appropriate Units to Conserve (UTC), including Demographically Independent Populations (DIP), Management Units (MU), Regional Management Units (RMUs), and Evolutionary Significant Units (ESU); see box 2.1. We review studies of behavior (multiple paternity) using genetic tools, and finally, we provide some thoughts and ideas about the future of genetic research in sea turtles, with an emphasis on implications for our understanding of leatherback turtle biology.

Phylogeny

That leatherback turtles are unique among sea turtles is obvious. Some studies suggested, on the basis of their morphological and behavioral

Box 2.1. Terms used in genetic studies of sea turtles.

Term	Definition
mtDNA	Mitochondrial DNA is maternally inherited DNA found in organelles called mitochondria within cells; in most multicellular organisms, mtDNA is closed circular double-stranded DNA. Mitochondrial DNA is frequently used for phylogenetic and phylogeographic analyses.
Microsatellites	Microsatellites or Short Sequence Repeats (SSR)/Short Tandem Repeats (STR) are repeating sequences typically of two to six base pairs of nuclear DNA, whose variation can be used to study population genetic structure, kinship, paternity, and so forth.
MU	Management Units (MUs) are populations or stocks that are functionally independent; animals lost from one are not likely to be replaced by individuals from another within ecologically relevant time frames. MUs are typically identified by significant differentiation in mtDNA haplotype or nuclear DNA allele frequencies (Moritz 1994). However, it is possible for populations to function as demographically independent, but with levels of genetic divergence that are not detected by genetic markers; an example is two sea turtle nesting populations that do not yet appear differentiated at the genetic markers due to recent colonization or due to low levels of gene flow from occasional migrants.
DIP	Demographically Independent Populations (DIPs) refer to such units that are functionally independent (as indicated by tagging or other demographic data), but not necessarily differentiated by genetic data (Taylor et al. 2010).
ESU	Evolutionary Significant Units (ESUs) broadly refer to subspecies level units needed to conserve the essential genetic variability for future evolutionary potential. ESUs are similar to Distinct Population Segments (DPS) under the US Endangered Species Act (Taylor et al. 2010). ESUs must be either geographically separate, show strong genetic differentiation from other groups (Moritz 1994), or have unique phenotypic traits as a consequence of selection.

differences, that leatherback turtles may be the sister taxon to all turtles (Rhodin et al. 1981; Rhodin 1985), although phylogenies from serological, immunological, and karyological data suggested that sea turtles are monophyletic (Chen et al. 1980; Chen and Mao 1981; Friar 1982; Bickham and Carr 1983). Several species of Dermochelyidae went extinct during the early Pleistocene (Wood et al. 1996), and *D. coriacea* is the sole surviving member of the family.

Understanding the degree of their uniqueness and relationship to other extant sea turtles improved with the development of molecular genetic techniques during the past two decades. In the first such study, Bowen et al. (1993) used mitochondrial DNA (mtDNA) cytochrome B sequence analysis to discover that *Dermochelys* is distantly related to all other extant sea turtles, diverging about 100 to 150 mya. Their results, however, did not support monophyly of all sea turtles.

Dutton et al. (1996) carried out sequence analysis of the ND4 LEU tRNA and control regions of mtDNA to build a molecular phylogeny of sea turtles. Their parsimony analysis supports the accepted distant and basal position of leatherback turtles in the systematics of sea turtles. Control region analysis suggests a shallow divergence of Atlantic and Pacific leatherback turtles, relative to green turtles (*Chelonia mydas*) (fig. 2.1).

More recently, Naro-Maciel et al. (2008) sequenced five nuclear genes (BDNF, Cmos, R35, Rag1, and Rag2) and two mitochondrial genes (12S and 16S) to reconstruct a sea turtle phylogeny. Their results support previous conclusions that sea turtles are monophyletic and that *Dermochelys* is basal to other sea turtles. At the core of this question lies the issue of adaptation to the sea environment: Did it arise once, or twice independently, in the evolutionary history of turtles? Results suggest that adaptation to the sea environment arose just once,

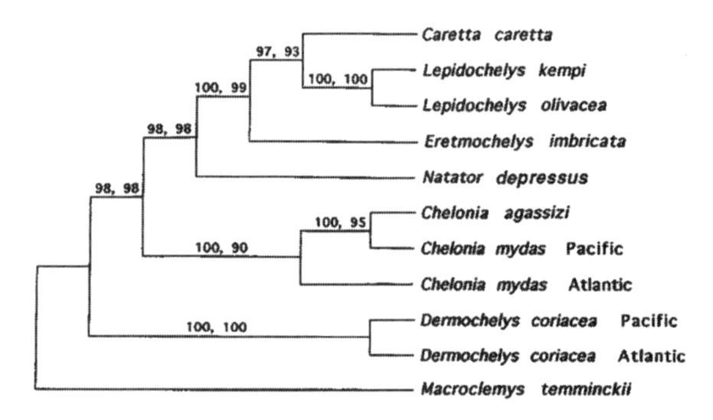

Fig. 2.1. Phylogenetic tree based on 1,433 bp mtDNA sequence data (control region, ND4-Leucine tRNA and cytochrome b) showing sea turtles are monophyletic. Bootstrap values for each analysis are reported on the corresponding branches as percentages of 1,000 replicates both with and without the control region data. From Dutton et al. 1996.

and that leatherback turtles were an early offshoot of the lineage.

In general, mean evolutionary rates in mtDNA are slower in Testudines than other reptiles, due either to their long generation times or slow metabolic rates (Avise et al. 1992). Within the mtDNA, the ND4-LEU and cytochrome b regions are more conserved than control regions. Even accounting for this, *Dermochelys* is characterized by low diversity and shallow mtDNA divergence relative to cheloniid sea turtles (Dutton et al. 1996; Naro-Maciel et al. 2008).

New technologies have allowed sequencing of the entire mtDNA genome of sea turtles (Frey et al. 2012; Morin 2012). These studies detected additional genetic variation in cheloniids (Shamblin et al. 2012) and provided new insights into evolutionary relationships among species (Duchene et al. 2012). Preliminary analysis of all 16,281 bp of each mitochondrial genome sequence for leatherbacks from representative rookeries around the world revealed surprisingly little additional variation (P. H. Dutton, unpublished data) and appeared to rule out the likelihood that there is greater diversity not detected in *Dermochelys* by the standard mtDNA genetic markers in earlier studies (Dutton et al. 1996; Dutton et al. 1999).

Phylogeography

Sea turtles are long-lived species and undertake extensive migrations as part of their life cycles, which suggests that there should be little genetic structure in the populations. On the other hand, females exhibit natal philopatry, and this leads to strong to moderate population structure in most species (Jensen et al. 2013).

Dutton et al. (1999) carried out a comprehensive global phylogeographic study with mtDNA control region sequences in 175 leatherback turtles from 10 nesting colonies. There were 11 haplotypes with mean sequence divergence ($p = 0.0058$) much lower than in other sea turtles. Haplotype A, which accounted for more than 50% of the sample, was fixed in the South African and several Atlantic populations, and it was also present in Indo-Pacific rookeries. Four rare haplotypes occurred in turtles from eastern Pacific rookeries, and the most divergent haplotypes occurred in turtles in that region and in western Atlantic rookeries.

Within the Indian Ocean, we sampled over 100 leatherbacks at a large rookery on Great Nicobar Island on beaches that were subsequently destroyed by the December 2004 tsunami (Shanker et al. 2011). In mitochondrial control regions, 43.5% of samples belonged to haplotype A, the dominant haplotype in the Atlantic and Indo-Pacific. Other haplotypes (D, E, I) occurred in Malaysia (D, E), Solomon Islands (D, E, I), Papua-Jamursba-Medi and Papua New Guinea (D, E, I), Papua-War Mon (I), and the East Pacific (Mexico and Costa Rica, D) (fig. 2.2).

Our group documented a number of new nesting sites in the last decade in the western Pacific (Dutton et al. 2007). The majority of clutches are laid at 4 nesting sites on the northwest coast of Papua-Barat, Indonesia; this portion of coast constitutes one of the most important nesting areas for the 5,000 to 9,000 nests built per year at 28 sites in the western Pacific and Indian Ocean (Benson et al. 2007; Hitipeuw et al. 2007). Genetic analysis of the control region of the mtDNA revealed six haplotypes (A, D, E, F, H, I), of which two

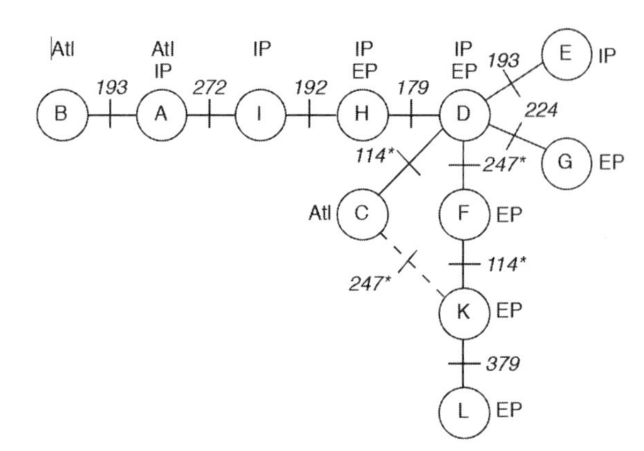

Fig. 2.2. Parsimony network describing relationships among the mtDNA haplotypes and ocean basins where they are found in rookeries; Atlantic (Atl), Indian-Pacific (IP), and eastern Pacific (EP). Mutation sites (base position) are shown on the branches and asterisks indicate assumed homoplasy. From Dutton et al. 1999, with permission from John Wiley and Sons.

(F, H) are also found in the eastern Pacific, while the others are Indo-Pacific or global haplotypes (Dutton et al. 2007).

Though leatherback turtles are part of an evolutionary lineage that originated at least 100 million years ago, intraspecific phylogeny recorded in mitochondrial lineages is less than a million years old (Dutton et al. 1999). Deepest nodes in other species—green, loggerhead (*Caretta caretta*), and hawksbill (*Eretmochelys imbricata*) turtles—are about 5–7% in control region sequences, which indicates a separation of about 2 to 4 million years and corresponds to closure of the isthmus of Panama (Dutton et al. 1996; Bowen and Karl 1997), or to climatic changes induced by the closure (Shanker et al. 2004). In the case of ridleys, this node represents the divergence between the Kemp's (*Lepidochleys kempii*) and olive (*Lepidochelys olivacea*) ridleys (Bowen et al. 1998; Shanker et al. 2004). Leatherback turtles, on the other hand, appear to be part of a global radiation of only a single mtDNA lineage, with the most divergent haplotypes separated by less than one million years. Leatherback turtles are able to withstand colder waters than other species, and this may explain their patterns of distribution. However, they have higher pivotal temperatures for sex determination (e.g. warmer temperatures are needed to produce females; chapter 8) compared to hard-shelled species and therefore need to nest on tropical beaches (Mrosovsky et al. 1984; Binckley et al. 1998). Hence, leatherback turtles may have been confined to narrow refugia during cold periods, followed by range expansion during warmer periods, leading to shallow contemporary mtDNA lineages. While mutation rates could be slower in leatherback turtles than for other species, data do not suggest that this is the case (Dutton et al. 1999; Duchene et al. 2012).

Gene genealogy and global distribution of mtDNA haplotypes indicated that leatherbacks may have radiated from a narrow refugium, possibly in the Indo Pacific, during the early Pleistocene glaciation (Dutton et al. 1999). Relatively higher diversity of haplotypes in the Indo-Pacific and the presence of the central D haplotype in Malaysia and Great Nicobar Island leatherbacks suggested that the Indo-Pacific may be the most recent source for global leatherback populations, as indicated by presence of putative ancestral haplotypes and the highest haplotype diversity (Dutton et al. 1999).

Effective population size (N_e; see box 2.2) is the size of an idealized population that would have the same amount of inbreeding or random gene frequency drift as an actual population under study. It provides a measure of the rate of evolutionary change in composition of alleles caused by genetic drift in a population.

Box 2.2. Explanation of terms for effective population size, N_e.

Effective population size (N_e) is a term that conceptualizes the evolutionary potential of a population by measuring the equivalent number of adults that contribute gametes or alleles (or in the case of mtDNA, haplotypes) to the next generation under a set of idealized assumptions (see Wright 1938; Kliman et al. 2008; and Charlesworth 2009). The N_e reflects the evolutionary history (and level of genetic variation) of a population and can often be much lower than the actual number of individuals (N_c) in a population at any given time. Genetic variation can be reduced by the loss of alleles through *genetic drift* in small populations. Genetic drift is a random change in gene (allele) frequencies and is most evident in small populations where rare alleles can be permanently lost each generation by sampling error, even if N_c subsequently increases. Genetic drift is greater in populations that remain small continuously over many generations, or that are drastically reduced by a sudden event, known as the *bottleneck effect*. A *founder effect* occurs when a population is initially established by just a few individuals, as is the case for sea turtles when a new beach is colonized by female nesters. Natal homing (philopatry to natal beaches) further establishes these new nesting populations as the offspring of these founders return to their natal beaches to breed. Although the population may subsequently grow in size and later consist of a large number of individuals, the gene pool of the population is derived from the genes present in the original founders. Genetic approaches using mtDNA markers (which quantify variation of female genetic lineages) allow estimation of N_e; this is of particular interest in the study of sea turtles, because many populations have declined dramatically, and new colonization of nesting beaches by females potentially creates founder effects. Nonrandom mating behavior and unequal sex ratios can contribute to the effects of genetic drift on genetic variation in a population and are also significant for sea turtle conservation genetics.

With the use of coalescent theory (Charlesworth 2009), we can calculate N_e and determine the time (in generations) elapsed from a common ancestor by tracing alleles to this progenitor; this model is based on observed genetic diversity and applies a molecular clock (Avise et al. 1988). Dutton et al. (1999) estimated that, based on a provisional evolutionary rate for mtDNA, the most divergent haplotypes detected in leatherbacks would coalesce to a common ancestor approximately 640,000 to 840,000 years BP, which corresponds to an N_e of 45,000 to 60,000, using a generation time of 14 years for leatherback turtles. This is consistent with the scenario in which leatherbacks went through at least one global population bottleneck during the Pleistocene and have since expanded, so that the current N_e reflects the past bottleneck and evolutionary processes that have occurred since then. Sea turtles are prone to genetic drift due to founder events, because new nesting colonies may be established by one or a few females (see box 2.2). In addition, leatherback turtles may show sweepstakes recruitment somewhat like marine fish, where reproductive success of a few nesters produce thousands of offspring that contribute disproportionally to population growth. On the other hand, they have subdivided populations due to female and possibly male philopatry to natal beaches. Founder events and high variance in reproductive success result in a lower N_e, but population subdivision favors retention of lineages and results in an increase in N_e as population expansion continues over evolutionary timescales.

Many major nesting colonies have declined precipitously in recent decades (Chan and Liew 1996; Spotila et al. 1996; Spotila et al. 2000; Sarti and Barragán 2011) with recent global population estimates of 23,000 to 33,000 adult females in 2010 (Wallace et al. 2013), but the decline appears to have been too recent to affect N_e, which is now higher than the actual population size. Evolutionary history of this species is one of local extinctions and recolonizations, and perhaps its ability to colonize rapidly (relative to other sea turtles) is one reason why it has persisted through previous population bottlenecks (Dutton et al. 1999). However, an apparent recent extinction of the Malaysian rookery (Liew 2011) combined with other persistent population declines in the Indo-Pacific may already have further reduced genetic diversity in this species. Then a lower N_e can be expected if population declines persist, because humans are essentially putting leatherbacks through another bottleneck (see box 2.2).

Olive ridley turtles follow a similar pattern, with divergence from the sister taxon (Kemp's ridley) three to four million years ago; however, the Atlantic and eastern Pacific populations of olive ridleys further diverged about 300,000 years ago from Indian Ocean populations of the same species (Bowen et al. 1998; Shanker et al. 2004). These results suggest the possibility of sources and sinks (sensu Pulliam 1988) at evolutionary time scales and global spatial scales for these two species (Shanker et al. 2004).

Population Genetics

Leatherback nesting populations are strongly structured at a global scale and within ocean basins (Dutton et al. 1999; Dutton et al. 2007). Within the Atlantic, significant differences in haplotype frequencies occur between mainland Caribbean populations and both the St. Croix (US Virgin Islands) population and Trinidad and Tobago population (Dutton et al. 1999). In the Pacific, leatherbacks from both the eastern and western Pacific nesting beaches converge on feeding areas in the northern and southeastern Pacific, and though they mix at feeding grounds, maternal lineages are distinct in the two nesting regions (Dutton, Frey, et al. 2000; Dutton et al. 2007).

While genetic results support the natal homing hypothesis for leatherback turtles, several proximal nesting populations cannot be distinguished, suggesting that natal homing may not be as precise as in other species of sea turtles. However, lack of genetic differentiation between the South African nesting population and those of the Caribbean most likely reflects the recent evolutionary connectedness between these populations, rather than contemporary ongoing gene flow; or it possibly reflects a lack of resolution of the genetic marker, rather than actual connectivity (Dutton 1995; Dutton et al. 1999). More recent studies using longer mtDNA sequences (763 bp) have now differentiated these rookeries and confirm that even these longer mtDNA sequences have low power to detect fine-scale (weak) structure (Dutton, Roden, et al. 2013). A comprehensive survey that includes larger sample sizes, additional rookeries, and analysis based on longer sequences in combination with an array of 17 microsatellite loci (Roden and Dutton 2011; Dutton, Roden, et al. 2013) resolves lingering questions on the level of population structure in Atlantic leatherbacks (Dutton, Roden, et al. 2013). With mtDNA data, this new analysis identifies seven distinct MUs in the Atlantic and Caribbean but also finds finer-scale demographic structuring (with more powerful microsatellite markers) among neighboring populations that identify nine rookeries as DIPs (fig. 2.3). Additional studies using a combination of these nuclear and mtDNA markers will continue to

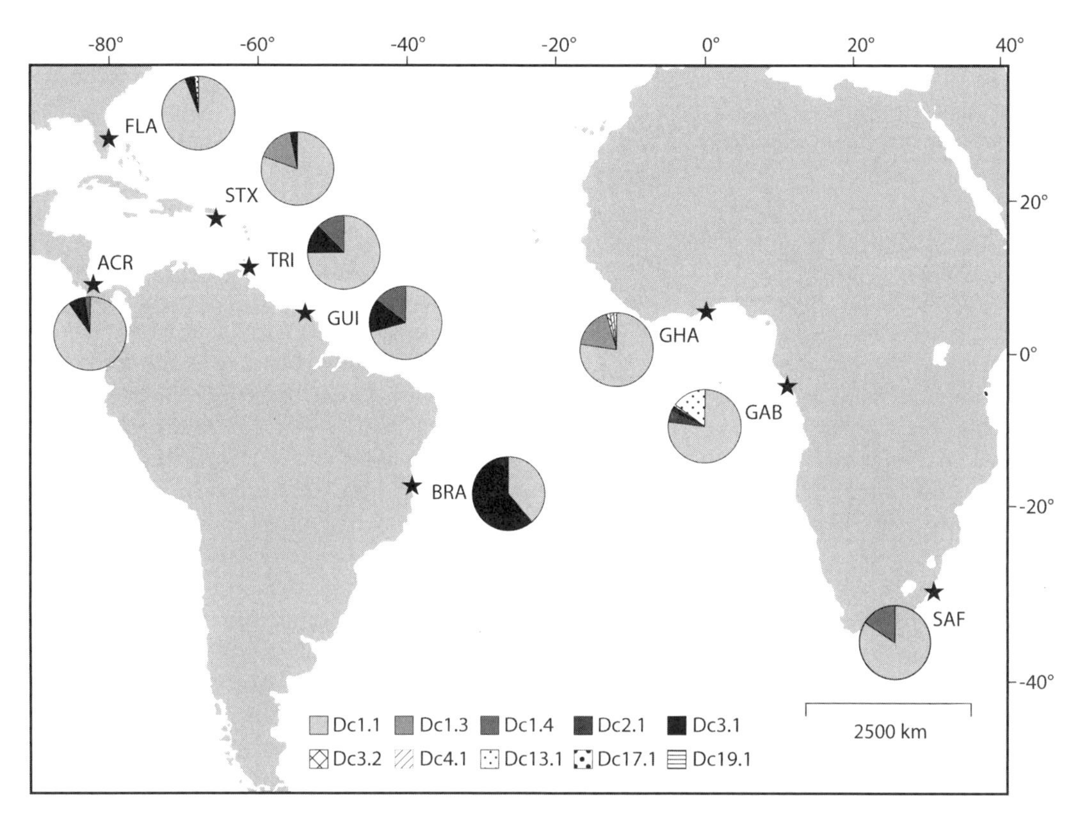

Fig. 2.3. Frequencies of 10 different haplotypes (based on 763 bp sequences of the mtDNA control region) at 9 leatherback rookeries (stars) in the Atlantic and Indian Oceans. Sampled nesting sites include northern Brazil (BRA), Atlantic Costa Rica (ACR), French Guiana and Suriname (GUI), Gabon (GAB), Ghana (GHA), Trinidad (TRI), Florida (FLA) in the United States, St. Croix (STX) in the US Virgin Islands, and Natal in South Africa (SAF). Based on mtDNA, pairwise comparisons indicated significant differentiation between all except FLA and ACR and TRI and GUI. However, all 9 rookeries were differentiated based on data from 17 nuclear microsatellite loci, indicating that they are all demographically independent populations. From Dutton, Roden, et al. 2013.

flesh out the finer-scale connectivity among rookeries and eventually provide comprehensive models of population structure for the species (Molfetti et al. 2013).

Leatherback populations have been increasing over the last decade in the Atlantic (Dutton et al. 2005; Turtle Expert Working Group 2007), and in Florida nesting has increased approximately 10% per year over the past 30 years (Stewart et al. 2011). This pattern is very similar to the increase that occurred at St. Croix (Dutton et al. 2005). However, preliminary analyses of genetic results did not provide any indication of immigration from St. Croix to Florida. There actually appeared to be genetic homogeneity between Costa Rica and Florida rookeries, suggesting that Costa Rica may be the source of founders for the Florida population, and the high connectivity between these two rookeries is either due to ongoing recruitment of nesters hatched in Costa Rica to the growing Florida breeding population, or to one or multiple recent colonization events from Costa Rica (Dutton, Roden, et al. 2013)

Leatherback turtles may also be constrained more by beach type than location; these turtles need deep access routes to the nesting beach and typically nest on open sand areas of steep, high-energy beaches (Pritchard 1976). These beaches are inherently unstable and turtles are likely to benefit from some flexibility in their nest site selection strategy. Genetic patterns similar to those in leatherbacks have been detected in olive ridley turtles that nest on very similar beach types (Shanker et al. 2004).

There is no difference in haplotype frequencies among four nesting sites in the western Pacific, including those on Papua Barat (Wermon and Jamursba-Medi beaches), Papua New Guinea, and the Solomon Islands. Haplotype frequencies at Terengganu, Malaysia, were different from those in the western Pacific rookeries

(Dutton et al. 2007). Preliminary results from reanalysis with longer mtDNA sequences appear to confirm that leatherback populations in the Pacific consist of two broad genetic populations: an eastern Pacific population that includes rookeries in Mexico and Central America, and a western Pacific population made up of rookeries in Papua-Barat (Indonesia), Papua New Guinea, and the Solomon Islands (Dutton et al. 2007; P. H. Dutton, unpublished data). Further investigation with microsatellites and more extensive sampling of these Pacific rookeries may yet resolve finer-scale population structuring; however, results are not yet available.

Given that many populations worldwide are threatened, it is necessary to identify genetic populations, and together with migratory patterns, define Evolutionary Significant Units or Regional Management Units for conservation. The MUs identified by Dutton, Roden, et al. (2013) and K. Shanker (unpublished) and summarized here provide baseline information with which to assess population composition of foraging aggregations and identify natal origins of turtles caught as fisheries bycatch. Oceanic distribution of animals from each MU can then be determined and population boundaries can be drawn. A recent attempt to define RMUs based on genetic studies, tag returns, satellite telemetry, and other data identified seven RMUs globally for leatherback turtles (Wallace et al. 2010).

Ongoing studies show that leatherbacks caught in longline and drift net fisheries in the North Pacific and foraging along the US West Coast belong to the western Pacific nesting population, indicating that this RMU extends all the way across the North Pacific (Dutton, Frey, et al. 2000; Dutton 2006a; Dutton et al. 2006; P. H. Dutton, unpublished data; LeRoux et al. 2007). Preliminary genetic results also indicate that the western Pacific RMU extends into the southeastern Pacific, with animals of western Pacific nesting population origin identified in the waters off Chile and Peru (Donoso and Dutton 2006; Dutton, LaCasella, et al. 2013). In the Atlantic, leatherbacks from Caribbean nesting populations have been identified at foraging areas in the North Atlantic while those at foraging areas off Argentina and Brazil in the South Atlantic appear to be from West Africa (L. Prosdocimi, unpublished data); Vargas et al. 2008; Dutton, Roden, et al. 2013; Stewart et al. 2013).

Behavior

Sea turtles are hard to observe, and behavior such as that involving mating systems is therefore hard to study. However, patterns of multiple paternity and parental genotypes may provide insights into reproductive behavior and strategies. Molecular techniques provide tools that can be used to examine these patterns. It is necessary to understand these components of life history to model population biology and dynamics, and to assess threats.

Sea turtles migrate from breeding to feeding grounds, and mate in offshore waters. The mating system is generally promiscuous, with many males mating with many females. Females are able to store sperm for a few months, suggesting that mechanisms for multiple paternity are present (Pearse and Avise 2001). Multiple paternity is believed to be beneficial to turtles, as they have limited parental care, and no opportunity to judge male fitness. Hence, it would be advantageous for females to mate with multiple males, while it would be advantageous for males to always mate with multiple females given the low cost of sperm production.

Multiple paternity has been examined using microsatellite markers in sea turtles, and has been documented in the loggerhead (Zbinden et al. 2007), olive ridley (Hoekert et al. 2002; Jensen et al. 2006), Kemp's ridley (Kichler et al. 1999), green (Ireland et al. 2003), leatherback (Dutton, Bixby, and Davis 2000; Crim et al. 2002) and flatback turtles (*Natator depressus*; Theissinger et al. 2009). However, several studies have found low or insignificant levels of multiple paternity (Fitzsimmons 1998; Dutton, Bixby, and Davis 2000; Curtis et al. 2000; see review in Jensen et al. 2013).

Curtis et al. (2000) examined 18 nests of leatherbacks in the Caribbean and detected multiple paternity in 10 of 11 nests. Crim et al. (2002) examined multiple clutches in 20 females at Playa Grande in Parque Nacional Marino Las Baulas, Costa Rica, and found single paternity in 12 of the females (31 of 50 clutches) and multiple paternity in 2 females (8 of 50 clutches). Similarly, no multiple paternity was detected in a study of 17 clutches (four females) at Sandy Point National Wildlife Refuge, St. Croix, US Virgin Islands (Dutton, Bixby, and Davis 2000).

Part of the problem in these studies is the issue of detection of multiple paternity. Crim et al. (2002) could not accept or reject single paternity in clutches from 6 of 20 females, due to the low resolution of the few markers used. However, a more comprehensive study examined over a thousand hatchlings from successive clutches of 12 known nesting female leatherback turtles at St. Croix, US Virgin Islands (Stewart and Dutton 2011, 2014). Seven microsatellite markers were used, which were developed for leatherback turtles and were polymorphic for the St. Croix population (Roden and Dutton 2011). Twelve mothers and 17 fathers were identified for 38 nests. While seven females had no sign of multiple paternity, five females mated with two males each. Multiple fathers did not contribute to clutches equally. However, primary and secondary fathers contributed to all clutches for a particular female, indicating that sperm storage did occur.

While there are many theoretical explanations for multiple paternity (increased offspring fitness, ensured fertilization, male coercion), Bowen and Karl (2007) suggested that it may be driven by male density and aggressive mating behavior. In fact, Jensen et al. (2006) found higher levels of multiple paternity in mass nesting populations (90%) than in solitary nesting populations (30%), indicating the role of density and/or adult sex ratio.

Population Vital Rates

Vital rates, such as age to maturity, survival, sex ratios, and population size (including the males), are still lacking for sea turtles and this makes it difficult to conduct meaningful population and risk assessments. Although vital rates are difficult to observe directly, genetic analysis provides a practical approach to understanding these processes and solving some old mysteries.

There has been much research focused on determining hatchling and juvenile sex ratios in sea turtles. Understanding the proportion of males to females in any population has important consequences for population demographic studies. In general, sex ratios of sea turtle hatchlings and juveniles are strongly female biased and there is some concern that under climate change scenarios, populations of turtles may become entirely feminized due to temperature-dependent sex determination in combination with warming temperatures of nesting beaches (Mrosovsky et al. 1984; Hawkes et al. 2007; chapter 16). To date, there have been no studies of sex ratios of breeding adult turtles, also known as the operational sex ratio (OSR; Hays et al. 2010). Using hatchling and maternal DNA fingerprints, one can deduce the paternal genotypes—from one to many fathers per clutch. Resulting genotypes represent individual males that are actively breeding in the population. This means that males can effectively be sampled without ever having been observed or caught in the field.

Stewart and Dutton (2014) assessed hatchlings from 46 female leatherbacks nesting on St. Croix and found that 47 different males had mated with those females, giving an estimated breeding sex ratio of 1.02 males for every female. One male had mated with 3 different females and several others had mated with 2 females. This finding has important implications, as there have been fears that there is a shortage of males and that breeding sex ratios may have been highly skewed toward females. However, these preliminary findings show that this is not the case; there are plenty of males actively breeding in the population and breeding males may actually outnumber breeding females within a given year. Stewart and Dutton (2011) identified one male that had been actively breeding in both 2009 and 2010 (with different females), indicating that males may be breeding yearly.

Age at first reproduction is one of the most important vital rates, especially for population modeling and predictions of future population fluctuations of long-lived species, yet it still remains an elusive holy grail of knowledge for sea turtle demographers and a continuing cause of controversy in the case of leatherbacks. Age at first reproduction of leatherback turtles, estimated from chondro-osseous morphology, skeletochronology, and growth-rate modeling, ranges from 3 to 29 years (Rhodin 1985; Zug and Parham 1996; Avens et al. 2009; Jones et al. 2011). Dutton et al. (2005) inferred age of first reproduction at around 12–15 years from analysis of demographic trend data, hatchling production data, and reproductive histories of nesting females monitored on St. Croix. They also used genetic fingerprinting to show that first-time nesters in the 1990s were closely related and possibly the genetic offspring of leatherbacks nesting in the 1980s; this was generally corroborated by the more recent estimates of 13–16 years at age of first reproduction proposed by Jones et al. (2011) (fig. 2.4). These indirect estimates need to be validated by direct capture-mark-recapture (CMR) of hatchlings to settle the mystery, and a mass-tagging experiment using genetic tags is now underway and is following a cohort of hatchlings to their adulthood.

The first phase of this long-term CMR study involves mass tagging of up to 100,000 leatherback hatchlings on St. Croix, using noninjurious sampling methods established for collecting hatchling DNA. This will be used to create a genetic fingerprint or "tag" to identify individual turtles throughout their lifetimes (Stewart and Dutton 2012; Dutton and Stewart 2013). All samples are stored and subsequently analyzed at the NMFS-SWFSC La Jolla Lab, in California. During the second phase of the study, genetic samples routinely collected from first-time nesters (neophytes) will be analyzed and compared to the stored hatchling genotypes to identify the individuals that were originally "tagged" at birth. Then, age at first reproduction and juvenile survival rates can be directly determined for this population by following a cohort of hatchlings to their adulthood. Other age-specific vital rates of adult females, such as birth and death rates, also can be estimated by moni-

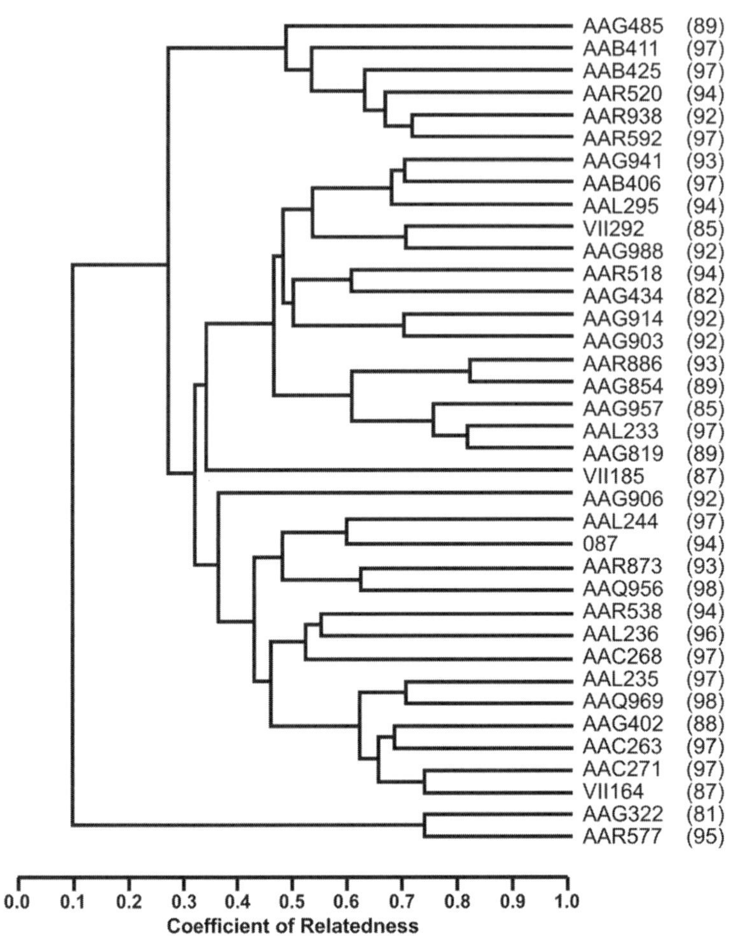

Fig. 2.4. Family groups identified among 37 St. Croix leatherback nesters based on relatedness determined with microsatellite genotyping and mtDNA sequencing (from Dutton et al. 2005). The year the turtle was first observed to nest is given in parentheses; old-timers, such as AAG322 (identified in 1981) and AAG434 (identified in 1982), are most likely mothers of recent (post-1993) first-time nesters, such as AAR577 (1995) and AAR518 (1994), respectively. From Dutton et al. 2005, with permission from Elsevier.

toring these cohorts through their lifetimes, providing crucial information for future studies of the species. Given rapid advances occurring in biotechnology and information management systems, it should be possible to expand the use of genetic fingerprinting in a broad range of CMR applications in the future.

The Future of Leatherbacks on a Changing Planet

Leatherbacks are prehistoric survivors—the message encoded in their DNA reveals an evolutionary history of extinction and recolonization in the genus (see Dutton et al. 1999). The patchy fossil record suggests that there were at least six species of leatherbacks (Wood et al. 1996) that disappeared during the mass extinctions of the early Pleistocene glaciations (chapter 1). However, this genus persists due to a small, tenacious group of one species (*coriacea*) that survived and has continued to recolonize and adapt through the most recent period of great climate change beginning 100,000 years ago. Perhaps its ability to colonize rapidly (relative to other sea turtles) is one reason why this species has persisted through previous population bottlenecks (Dutton et al. 1999). Could the catastrophic declines in the eastern Pacific and loss of the Malaysian nesting population now signal the beginning of another great upheaval for leatherbacks as we enter another period of global climate change?

Evolutionary history gleaned from phylogenetic analysis suggests that the eastern Pacific population is relatively new and has undergone recent recolonization (Dutton et al. 1999). The southeastern Pacific is also one of the most unstable ocean ecosystems due to boom and bust ENSO (El Niño Southern Oscillation) cycles that affect food availability and survival of leatherbacks (Saba et al. 2007). The East Pacific population is particularly vulnerable to these environmental impacts since it appears to lack the complex metapopulation structure that characterizes western Pacific and Atlantic populations (Dutton 2006b; Benson et al. 2011; Bailey et al. 2012).

With a warming planet, as some beaches grow warmer and drier, they become inhospitable to embryonic development, while others that were previously too wet and cold become potentially viable nesting habitat. The former appears to be happening in the western Pacific, where a continuing, long-term decline in the population nesting during the austral summer at Jamursba-Medi may be due in part to hatching failure associated with warmer and drier sand (Tapilatu and Tiwari 2007). At the same time, the winter nesting population unit at neighboring Wermon appears to be increasing. Components of the metapopulation may be blinking on and off, and while they are demographically distinct, they are genetically homogenous (with high connectivity).

A more dramatic example is Baja California, Mexico, where a few nests usually observed every year are laid on beaches that are too cold for the embryos to develop properly. This region represents the northern frontier of the nesting range for the eastern Pacific leatherback population, and if these beaches continue to warm, additional viable rookeries are likely to be established by the offspring of these pioneer nesters.

The level of regional connectivity derived from genetic results suggests that natal homing instincts may not be as rigid in leatherbacks as in other sea turtles. Perhaps this bodes well for conservation of this species, as it allows more flexibility to exploit new reproductive habitat. The increased number of leatherbacks nesting in Florida over the last few years, from less than 100 nests per year in the 1980s to nearly 1,000 nests per year in the 2000s (Stewart et al. 2011) may represent such colonization by migrants from the Costa Rican population (Dutton, Roden, et al. 2013). This is a good sign for a taxon that has persisted through previous global climate change, as long as conservation efforts can help preserve the viability of the metapopulations in each of the main ocean basins (Dutton 2006b).

LITERATURE CITED

Avens, L., J. C. Taylor, L. R. Goshe, T. T. Jones, and M. Hastings. 2009. Use of skeletochronological analysis to estimate the age of leatherback sea turtles *Dermochelys coriacea* in the western North Atlantic. Endangered Species Research 8: 165–177.

Avise, J. C., R. M. Ball, and J. Arnold. 1988. Current versus historical population sizes in vertebrate species with high gene flow: a comparison based on mitochondrial DNA lineages and inbreeding theory for neutral mutations. Molecular Biology and Evolution 5: 331–344.

Avise, J. C., B. W. Bowen, T. Lamb, A. B. Meylan, and E. Bermingham. 1992. Mitochondrial DNA evolution at a turtle's pace: evidence for low genetic variability and reduced microevolutionary rate in the testudines. Molecular Biology and Evolution 9: 457–473.

Bailey, H., S. R. Benson, G. L. Shillinger, S. J. Bograd, P. H. Dutton, S. A. Eckert, S. J. Morreale, et al. 2012. Identification of distinct migration and foraging patterns in Pacific leatherback turtle populations influenced by ocean conditions. Ecological Applications 22: 735–747.

Benson, S. R., T. Eguchi, D. G. Foley, K. A. Forney, H. Bailey, C. Hitipeuw, B. P. Samber, et al. 2011. Large-scale movements and high-use areas of western Pacific leatherback turtles, *Dermochelys coriacea*. Ecosphere 2(7): art84. doi:10.1890/ES11-00053.1.

Benson, S. R., K. M. Kisokau, L. Ambio, V. Rei, P. H. Dutton, and D. Parker. 2007. Beach use, internesting movement, and migration of leatherback turtles, *Dermochelys coriacea*, nesting on the north coast of Papua New Guinea. Chelonian Conservation and Biology 6: 17–14.

Bickham, J. W., and J. L. Carr. 1983. Taxonomy and phylogeny of the higher categories of cryptodiran turtles based on a cladistic analysis of chromosomal data. Copeia 1984: 918–932.

Binckley, C. A., J. R. Spotila, K. S. Wilson, and F. V. Paladino. 1998. Sex determination and sex ratios of Pacific leatherback turtles, *Dermochelys coriacea*. Copeia 1998: 291–300.

Bowen B. W., A. M. Clark, F. A. Abreu-Grobois, A. Chaves, H. A. Reichert, and R. J. Ferl. 1998. Global phylogeography of the ridley sea turtles (*Lepidochelys spp*) as inferred from mitochondrial DNA sequences. Genetica 101: 179–189.

Bowen, B. W., and S. A. Karl. 1997. Population genetics, phylogeography and molecular evolution. In: P. L. Lutz and J. A. Musick (eds.), The biology of sea turtles. CRC Press, Boca Raton, FL, USA, pp. 29–50.

Bowen, B. W., and S. A. Karl. 2007. Population genetics and phylogeography of sea turtles. Molecular Ecology 16: 4886–4907.

Bowen, B. W., W. S. Nelson, and J. C. Avise. 1993. A molecular phylogeny for marine turtles: trait mapping, rate assessment and conservation relevance. Proceedings of the National Academy of Science United States of America 90: 5574–5577.

Chan, E. H., and H. C. Liew. 1996. Decline of the leatherback population in Terengganu, Malaysia, 1956–1995. Chelonian Conservation and Biology 2: 196–203.

Charlesworth, B. 2009. Effective population size and patterns of molecular evolution and variation. Nature Reviews Genetics 10:195–205.

Chen, B.-Y., and S.-H. Mao. 1981. Hemoglobin fingerprint correspondence and relationships of turtles. Comparative Biochemistry and Physiology B 68: 497–503.

Chen, B.-Y., S.-H. Mao, and Y.-H. Ling. 1980. Evolutionary relationships of turtles suggested by immunological cross-reactivity of albumins. Comparative Biochemistry and Physiology B 66: 421–425.

Crim, J. L., L. D. Spotila, J. R. Spotila, M. O'Connor, R. Reina, C. J. Williams, and F. V. Paladino. 2002. The leatherback turtle, *Dermochelys coriacea*, exhibits both polyandry and polygyny. Molecular Ecology 11: 2097–2106.

Curtis, C., C. J. Williams, and J. R. Spotila. 2000. Mating system of Caribbean leatherback turtles as indicated by analysis of microsatellite DNA from hatchlings and adult females. Abstract. In: F. A. Abreu-Grobois, R. Briseño-Dueñas, R. Márquez, and L. Sarti (compilers), Proceedings of the Eighteenth International Symposium on Sea Turtle Biology and Conservation. Technical Memorandum NMFS-SEFSC-436. NOAA, Miami, FL, USA, p. 155.

Donoso, M., and P. H. Dutton. 2006. Distribution and stock origin of sea turtles caught incidentally in the Chilean longline fishery for swordfish, 2001–2004. Abstract. In: M. Frick, A. Panagopoulou, F. Rees, and K. Williams (compilers), Abstracts of the Twenty-sixth International Symposium on Sea Turtle Biology and Conservation. International Sea Turtle Society, Athens, Greece, pp. 242–243.

Duchene, S., A. Frey, A. Alfaro-Núñez, P. H. Dutton, M. Thomas, P. Gilbert, and P. A. Morin. 2012. Marine turtle mitogenome phylogenetics and evolution. Molecular Phylogenetics and Evolution 65: 241–250.

Dutton, D. L., P. H. Dutton, M. Chaloupka, and R. H. Boulon. 2005. Increase of a Caribbean leatherback turtle *Dermochelys coriacea* nesting population linked to long-term nest protection. Biological Conservation 126: 186–194.

Dutton, P. H. 1995. Molecular evolution of the sea turtles with special reference to the leatherback, *Dermochelys coriacea*. PhD diss., Texas A&M University, College Station, TX, USA.

Dutton, P. H. 2006a. Building our knowledge of the leatherback stock structure. State of the World's Sea Turtles Report 1: 10–11.

Dutton, P. H. 2006b. Multiple foraging strategies in leatherbacks: A hedge against catastrophe? Abstract. In: M. Frick, A. Panagopoulou, F. Rees, and K. Williams (compilers), Abstracts of the Twenty-sixth International Symposium on Sea Turtle Biology and Conservation. International Sea Turtle Society, Athens, Greece, p. 189.

Dutton, P. H., S. R. Benson, and S. A. Eckert. 2006. Identifying origins of leatherback turtles from Pacific foraging grounds off central California, USA. Abstract. In: N. J. Pilcher (compiler), Proceedings of the Twenty-third International Symposium on Sea Turtle Biology and Conservation. NOAA Technical Memorandum NMFS-SEFSC-536. NOAA, Miami, FL, USA, p. 228.

Dutton, P. H., E. Bixby, and S. K. Davis. 2000. Tendency towards single paternity in leatherbacks detected with microsatellites. Abstract. In: F. A. Abreu-Grobois, R. Briseno-Duenas, R. Marquez-Millian, and L. Sarti-Martínez (compilers), Proceedings of the Eighteenth International Symposium on Sea Turtle Biology and Conservation. NOAA Technical Memorandum NMFS-SEFSC-436. NOAA, MIAMI, FL, USA, p 39.

Dutton, P. H., B. W. Bowen, D. W. Owens, A. Barragán, and S. Davis. 1999. Global phylogeography of the leatherback turtle (*Dermochelys coriacea*). Journal of Zoology 248: 397–409.

Dutton, P. H., S. K. Davis, T. Guerra, and D. Owens. 1996. Molecular phylogeny for marine turtles based on sequences of the ND4-Leucine tRNA and control region of mitochondrial DNA. Molecular Phylogenetics and Evolution 5: 511–521.

Dutton, P. H., A. Frey, R. A. LeRoux, and G. Balazs. 2000. Molecular ecology of leatherback turtles in the Pacific. In: N. Pilcher and G. Ismail (eds.), Sea turtles of the Indo-Pacific: Research, management and conservation. ASEAN Academic Press, London, UK, pp. 248–253.

Dutton, P. H., C. Hitipeuw, M. Zein, S. R. Benson, G. Petro, J. Pita, V. Rei, et al. 2007. Status and genetic structure of nesting populations of leatherback turtles (*Dermochelys coriacea*) in the western Pacific. Chelonian Conservation and Biology 6: 47–53.

Dutton, P. H., E. L. LaCasella, J. Alfaro-Shigueto, and M. Donoso. 2013. Stock origin of leatherbacks (*Dermochelys coriacea*) foraging in the southeastern Pacific. Abstract. In: J. Blumenthal, A. Panagopoulou, and A. F. Rees (compilers), Proceedings of the Thirtieth International Symposium on Sea Turtle Biology and Conservation. NOAA Technical Memorandum NMFS-SEFSC-640. NOAA, Miami, FL, USA, p. 91.

Dutton, P. H., S. E. Roden, K. R. Stewart, E. LaCasella, M. Tiwari, A. Formia, J. C. Thomé, et al. 2013. Population structure of leatherback turtles (*Dermochelys coriacea*) in the Atlantic revealed using mtDNA and microsatellite markers. Conservation Genetics 14: 625–636.

Dutton, P. H., and K. R. Stewart. 2013. A method for sampling hatchling sea turtles for the development of a genetic tag. Marine Turtle Newsletter 138: 3–7.

FitzSimmons, N. N. 1998. Single paternity of clutches and sperm storage in the promiscuous green turtle (*Chelonia mydas*). Molecular Ecology 7: 575–584

Frey, A., P. H. Dutton, and P. Morin. 2012. Whole mitogenomic sequences for further resolution of ubiquitous dloop haplotypes in Pacific green turtles. Abstract. In: T. T. Jones and B. P. Wallace (compilers), Proceedings of the Thirty-first International Symposium on Sea Turtle Biology and Conservation. NOAA Technical Memorandum NMFS-SEFSC-631. NOAA, MIAMI, FL, USA, p. 27.

Friar, W. 1982. Serum electrophoresis and sea turtle classification. Comparative Biochemistry and Physiology B 72: 1–5.

Gaffney, E. S., and P. A. Meylan. 1988. A phylogeny of marine turtles. In: M. J. Benton (ed.), The phylogeny and classification of the tetrapods, Vol. 1. Clarendon, Oxford, UK, pp. 157–219.

Hawkes, L. A., A. C. Broderick, M. H. Godfrey, and B. J. Godley. 2007. Investigating the potential impacts of climate change on a marine turtle population. Global Change Biology 13: 923–932.

Hays, G. C., S. Fossette, K. A. Katselidis, G. Schofield, and M. B. Gravenor. 2010. Breeding periodicity for male sea turtles, operational sex ratios, and implications in the face of climate change. Conservation Biology 24: 1636–1643.

Hitipeuw, C., P. H. Dutton, S. R. Benson, J. Thebu, and J. Bakarbessy. 2007. Population status and internesting movement of leatherback turtles, *Dermochelys coriacea*, nesting on the northwest coast of Papua, Indonesia. Chelonian Conservation and Biology 6: 128–36.

Hoekert, W.E.J., H. Neuféglise, A. D. Schouten, and S.B.J. Menken. 2002. Multiple paternity and female-biased mutation at a microsatellite locus in the olive ridley sea turtle (*Lepidochelys olivacea*). Journal of Heredity 89: 107–113.

Ireland, J. S., A. C. Broderick, F. Glen, B. J. Godley, G. C. Hays, P.L.M. Lee, and D.O.F. Skibinski. 2003. Multiple paternity assessed using micro-satellite markers, in green turtle *Chelonia mydas* (Linnaeus, 1758) of Ascension Island, South Atlantic. Journal of Experimental Marine Biology and Ecology 291: 149–160.

Jensen, M. P., F. A. Abreu-Grobois, J. Frydenberg, and V. Loeschcke. 2006. Microsatellites provide insight into contrasting mating patterns in arribada vs. non-arribada olive ridley sea turtle rookeries. Molecular Ecology 15: 2567–2575

Jensen, M. P., N. N. FitzSimmons, and P. H. Dutton. 2013. Molecular genetics of sea turtles. In: J. Musick, K. Lohman, and J. Wyneken (eds.), Biology of the sea turtles, Vol. 3. CRC Press, Boca Raton, FL, USA, pp. 135–154.

Jones, T. T., M. D. Hastings, B. L. Bostrom, D. Pauly, and D. R. Jones. 2011. Growth of captive leatherback turtles, *Dermochelys coriacea*, with inferences on growth in the wild: implications for population decline and recovery. Journal of Experimental Marine Biology and Ecology 399: 84–92.

Kichler, K., M. T. Holder, S. K. Davis, M. R. Marquez, and D. W. Owens. 1999. Detection of multiple paternity in Kemp's ridley sea turtle with limited sampling. Molecular Ecology 8: 819–830.

Kliman, R., B. Sheehy, and J. Schultz. 2008. Genetic drift and effective population size. Nature Education 1: 3.

LeRoux, R. A., G. H. Balaz, and P. H. Dutton. 2007. Genetic composition of sea turtles caught in the Hawaii-based longline fishery using mtDNA genetic analysis. Abstract. In: R. B. Mast, B. J. Hutchinson, and A. H. Hutchinson (compilers), Proceedings of the Twenty-fourth International Symposium on Sea Turtle Biology and Conservation. NOAA Technical Memorandum NMFS-SEFSC-567. NOAA, MIAMI, FL, USA, p. 136.

Liew, H.-C. 2011. Tragedy of the Malaysian leatherback population: what went wrong. In: P. H. Dutton, D. Squires, and A. Mahfuzuddin (eds.), Conservation and sustainable management of sea turtles in the Pacific Ocean. University of Hawaii Press, Oahu, HI, USA, pp. 97–107.

Molfetti É., S. Torres Vilaça, J.-Y. Georges, V. Plot, E. Delcroix, R. Le Scao, A. Lavergne, et al. 2013. Recent demographic history and present fine-scale structure in the Northwest Atlantic leatherback (*Dermochelys coriacea*) turtle population. PLoS ONE 8(3): e58061. doi:10.1371/journal.pone.0058061.

Morin, P. 2012. Applications of "next generation" sequencing and SNP genotyping for population genetics and phylogeography studies. Abstract. In: T. T. Jones and B. P. Wallace (compilers), Proceedings of the Thirty-first International Symposium on Sea Turtle Biology and Conservation. NOAA Technical Memorandum NMFS-SEFSC-631. NOAA, MIAMI, FL, USA, p. 23.

Moritz, C. 1994. Defining "evolutionarily significant units" for conservation. Trends in Ecology and Evolution 9: 373–375.

Mrosovsky, N., P. H. Dutton, and C. P. Whitmore. 1984. Sex ratios of two species of sea turtle nesting in Suriname. Canadian Journal of Zoology 62: 2227–2239.

Naro-Maciel, E., M. Le, N. N. FitzSimmons, and G. Amato. 2008. Evolutionary relationships of marine turtles: a molecular phylogeny based on nuclear and mitochondrial genes. Molecular Phylogenetics and Evolution 49: 659–662.

Pearse, D., and J. C. Avise. 2001. Turtle mating systems: behavior, sperm storage, and genetic paternity. Journal of Heredity 92: 206–211.

Pritchard, P.C.H. 1976. Post-nesting movements of marine turtles (Cheloniidae and Dermochelyidae) tagged in the Guianas. Copeia 1976: 749–754.

Pritchard, P.C.H. 1997. Evolution, phylogeny and current status. In: P. L. Lutz and J. A. Musick (eds.), The biology of sea turtles. CRC Press, Boca Raton, FL, USA, pp. 1–28.

Pulliam, H. R. 1988. Sources, sinks and population regulation. American Naturalist 132: 652–661.

Rhodin, A.G.J. 1985. Comparative chondro-osseous development and growth in marine turtles. Copeia 1985: 752–771.

Rhodin, A.G.J., J. A. Ogden, and G. J. Conlogue. 1981. Chondro-osseous morphology of *Dermochelys coriacea*, a marine reptile with mammalian skeletal features. Nature 290: 244–246.

Roden, S. E., and P. H. Dutton. 2011. Isolation and characterization of 14 polymorphic microsatellite loci in the leatherback turtle (*Dermochelys coriacea*) and cross-species amplification. Conservation Genetics Resources 3: 49–52.

Saba, V. S., P. Santidrián-Tomillo, R. D. Reina, J. R. Spotila, J. A. Musick, D. A. Evans, and F. V. Paladino. 2007. The effect of the El Niño Southern Oscillation on the reproductive frequency of eastern Pacific leatherback turtles. Journal of Applied Ecology 44: 395–404.

Sarti, L., and A. R. Barragán. 2011. Importance of networks for conservation of the Pacific leatherback turtle: the case of "Proyecto Laúd" in Mexico. In: P. H. Dutton, D. Squires, and A. Mahfuzuddin (eds.), Conservation and sustainable management of sea turtles in the Pacific Ocean. University of Hawaii Press, Oahu, HI, USA, pp. 120–131.

Shamblin, B. M., K. A. Bjorndal, A. B. Bolten, et al. 2012. Mitogenomic sequences better resolve stock structure of southern greater Caribbean green turtle rookeries. Molecular Ecology 21: 2330–2340.

Shanker, K., B. C. Choudhury, and R. K. Aggarwal. 2011. Conservation genetics of marine turtles on the mainland coast of India and offshore islands. Final Project Report. Wildlife Institute of India, Dehradun and Centre for Cellular and Molecular Biology, Hyderabad, India.

Shanker, K., J. Rama Devi, B. C. Choudhury, L. Singh, and R. K. Aggarwal. 2004. Phylogeography of olive ridley turtles (*Lepidochelys olivacea*) on the east coast of India:

implications for conservation theory. Molecular Ecology 13: 1899–1909.

Spotila, J. R., A. E. Dunham, A. J. Leslie, A. C. Steyermark, P. T. Plotkin, and F. V. Paladino. 1996. Worldwide population decline of *Dermochelys coriacea*: Are leatherbacks going extinct? Chelonian Conservation Biology 2: 209–222.

Spotila, J. R., R. D. Reina, A. C. Steyermark, P. T. Plotkin, and F. V. Paladino. 2000. Pacific leatherback turtles face extinction. Nature 405: 529–530.

Stewart, K. R., and P. H. Dutton. 2011. Paternal genotype reconstruction reveals multiple paternity and sex ratios in a breeding population of leatherback turtles (*Dermochelys coriacea*). Conservation Genetics 12: 1101–1113.

Stewart, K. R., and P. H. Dutton. 2012. Sea turtle CSI: it's all in the genes. State of the World's Sea Turtles Report 7: 12–13.

Stewart, K. R., and P. H. Dutton. 2014. Breeding sex ratios in adult leatherback turtles (*Dermochelys coriacea*) may compensate for female-biased hatchling sex ratios. PLoS ONE 9(2): e88138. doi:10.1371/journal.pone.0088138.

Stewart, K. R., S. Roden, M. C. James, and P. H. Dutton. 2013. Assignment tests, telemetry, and tag-recapture data converge to identify natal origins of leatherback turtles, *Dermochelys coriacea*, foraging in Canadian waters. Journal of Animal Ecology 82: 791–803.

Stewart, K., M. Sims, A. Meylan, B. Witherington, B. Brost, and L. Crowder. 2011. Leatherback nests increasing significantly in Florida, USA; trends assessed over 30 years using multi-level modeling. Ecological Applications 21: 263–273.

Tapilatu, R. F., and M. Tiwari. 2007. Leatherback turtle, *Dermochelys coriacea*, hatching success at Jamursba-Medi and Wermon Beaches in Papua, Indonesia. Chelonian Conservation and Biology 6: 154–158.

Taylor B. L., K. Martien, and P. Morin. 2010. Identifying units to conserve using genetic data. In: I. L. Boyd, W. D. Bowen, and S. J. Iverson (eds.), Marine mammal ecology and conservation—A handbook of techniques. Oxford University Press, Oxford, UK, pp. 306–344.

Theissinger, K., N. N. FitzSimmons, C. J. Limpus, C. J. Parmenter, and A. D. Phillott. 2009. Mating system, multiple paternity and effective population size in the endemic flatback turtle (*Natator depressus*) in Australia. Conservation Genetics 10: 329–346.

Turtle Expert Working Group (TEWG). 2007. An assessment of the leatherback turtle population in the Atlantic Ocean. NOAA Technical Memorandum NMFS-SEFSC-555. NOAA, MIAMI, FL, USA.

Vargas, S. M., F.C.F. Araujo, D. S. Monteiro, S. C. Estima, A. P. Almeida, L. S. Soares, and F. R. Santos. 2008. Genetic diversity and origin of leatherback turtles (*Dermochelys coriacea*) from the Brazilian coast. Journal of Heredity 99: 215–220.

Wallace B. P., A. D. DiMatteo, B. J. Hurley, E. M. Finkbeiner, A. B. Bolten, M. Y. Chaloupka, B. J. Hutchinson, et al. 2010. Regional management units for marine turtles: a novel framework for prioritizing conservation and research across multiple scales. PLoS ONE 5(12): e15465. doi:10.1371/journal.pone.0015465.

Wallace, B. P., M. Tiwari, and M. Girondot. 2013. "*Dermochelys coriacea*." In: IUCN 2013. IUCN Red List of Threatened Species. Version 2013.2. www.iucnredlist.org. Accessed 27 February 2014.

Wood, R. C., J. Johnson-Gove, E. S. Gaffney, and K. F. Maley. 1996. Evolution and phylogeny of leatherback turtles (Dermochelyidae), with description of new fossil taxa. Chelonian Conservation and Biology 2: 266–286.

Wright, S. 1938. Size of a population and breeding structure in relation to evolution. Science 87: 430–431.

Zbinden, J. A., C. R. Largaider, F. Leippert, D. Margaritoulis, and R. Arlettaz. 2007. High frequency of multiple paternity in the largest rookery of Mediterranean loggerhead sea turtles. Molecular Ecology 16: 3703–3711.

Zug, G. R., and J. F. Parham. 1996. Age and growth in leatherback turtles, *Dermochelys coriacea*: A skeletochronological analysis. Chelonian Conservation and Biology 2: 244–249.

3

Diving Behavior and Physiology of the Leatherback Turtle

NATHAN J. ROBINSON AND
FRANK V. PALADINO

Archie Carr was particularly insightful when he titled his classic book on the natural history of sea turtles as *So excellent a fishe* after a reference in a 1620 law protecting sea turtles in Bermuda (Carr 1967). Able to remain submerged for prolonged periods of time, sea turtles spend the majority of their lives deep below the water's surface. Here they forage, mate, and even sleep, only ascending to the surface to breathe. Perhaps it is even a misnomer to refer to sea turtles as "divers"; a more apt term may be "surfacers" (Kramer 1987).

Of all the sea turtles, the leatherback turtle *Dermochelys coriacea* is the largest and most uniquely adapted to a life plumbing the depths of the world's oceans. Leatherbacks spend most of their time in the top 300 m of the water column in dives of less than 20–30 minutes (James et al. 2005; Houghton et al. 2008; Fossette, Gleiss, et al. 2010; Shillinger et al. 2011). However, they can dive to depths of over 1 km (Doyle et al. 2008) and remain submerged for almost 90 minutes between breaths (López-Mendilaharsu et al. 2009). They are similar in many ways to marine mammals (Paladino et al. 1996; Lutcavage and Lutz 1997). To achieve such feats, leatherback turtles have an array of adaptations that enable them to withstand extreme changes in temperature and pressure as well as survive for prolonged periods of time on a single breath.

In this chapter we summarize current knowledge of leatherback turtle diving behavior and physiology. We begin by outlining use of bio-logging devices for studying at-sea behavior. Subsequently, we describe diving behavior during different life stages. Last, we overview the physiological adaptations to deep prolonged dives and discuss how these adaptations compare to those in other air-breathing, diving vertebrates.

Use of Bio-loggers for Studying Diving Behavior

In the 1960s, the study of free-diving behavior in marine animals was revolutionized by the invention of small, waterproof bio-loggers (Naito 2004). Able to record depth every few seconds, these devices provided a uniquely quantitative view of how marine species travel through the water column.

Eckert et al. (1986) were the first to attach bio-loggers to leatherback turtles. They focused on inter-nesting females because they were predictably reencountered in subsequent nesting events, making it possible to recover the device along with the stored data. As many modern devices still require manual recovery, most studies on diving behavior are similarly conducted on inter-nesting females (Hochscheid 2014). Yet technological advances now allow for data recovery via satellite, thus enabling the study of sea turtle diving behavior as they travel far from nesting locations (Hays et al. 2004). In addition, other technological innovations have augmented the types of sensors that can be fitted to bio-loggers. Bio-loggers are no longer limited to coarse-resolution measures of depth. They are now able to record an ever-widening array of variables, including temperature, blood respiratory gas concentrations, movements of specific body parts (beak or flippers), heart rates, and even 3D accelerometry.

As new bio-logging technologies become available, further insights are gained into the underwater life of leatherback turtles. However, before deploying any bio-logger, consideration must be given to the impacts such devices may have on the study animal. Depending on the size and shape of the bio-logger and the design of its attachment mechanism, it may impede the animal's natural movements as well as cause immediate or sustained injury (Troëng et al. 2006). More subtly, external bio-loggers can disturb the hydrodynamic patterns of water flowing over an animal's body and, in turn, alter its drag coefficient (Jones et al. 2013).

There are obvious ethical reasons for minimizing the impacts of bio-loggers, but there are also implications for the utility of the data collected. If the attachment of a bio-logger significantly alters an animal's behavior, then the data are not comparable to observations of individuals unencumbered by such devices. Minimizing the long-term impacts of bio-loggers is therefore necessary if we are to use these devices to understand the natural behavior of free-diving animals.

Leatherback Turtle Diving Behavior

Juvenile Stages

Our knowledge of the diving behavior of leatherback turtles during their early life stages remains limited because leatherbacks are difficult to rear in captivity (Jones et al. 2000) and juveniles are rarely encountered in the wild (Eckert 2002a). Yet by relating the findings from the few available studies to what we know about how body size affects the diving capacity of air-breathing vertebrates, we can begin to predict ontogenetic patterns in diving behavior for leatherback turtles during their first few years of life.

By the time leatherback hatchlings emerge from the nest, they already exhibit the respiratory characteristics of actively diving adults: deep breaths, punctuated by long pauses (Price et al. 2007). Although hatchlings are able to dive, they can descend to only a few meters and for short durations (< 3 minutes; Wyneken and Salmon 1992). These small dives probably help hatchlings avoid surface waves that would push them back toward the nesting beach. Hatchlings that survived their first day soon developed diel patterns in swimming activity and were less active at night (Wyneken and Salmon 1992). Although no studies have separated diel patterns in leatherback hatchling swimming activity from diving behavior, it is expected that fewer dives also occur at night and more time is spent resting at the surface.

Three to five days after hatching, leatherback turtles began foraging and growing (Jones et al. 2011). At the same time, their diving abilities increase. Larger individuals tend to have lower metabolic rates and greater oxygen stores per unit body mass, enabling them to hold their breath for longer periods (Schreer and Kovacs 1997; Hochscheid et al. 2007). Larger individuals also are often stronger swimmers and able to descend to depth more quickly. After being reared in captivity for a period of 10 weeks, leatherback turtle juveniles (straight carapace length: ~72.5 mm ± 5.0 SD) were able to remain submerged for almost 6 minutes at a time, reaching depths of 17.1 m (Salmon et al. 2004).

Although leatherback hatchlings possess some diving ability, they lack some of the necessary physiological adaptations seen in adults for deep diving. Specifically, hatchlings do not have a trachea that can readily collapse or reinflate as turtles descend or ascend, respectively (Murphy et al. 2012). This adaptation is still not fully present in juvenile leatherback turtles (curved carapace length: 70.9–87.3 cm) and mid-sized individuals are therefore probably incapable of reaching the deepest depths recorded for this species (Davenport et al. 2009). Another limiting factor might be the capacity for juvenile leatherback turtles to maintain elevated body temperatures relative to their environment. Smaller leatherback turtles are generally found at warmer, equatorial latitudes than their larger counterparts (Witt et al. 2007), probably because larger individuals are better suited to retain body heat (Paladino et al. 1990). The lower thermoregulatory capacity of juvenile leatherback turtles could therefore constrain them from diving into particularly deep, and thus colder, waters.

By the time leatherback turtles reach subadult status, around 120 cm curved carapace length, their diving pat-

Table 3.1. Dive parameters for foraging or transiting leatherback turtles. In those studies that separated transiting (rapid movements in a relatively straight direction) from foraging behaviors (slower movements with high turn angles), we provide dive parameters for each behavioral phase individually. For those studies that did not separate transiting from foraging behavior, dive parameters for both behavioral phases are combined.

				Dive depth (m)		Dive duration (min)		Dive frequency (dives hr $^{-1}$)	
Behavior	Location	Number of turtles	Dive criteria	Mean	Maximum	Mean	Maximum	Mean	Source
Transiting and foraging	North Atlantic (oceanic)	2	>10 m		1280		68.5		Doyle et al. 2008
Transiting	North Atlantic (oceanic)	9	>10 m	95.0		25.3			Fossette, Girard, et al. 2010
Foraging	North Atlantic (neritic)	8	>10 m	39.5		17.0			Fossette, Girard, et al. 2010
Foraging	North Atlantic (oceanic)	9	>10 m	59.0		23.8			Fossette, Girard, et al. 2010
Transiting	North Atlantic (oceanic)	2	>10 m	76.7	626	28.9	54.0		Hays et al. 2004
Transiting	North Atlantic (Caribbean)	2	>10 m	51.7		11.0			Hays et al. 2004
Transiting and foraging	North Atlantic (oceanic)	1	N/A		1186		86.5		López-Mendilaharsu et al. 2009
Transisting and foraging	Western Indian Ocean and South Atlantic (oceanic)	4	>2 m for over 30 sec		940	8.6	82.3	10.3	Sale et al. 2006
Transiting	Eastern Pacific (oceanic)	46	>10 m	45.0		23.6			Shillinger et al. 2011
Foraging	Eastern Pacific (oceanic)	46	>10 m	56.7		26.4			Shillinger et al. 2011

terns are largely similar to those observed for sexually mature individuals (Standora et al. 1984; James et al. 2006). In addition, leatherback turtle growth rates diminish rapidly upon reaching sexual maturity (Price et al. 2004), so further improvements to diving abilities after reaching reproductive maturity are unlikely.

Migrating and Foraging

Foraging migrations of adult leatherback turtles take them across entire ocean basins in search of aggregations of gelatinous zooplankton (Shillinger et al. 2010; Witt et al. 2011). While migrating from their nesting beaches to foraging areas, leatherbacks dive to mean depths between 40.0 and 95.0 m with mean durations between 11.0 to 28.9 minutes (table 3.1; Hays et al. 2004; Fossette, Girard, et al. 2010; Shillinger et al. 2011), although individuals occasionally make extraordinarily long and deep dives (Houghton et al. 2008). The deepest dive ever recorded for a leatherback turtle was made off the coast of Cape Verde by a male that dove to a depth of 1,280 m (Doyle et al. 2008). The longest dive was made off the coast of Brazil by a female that remained submerged for 86 min (López-Mendilaharsu et al. 2009). Though impressive, these dives reflect exceptional events, whereas less than 1% of all dives actually reach depths below 300 m, or last longer than 60 minutes (Shillinger et al. 2011). It has been hypothesized that these deep, infrequent dives serve to assess the water column for the distribution and abundance of diel, vertically migrating zooplankton (Houghton et al. 2008). If high densities of food are encountered, the turtle may remain in temporary residence at that location until resources are depleted.

On arrival at a foraging patch, leatherback turtles shift their diving patterns depending on the distribution and abundance of their prey. At particularly high latitudes, for example, dives were generally shallow and diel activity patterns were not observed (James et al. 2006). It has been argued that this is because leatherback turtles at high latitudes are able to feed on ample amounts of shallow-living gelatinous zooplankton during both day and night (James et al. 2006). In contrast, when turtles are at low-productivity foraging grounds, dives were

Table 3.2. Dive parameters for inter-nesting leatherback turtles.

				Dive depth (m)		Dive duration (min)		Dive frequency (dives hr^{-1})	
Behavior	Location	Number of turtles	Dive criteria	Mean	Maximum	Mean	Maximum	Mean	Source
Inter-nesting	North Atlantic (St. Croix, US Virgin Islands)	8	>3 m	90.0		12.4		3.6	Casey et al. 2010
Inter-nesting	North Atlantic (St. Croix, US Virgin Islands)	2	Not specified	82.8	475	12.1	37.4	3.0	Eckert et al. 1986
Inter-nesting	North Atlantic (St. Croix, US Virgin Islands)	6	Not specified	61.6	>1,000	9.9	37.0		Eckert et al. 1989
Inter-nesting	North Atlantic (French Guiana / Suriname)	23	>2 m	9.4	84	4.4	28.2	10.7	Fossette et al. 2007
Inter-nesting	North Atlantic (French Guiana / Suriname)	4	>3 m	11.8		6.6			Fossette et al. 2008
Inter-nesting	North Atlantic (Grenada)	11	>10 m	54.7		11.1			Myers and Hays 2006
Inter-nesting	East Pacific (Costa Rica)	46	>10 m	23.4	314	11.6	51.7		Shillinger et al. 2010
Inter-nesting	East Pacific (Costa Rica)	6	>4 m	19.4	124	7.4	67.3	4.6	Southwood et al. 1999
Inter-nesting	East Pacific (Costa Rica)	18	>3 m	22.6	200	7.8	44.9	5.4	Wallace et al. 2005

consistently deeper and longer during the day than at night (Shillinger et al. 2011). Many prey items for leatherback turtles show similar diel movement patterns and ascend up the water column at sunset and descend at sunrise (Hays 2003). If sufficient food is not available at the surface during the day, leatherback turtles may dive deeper in search of prey. At nighttime, individuals are either resting at shallow depths or foraging on prey that have risen up the water column.

The majority of data on leatherback turtle diving behavior while migrating or foraging are from females. However, the few studies available for males suggest that diving patterns are similar between sexes (James et al. 2005).

Inter-nesting

Gravid leatherback turtles nest every 8 to 14 days (Reina et al. 2002). During this inter-nesting interval, females swim almost continuously (Eckert 2002b) and can travel over 800 km (Witt et al. 2008). At locations such as St. Croix, US Virgin Islands, and Grenada, where the continental shelf is close to the nesting beach and drops off in a steep decline to great depths, inter-nesting turtles move rapidly into deep waters just offshore (Eckert et al. 1986; Georges et al. 2007). At these sites, mean dive depths of about 80 m are commonly recorded for inter-nesting turtles (table 3.2; Eckert et al. 1986; Myers and Hays 2006). In contrast, at locations including Playa Grande, Costa Rica, or French Guiana, inter-nesting habitats are primarily found in shallow waters and mean dive depths are generally closer to approximately 20 m (Fossette et al. 2007; Shillinger et al. 2010).

At both shallow- and deep-water sites, inter-nesting dives generally last between 5 and 30 minutes (Fossette et al. 2008; Shillinger et al. 2010). While submerged, turtles exhibit brief periods of low activity lasting around 5 to 10 minutes (Eckert 2002b; Reina et al. 2005). This behavior, which is indicative of resting, is more common in the first half of an inter-nesting interval (Southwood et al. 2005). Dives are also longer and deeper during the first half of the inter-nesting interval than they are in the few days prior to nesting (Southwood et al. 1999; Myers and Hays 2006).

The combination of deeper dives and periods of reduced activity early in the inter-nesting interval may be indicative of behavioral thermoregulation. Sea surface temperatures adjacent to leatherback nesting beaches are often about 27°C, yet water temperatures at depth can be many degrees cooler even at shallow water sites (Fossette et al. 2009; Shillinger et al. 2010). As the body temperatures of leatherback turtles are influenced by ambient water temperature (Southwood et al. 2005), the deeper and longer dives recorded at the beginning of the inter-nesting interval could help females shed excess heat generated and stored during the nesting process (Wallace et al. 2005). Furthermore, this could explain why leatherback turtles ingest large quantities of cooling seawater immediately after nesting, although this would also help to recover water lost in egg production and while the animal was respiring on the beach (Southwood et al. 2005; Casey et al. 2010).

Inter-nesting leatherback turtles generally exhibit diel movement patterns similar to those of foraging individuals, with longer and deeper dives occurring during the day (Myers and Hays 2006; Fossette et al. 2007; Shillinger et al. 2010). These data have been used to argue that leatherback turtles forage during the nesting period (Eckert et al. 1986). Other data that support this hypothesis have come from studies tracking mass changes in individuals over the nesting season (Eckert et al. 1989) and the use of beak movement sensors (Myers and Hays 2006; Fossette et al. 2008). Yet a similar number of studies also suggest that leatherback turtles do not feed during the breeding season; these studies include biochemical assays (Plot et al. 2013), the use of video cameras mounted on the carapace of leatherbacks (Reina et al. 2005), and metabolic rate experiments (Wallace et al. 2005). Such conflicting results indicate that some level of behavioral plasticity may exist among different leatherback turtle populations, with inter-nesting individuals foraging at some locations where food is locally abundant and not where food availability is lower. Such behavioral plasticity is also observed in green turtles *Chelonia mydas* (Hays et al. 2002).

Diving Physiology

The respiratory system of air-breathing species is essentially a closed system while diving. As a result, the capacity for leatherbacks to withstand prolonged periods of apnea (the suspension of external breathing) is dependent on three major factors: the quantity of oxygen that can be stored in the respiratory system and other tissues, the rate at which these oxygen stores are depleted, and the ability to withstand hypoxia and the accumulation of respiratory waste products, such as lactic acid. A separate adaptation for deep diving is a tolerance to crushing hydrostatic pressures and near-freezing temperatures. Here, we will cover all these adaptations present in leatherback turtles that enable them to conduct prolonged and deep dives, with the exception of cold tolerance; this last is covered elsewhere in this book (chapter 13).

Oxygen Stores

Air-breathing vertebrates store oxygen in their respiratory system, blood, muscle tissue, and organs such as the kidneys, liver, and spleen. This oxygen is (1) attached to respiratory pigments like hemoglobin in blood cells and myoglobin in muscles; (2) dissolved in body fluids like plasma, cerebral spinal fluid, or cytoplasm; and (3) stored as a gas in the respiratory passageways and lungs.

Leatherback turtles have the largest oxygen storage capacity of all the sea turtle species at 27 O_2 kg^{-1} (Lutcavage et al. 1992). They have greater amounts of muscle myoglobin (Lutz and Hochachka 1993), higher levels of hemoglobin, and a higher hematocrit than other sea turtles (Lutcavage et al. 1992). The largest proportion of oxygen is stored in the blood (50.4%), closely followed by the lungs (44.5%), with the smallest percentage (5.1%) in the muscles and other tissues (Lutcavage et al. 1992). Although leatherback turtles have proportionately smaller lungs than other sea turtle species (Lutcavage and Lutz 1997), the proportion of oxygen stored in the lungs of leatherback turtles greatly exceeds that of most marine mammals (fig. 3.1). Leatherback sea turtles store a greater proportion of oxygen in myoglobin than other sea turtles, but this still only constitutes a relatively minor oxygen store, especially when compared to most marine mammals.

LUNG OXYGEN STORES

Leatherback turtles inhale prior to diving (Reina et al. 2005) and may even regulate the amount of air inhaled, depending on the intended dive duration (Fossette, Gleiss, et al. 2010). In this manner, leatherback turtles could control the depth at which they reach neutral buoyancy, potentially reducing the energetic costs of swimming to, or maintaining position at, a particular depth. An additional benefit of inhaling before a dive is that the lung can function as an oxygen store; however, this would only be the case during shallow dives. Berkson (1967) determined that the lungs of green turtles collapse at depths between 90 and 190 m, and that the lungs of leatherbacks may collapse at similar depths. A well-developed sphincter valve in the pulmonary artery

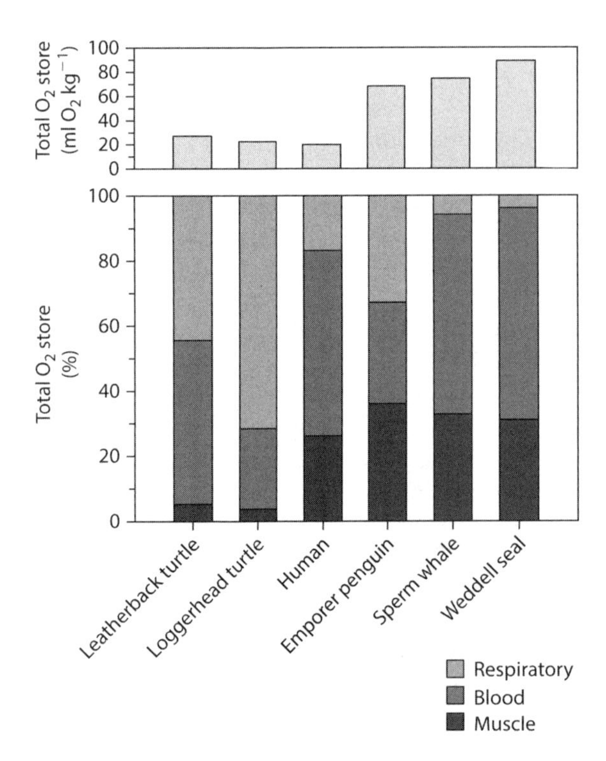

Fig. 3.1. *Top:* Total O_2 stores in a variety of air-breathing organisms. *Bottom:* Distribution of total O_2 stores in a variety of organisms. Data are from Lutcavage et al. (1990) for leatherback turtles; Lutz and Bentley (1985) for loggerhead turtles; Kooyman (1989) for humans; and Ponganis et al. (2011) for emperor penguins, sperm whales, and Weddell seals.

suggests intracardiac shunting of blood away from collapsed lungs during a dive (Wyneken and Rhodin 1999). The trachea of adult leatherbacks would collapse easily in the upper water column (Davenport et al. 2009). With the lungs collapsed, the oxygen stored within becomes essentially unavailable. On deep dives, animals must consequently rely on oxygen held in hemoglobin and myoglobin stores (Scholander 1940).

HEMOGLOBIN OXYGEN STORES

While diving, leatherback turtles store the majority of their O_2 in their blood (Lutcavage et al. 1992). This high level of storage is enabled by leatherbacks' notably high hemoglobin concentration of approximately 15.6 dL^{-1}, along with a hematocrit of about 39 %, yielding a mean oxygen carrying capacity of nearly 21 mL dL^{-1} (Lutcavage et al. 1990). Thus, the oxygen storing capacity of leatherback blood is higher than other sea turtles or diving reptiles, but still lower than most diving mammals.

In comparison to hard-shelled sea turtles, leatherbacks have intermediate blood O_2 binding curves at high PO_2; however, at low PO_2 the shape of the O_2 binding curves differs notably (Lutcavage et al. 1990). Hard-shelled sea turtles retain a steep incline in binding curves at low PO_2, while leatherbacks have the more traditional sigmoidal curve (Lutcavage et al. 1990). This means that hard-shelled turtles are able to extract oxygen from the lungs at lower blood PO_2, while leatherback blood is better suited to release O_2 to the tissues from the blood at low blood PO_2. The logical reason for these different O_2 binding curves is that the largest O_2 store for leatherbacks is in the blood, while in other hard-shelled turtles it is in the lungs (Berkson 1966).

MYOGLOBIN OXYGEN STORES

Myoglobin concentrations in leatherback turtles are 4.9 mg g^{-1}, which are the highest recorded among reptiles, but are still an order of magnitude lower than deep-diving marine mammals (Lutcavage et al. 1990). To date, no studies have focused on identifying the structural properties of leatherback myoglobin.

Oxygen Consumption Rates

At 27 ml O_2 kg^{-1}, leatherback turtles have larger total oxygen stores than hard-shelled turtles, yet far lower than those recorded for deep diving mammals. Comparatively, sperm whales *Physeter macrocephalus* and Weddell seals *Leptonychotes weddellii* have total oxygen stores of 74 and 89 ml O_2 kg^{-1}, respectively (table 3.2; Snyder 1983). In fact, the oxygen stores in leatherback turtles are surprisingly similar to those for humans, which are at 24 ml O_2 kg^{-1} (Kooyman 1989). However, the longest dives of leatherback turtles and Weddell seals remain roughly equivalent at 86 and 82 min respectively (Castellini et al. 1991; López-Mendilaharsu et al. 2009). While both leatherback turtles and Weddell seals have numerous adaptations that enable them to remain submerged for extended periods of time, the metabolic rates of leatherbacks are only half that predicted for mammals of equivalent size (Paladino et al. 1990). Thus, the rate at which oxygen is depleted during a dive should be much slower in leatherback turtles than for similarly sized marine mammals, enabling the turtles to stay submerged for equivalent periods of time despite lower total oxygen stores.

Regional Blood Flow

Forced diving in the freshwater turtle *Chrysemys scripta* prompts a reduction in blood flow to the kidneys, stomach, pancreas, liver, and small intestine; this leads to highly suppressed metabolic rates when submerged (Herbert and Jackson 1985; Davies 1989). Leatherback turtles may similarly reduce their diving metabolic rates by reducing blood flow to, and thus oxygen consumption of, organs that are superfluous while diving. When at the surface, circulation would then resume to those organs for which blood flow was previously suppressed. Indeed, heart rates of diving leatherback turtles are lower than those breathing at the surface (Southwood et al. 1999), similar to the reflex bradycardia of other diving animals (Kooyman 1989). During ascent, heart rate increases and is highest at the surface; this ensures rapid and efficient dumping of CO_2 and uptake of O_2 to facilitate continued diving (Southwood et al. 1999; Paladino et al. 1996).

Aerobic Dive Limits and Calculated Aerobic Dive Limits

While diving, an animal's oxygen stores become depleted. This requires the animal to depend increasingly on anaerobic respiration, but anaerobic respiration also leads to the accumulation of lactic acid in the blood and muscles. The time it takes for blood lactate concentration to increase beyond pre-dive levels is termed the aerobic dive limit (ADL). This is a useful parameter for understanding the physiological and energetic constraints of diving (Kooyman et al. 1983). Due to the difficulties of working with free-diving animals, however, the ADL is generally calculated indirectly by dividing an animal's total oxygen stores by its diving metabolic rate (Kooyman et al. 1980; Kooyman et al. 1983). The resulting value is termed the calculated aerobic dive limit (ADLc). Using this method, Wallace et al. (2005) estimated the ADLc for adult leatherback turtles to be between 11.7 and 44.3 minutes. However, this value is far shorter than the maximum observed dive durations for this species and it is unlikely that an animal would ever dive to the point at which its oxygen stores are on the verge of exhaustion. Consequently, Wallace et al. (2005) must have underestimated the ADLc. This is probably because their estimate of metabolic rate was generated from leatherbacks over periods of time when they were both diving, where metabolic rates were likely suppressed, and at the surface, where metabolic rates were likely elevated.

It should also be noted that the ADLc is an estimate of the time it takes for an organism to deplete all their useable oxygen stores and this is not the same as the ADL. This is because anaerobic metabolism begins before complete oxygen depletion. Consequently, beyond the ADL there is a significant increase in anaerobic metabolism as well as continuation of aerobic metabolism (Ponganis et al. 2011; Williams et al. 2011). To this extent, an alternate method to determine the ADL indirectly from diving behavior data was proposed by Hays et al. (2004) and then applied to leatherback turtles by Bradshaw et al. (2007). This method uses the asymptotic increase in maximum dive duration with depth, along with information on total oxygen stores, to estimate both the ADL and metabolic rate of a diving leatherback turtle. From this study, Bradshaw et al. (2007) concluded that the metabolic rates of leatherback turtles while diving were lower than previous estimates by both Paladino et al. (1990) and Wallace et al. (2005); the former estimate was based on nesting individuals, while the latter was based on individuals over their entire inter-nesting periods. This provides further support that leatherback turtles may suppress their metabolic rates while diving.

Withstanding Hypoxia

Turtles, like all vertebrates, are aerobic animals that require oxygen to survive. However, turtles also have traits that enable them to maintain function with little or no available oxygen. The mechanisms for anoxia tolerance have been studied extensively in freshwater turtles (Bickler and Buck 2007 and references within) and are divisible into five categories: (1) suppression of metabolic rates, facilitated by controls in regional blood flow; (2) large reserves of fermentable fuels, such as reserves of glycogen found in the liver; (3) mechanisms for buffering or excreting metabolic end products (e.g. utilizing carbonates released from the bone and shell to help buffer the effects of lactic acid build up); (4) antioxidant mechanisms to reduce oxidative stress when oxygen is reintroduced; and (5) production of proteins that aid anoxia survival. These mechanisms observed in freshwater turtles are probably similar to those in sea turtles, though this is still unconfirmed.

Adaptations to Hydrostatic Pressures

As a diving animal descends through the water column it is subjected to increasing hydrostatic pressure at a rate of 1 atmosphere (101.3 kPA) for every 10 m of depth. During descent, gases decrease in volume, in accordance with Boyle's Law, placing a large compressive force on the respiratory system. To equalize these pressure differentials (termed the "thoracic squeeze"), sea turtles' lungs collapse at depths between 90 and 190 m (Berkson 1967). Following collapse of the lungs, the trachea also collapses, shunting most of the air into the laryngeal region as the animal continues to descend (Davenport et al. 2009; Murphy et al. 2012). Also, at this time the tongue of the leatherback turtle functions as an epiglottis, stopping water flooding into the larynx (Fraher et al. 2010).

The carapace of a leatherback turtle can accommodate the thoracic squeeze by expanding or "collapsing" when necessary. The interlocking cartilaginous ostraderm ossicles that form the leatherbacks' carapace and plastron allow for significant flexibility beyond that of the bone and shell scutes of hard-shelled turtles (Rhodin et al. 1980). Indeed, the entire shell of foraging leatherbacks at high latitudes expands to accommodate the increasing bulk of the animal as they become seasonally engorged with high volumes of food (Davenport et al. 2011).

Alongside structural issues, a leatherback turtle faces difficulties concerning the solubility of gases at high pressures. As gases in the lungs are pressurized, they become increasingly soluble and more readily dissolve into the body's tissues. The increased concentration of dissolved nitrogen can lead to nitrogen narcosis. Furthermore, as the organism ascends the nitrogen that was dissolved comes out of solution. If this occurs too quickly gas bubbles can form in the blood or other tissues and this leads to decompression sickness ("the bends"). The detrimental effects are, however, minimized by the collapse of lungs and trachea at depth. In the larynx, the potential for gases to dissolve into the blood stream at depth is lower as its surface is relatively impermeable to gaseous diffusion and it is not as well supplied with blood. In addition, leatherback turtles modify time spent at different depths and tend to ascend relatively slowly from deep dives, reducing the rate at which nitrogen bubbles out of solution (Fossette, Gleiss, et al. 2010). As a result of these physiological and behavioral adaptations, indications of decompression sickness, which is identifiable through osteonecrosis or small holes in the bones, is rarely recorded in all extant sea turtle lineages (Rothschild 1991).

Summary

While much attention has been paid to the epic migrations that leatherback turtles take between their feeding and breeding grounds, their routine ventures into the ocean depths are no less impressive. This species moves through the ocean in a truly three-dimensional manner and, thus, to understand leatherback turtle behavior we must consider both horizontal and vertical movement patterns.

A suite of physiological adaptations enables leatherbacks to conduct some of deepest and longest dives in the animal kingdom. Many of these adaptations are similar to those found in other diving reptiles; however, others are more comparable to those found in distantly related diving mammals. These adaptations reveal the power of evolution to facilitate a small handful of polyphyletic vertebrates to inhabit the deep sea, arguably one of the least hospitable environments on Earth for air-breathing organisms.

ACKNOWLEDGMENTS

We would like to extend our thanks to the editors, J. R. Spotila and P. Santidrián Tomillo, for making this book a reality and inviting us to write this chapter. We are also grateful for support from G. Goldring, L. Gund, the Schrey Distinguished Professorship, the Goldring-Gund Marine Biology Station, and the Leatherback Trust.

LITERATURE CITED

Berkson, H. 1966. Physiological adjustments to prolonged diving in the Pacific green turtle (*Chelonia mydas agassizii*). Comparative Biochemistry and Physiology A 18: 101–119.

Berkson, H. 1967. Physiological adjustments to deep diving in the Pacific green turtle (*Chelonia mydas agassizii*). Comparative Biochemistry and Physiology A 21: 507–524.

Bickler, P. E., and L. T. Buck. 2007. Hypoxia tolerance in reptiles, amphibians, and fishes: life with variable oxygen availability. Annual Review of Physiology 69: 145–170.

Bradshaw, C.J.A., C. R. McMahon, and G. C. Hays. 2007. Behavioral inference of diving metabolic rate in free-ranging leatherback turtles. Physiological and Biochemical Zoology 80: 209–219.

Carr, A. 1967. So excellent a fishe. Natural History Press, Garden City, New York, NY, USA.

Casey, J., J. Garner, G. Steve, and A. Southwood Williard. 2010. Diel foraging behavior of gravid leatherback sea turtles in deep waters of the Caribbean Sea. Journal of Experimental Biology 213: 3961–3971.

Castellini, M. A., R. W. Davis, and G. L. Kooyman. 1991. Annual cycles of diving behavior and ecology of the Weddell seal. University of California Press, Berkeley, CA, USA.

Davenport, J., J. Fraher, E. Fitzgerald, P. McLaughlin, T. Doyle, L. Harman, T. Cuffe, and P. Dockery. 2009. Ontogenetic changes in tracheal structure facilitate deep dives and cold water foraging in adult leatherback sea turtles. Journal of Experimental Biology 212: 3440–3447.

Davenport, J., V. Plot, J.-Y. Georges, T. K. Doyle, and M. C. James. 2011. Pleated turtle escapes the box-shape changes in *Dermochelys coriacea*. Journal of Experimental Biology 214: 3474–3479.

Davies, D. G. 1989. Distribution of systemic blood flow during anoxia in the turtle, *Chrysemys scripta*. Respiration Physiology 78: 383–389.

Doyle, T. K., J.D.R. Houghton, P. F. O'Súilleabháin, V. J. Hobson, F. Marnell, J. Davenport, and G. C. Hays. 2008. Leatherback turtles satellite-tagged in European waters. Endangered Species Research 4: 23–31.

Eckert, S. A. 2002a. Distribution of juvenile leatherback sea turtle (*Dermochelys coriacea*) sightings. Marine Ecology Progress Series 230: 289–293.

Eckert, S. A. 2002b. Swim speed and movement patterns of gravid leatherbacks at St. Croix, U.S. Virgin Islands. Journal of Experimental Biology 205: 3689–3697.

Eckert, S. A., K. L. Eckert, P. Ponganis, and G. L. Kooyman. 1989. Diving and foraging behavior by leatherback sea turtles (*Dermochelys coriacea*). Canadian Journal of Zoology 67: 2834–2840.

Eckert, S. A., D. W. Nellis, K. L. Eckert, and G. L. Kooyman. 1986. Diving patterns of two leatherback sea turtles (*Dermochelys coriacea*) during interesting intervals at Sandy Point, St. Croix, U.S. Virgin Islands. Herpetologica 42: 381–388.

Fossette, S., S. Ferraroli, H. Tanaka, Y. Ropert-Coudert, N. Arai, K. Sato, Y. Naito, et al. 2007. Dispersal and dive patterns in gravid leatherback turtles during the nesting season in French Guiana. Marine Ecology Progress Series 338: 233–247.

Fossette, S., P. Gaspar, Y. Handrich, Y. Le Maho, and J.-Y. Georges. 2008. Dive and beak movement patterns in leatherback turtles *Dermochelys coriacea* during internesting intervals in French Guiana. Journal of Animal Ecology 77: 236–246.

Fossette, S., C. Girard, T. Bastian, B. Calmettes, S. Ferraroli, P. Vendeville, F. Blanchard, and J.-Y. Georges. 2009. Thermal and trophic habitats of the leatherback turtle during the nesting season in French Guiana. Journal of Experimental Marine Biology and Ecology 378: 8–14.

Fossette, S., C. Girard, M. López-Mendilaharsu, P. Miller, A. Domingo, D. Evans, L. Kelle, et al. 2010. Atlantic leatherback migratory paths and temporary residence areas. PLoS ONE 5(11): e13908. doi: 10.1371.

Fossette, S., A. C. Gleiss, A. E. Myers, S. Garner, N. Liebsch, N. M. Whitney, G. C. Hays, et al. 2010. Behaviour and buoyancy regulation in the deepest-diving reptile: the leatherback turtle. Journal of Experimental Biology 213: 4074–4083.

Fraher, J., J. Davenport, E. Fitzgerald, P. McLaughlin, T. Doyle, L. Harman, and T. Cuffe. 2010. Opening and closing mechanisms of the leatherback sea turtle larynx: a crucial role for the tongue. Journal of Experimental Biology 213: 4137–4145.

Georges, J.-Y., S. Fossette, A. Billes, S. Ferraroli, J. Fretey, D. Grémillet, Y. Le Maho, et al. 2007. Meta-analysis of movements in Atlantic leatherback turtles during the nesting season: conservation implications. Marine Ecology Progress Series 338: 225–232.

Hays, G. C. 2003. A review of the adaptive significance and ecosystem consequences of zooplankton diel vertical migrations. Hydrobiologia 503: 163–170.

Hays, G. C., F. Glen, A. C. Broderick, B. J. Godley, and J. D. Metcalfe. 2002. Behavioural plasticity in a large marine herbivore: contrasting patterns of depth utilization between two green turtle (*Chelonia mydas*) populations. Marine Biology 141: 985–990.

Hays, G. C., J.D.R. Houghton, C. Isaacs, R. S. King, C. Lloyd, and P. Lovell. 2004. First records of oceanic dive profiles for leatherback turtles, *Dermochelys coriacea*, indicate behavioral plasticity associated with long-distance migration. Animal Behavior 67: 733–743.

Herbert, C. V., and D. C. Jackson. 1985. Temperature effects on the responses to prolonged submergence in the turtle *Chrysemys picta bellii*, II. Metabolic rate, blood acid-base and ionic changes, and cardiovascular function in aerated and anoxic water. Physiological Zoology 58: 670–681.

Hochscheid, S. 2014. Why we mind sea turtles' underwater business: a review on the study of diving behavior. Journal of Experimental Marine Biology and Ecology 450: 118–136.

Hochscheid, S., C. R. McMahon, C.J.A. Bradshaw, F. Maffucci, F. Bentivegna, and G. C. Hays. 2007. Allometric scaling of lung volume and its consequences for marine turtle diving performance. Comparative Biochemistry and Physiology A 148: 360–367.

Houghton, J.D.R., T. K. Doyle, J. Davenport, R. P. Wilson, and G. C. Hays. 2008. The role of infrequent and extraordinary deep dives in leatherback turtles (*Dermochelys coriacea*). Journal of Experimental Biology 211: 2566–2575.

James, M. C., S. A. Eckert, and R. A. Myers. 2005. Migratory and reproductive movements of male leatherback turtles (*Dermochelys coriacea*). Marine Biology 147: 845–853.

James, M. C., C. A. Ottensmeyer, S. A. Eckert, and R. A. Myers. 2006. Changes in diel diving patterns accompany shifts between northern foraging and southward migration in leatherback turtles. Canadian Journal of Zoology 84: 754–765.

Jones, T. T., M. D. Hastings, B. L. Bostrom, D. Pauly, and D. R. Jones. 2011. Growth of captive leatherback turtles, *Dermochelys coriacea*, with inferences on growth in the wild: implications for population decline and recovery. Journal of Experimental Marine Biology and Ecology 399: 84–92.

Jones, T. T., M. Salmon, J. Wyneken, and C. Johnson. 2000. Rearing leatherback hatchlings: protocols, growth and survival. Marine Turtle Newsletter 90: 3–6.

Jones, T. T., K. S. Van Houtan, B. L. Bostrom, P. Ostafichuk, J. Mikkelsen, E. Tezcan, M. Carey, et al. 2013. Calculating the ecological impacts of animal-borne instruments on aquatic organisms. Methods in Ecology and Evolution 4: 1178–1186.

Kooyman, G. L. 1989. Diverse divers: Physiology and behavior. Springer, Berlin, Germany.

Kooyman, G. L., M. A. Castellini, R. W. Davis, and R. A. Maue. 1983. Aerobic diving limits of immature Weddell seal. Journal of Comparative Physiology B 151: 171–174.

Kooyman, G. L., E. A. Wahrenbrock, M. A. Castellini, R. W. Davis, and E. E. Sinnett. 1980. Aerobic and anaerobic metabolism during voluntary diving in Weddell seals:

evidence of preferred pathways from blood chemistry and behavior. Journal of Comparative Physiology B 138: 335–346.

Kramer, D. L. 1987. The behavioral ecology of air-breathing by aquatic animals. Canadian Journal of Zoology 66: 89–94.

López-Mendilaharsu, M., C.F.D. Rocha, A. Domingo, B. P. Wallace, and P. Miller. 2009. Prolonged, deep dives by the leatherback turtle *Dermochelys coriacea*: pushing their aerobic dive limits. Marine Biodiversity Records 2: e35. doi: 10.1017/S1755267208000390.

Lutcavage, M. E., P. G. Bushnell, and D. R. Jones. 1990. Oxygen transport in the leatherback sea turtle *Dermochelys coriacea*. Physiological Zoology 63: 1012–1024.

Lutcavage, M. E., P. G. Bushnell, and D. R. Jones. 1992. Oxygen stores and aerobic metabolism in the leatherback sea turtle. Canadian Journal of Zoology 70: 348–351.

Lutcavage, M. E., and P. L. Lutz. 1997. Diving physiology. In: P. L. Lutz and J. A. Musick (eds.), The biology of sea turtles. CRC Press, Boca Raton, FL, USA, pp. 277–296.

Lutz, P. L., and T. B. Bentley. 1985. Respiratory physiology of diving in the sea turtle. Copeia 1985: 671–679.

Lutz, P. L., and P. W. Hochachka. 1993. Hypoxic defense mechanisms: a comparison between diving reptiles and mammals. In: P. W. Hochachka, P. L. Lutz, T. Sick, M. Rosenthal, and G. Van den Thillart (eds.), Surviving hypoxia: Mechanisms of control and adaptation. CRC Press, Boca Raton, FL, USA, pp. 459–472.

Murphy, C., D. Kelliher, and J. Davenport. 2012. Shape and material characteristics of the trachea in the leatherback sea turtle promote progressive collapse and reinflation during dives. Journal of Experimental Biology 215: 3064–3071.

Myers, A. E., and G. C. Hays. 2006. Do leatherback turtles *Dermochelys coriacea* forage during the breeding season? A combination of data-logging devices provide new insights. Marine Ecology Progress Series 322: 259–267.

Naito, Y. 2004. New steps in biologging science. Memoirs of the National Institute of Polar Research SI 58: 50–57.

Paladino, F. V., M. O. O'Connor, and J. R. Spotila. 1990. Metabolism of leatherback turtles, gigantothermy, and thermoregulation of dinosaurs. Nature 344: 858–860.

Paladino, F. V., J. R. Spotila, M. P. O'Connor, and R. E. Gatten Jr. 1996. Respiratory physiology of adult leatherback turtles (*Dermochelys coriacea*) while nesting on land. Chelonian Conservation and Biology 2: 223–229.

Plot, V., T. Jenkins, J.-P. Robin, S. Fossette, and J.-Y. Georges. 2013. Leatherback turtles are capital breeders: morphometric and physiological evidence from longitudinal monitoring. Physiological and Biochemical Zoology 86: 385–397.

Ponganis, P. J., J. U. Meir, and C. L. Williams. 2011. In pursuit of Irving and Scholander: a review of oxygen store management in seals and penguins. Journal of Experimental Biology 214: 3325–3339.

Price, E. R., F. V. Paladino, K. P. Strohl, P. Santidrián Tomillo, K. Klann, and J. R. Spotila. 2007. Respiration in neonate sea turtles. Comparative Biochemistry and Physiology A 146: 422–428.

Price, E. R., B. P. Wallace, R. D. Reina, J. R. Spotila, F. V. Paladino, R. Piedra, and E. Vélez. 2004. Size, growth, and reproductive output of adult female leatherback turtles *Dermochelys coriacea*. Endangered Species Research 5: 1–8.

Reina, R. D., K. J. Abernathy, G. J. Marshall, and J. R. Spotila. 2005. Respiratory frequency, dive behaviour and social interactions of leatherback turtles, *Dermochelys coriacea* during the inter-nesting interval. Journal of Experimental Marine Biology and Ecology 316: 1–16.

Reina, R. D., P. A. Mayor, J. R. Spotila, R. Piedra, and F. V. Paladino. 2002. Nesting ecology of leatherback turtles, *Dermochelys coriacea*, at Parque Nacional Marino Las Baulas, Costa Rica: 1988–1989 to 1999–2000. Copeia 2002: 653–664.

Rhodin, A.G.J., J. A. Ogden, and G. J. Conlogue. 1980. Preliminary studies on skeletal morphology of the leatherback turtle. Marine Turtle Newsletter 16: 7–9.

Rothschild, B. M. 1991. Stratophenetic analysis of avascular necrosis in turtles: affirmation of the decompression syndrome hypothesis. Comparative Biochemistry and Physiology A 100: 529–535.

Sale, A., P. Luschi, R. Mencacci, P. Lambardi, G. R. Hughes, G. C. Hays, S. Benvenuti, and F. Papi. 2006. Long-term monitoring of leatherback turtle diving behaviour during oceanic movements. Journal of Experimental Marine Biology and Ecology 328: 197–210.

Salmon, M., T. T. Jones, and K. W. Horch. 2004. Ontogeny of diving and feeding behavior in juvenile sea turtles: leatherback sea turtles (*Dermochelys coriacea* L) and green sea turtles (*Chelonia mydas* L) in the Florida Current. Journal of Herpetology 38: 3–43.

Scholander, P. F. 1940. Experimental investigatory function in diving mammals and birds. Hvalradets Skrifter 22: 1–131.

Schreer, J. F., and K. M. Kovacs. 1997. Allometry of diving capacity in air-breathing vertebrates. Canadian Journal of Zoology 75: 339–357.

Shillinger, G. L., A. M. Swithenbank, S. J. Bograd, H. Bailey, M. R. Castelton, B. P. Wallace, J. R. Spotila, et al. 2010. Identification of internesting habitats for eastern Pacific leatherback turtles: role of the environment and implications for conservation. Endangered Species Research 10: 215–232.

Shillinger, G. L., A. M. Swithenbank, S. J. Bograd, H. Bailey, M. R. Castelton, B. P. Wallace, J. R. Spotila, et al. 2011. Vertical and horizontal habitat preferences of post-nesting leatherback turtles in the South Pacific Ocean. Marine Ecology Progress Series 422: 275–289.

Snyder, G. K. 1983. Respiratory adaptations in diving mammals. Respiration Physiology 54: 269–294.

Southwood, A. L., R. D. Andrews, M. E. Lutcavage, F. V. Paladino, N. H. West, R. H. George, and D. R. Jones. 1999. Heart rates and diving behavior of leatherback sea turtles in the eastern Pacific Ocean. Journal of Experimental Biology 202: 1115–1125.

Southwood, A. L., R. D. Andrews, F. V. Paladino, and D. R. Jones. 2005. Effects of diving and swimming behavior on body temperatures of Pacific leatherback turtles in tropical seas. Physiological and Biochemical Zoology 78: 285–297.

Standora, E. A., J. R. Spotila, J. A. Keinath, and C. R. Shoop. 1984. Body temperatures, diving cycles, and movement of a subadult leatherback turtle, *Dermochelys coriacea*. Herpetologica 40: 169–176.

Troëng, S., R. Solano, A. Díaz-Merry, J. Ordoñez, J. Taylor, D. R. Evans, D. Godfrey, et al. 2006. Report on long-term transmitter harness retention by a leatherback turtle. Marine Turtle Newsletter 111: 6–7.

Wallace, B. P., C. L. Williams, F. V. Paladino, S. J. Morreale, R. T. Lindstrom, and J. R. Spotila. 2005. Bioenergetics and diving activity of internesting leatherback turtles *Dermochelys coriacea* at Parque Nacional Marino Las Baulas, Costa Rica. Journal of Experimental Biology 208: 3873–3884.

Williams, C. L., J. U. Meir, and P. J. Ponganis. 2011. What triggers the aerobic dive limit? Patterns of muscle oxygen depletion during dives of emperor penguins. Journal of Experimental Biology 214: 1802–1812.

Witt, M. J., E. Augowet Bonguno, A. C. Broderick, M. S. Coyne, A. Formia, A. Gibudi, G. A. Mounguengui Mounguengui, et al. 2011. Tracking leatherback turtles from the world's largest rookery: assessing threats across the South Atlantic. Proceedings of the Royal Society B 278: 2338–2347.

Witt, M. J., A. C. Broderick, M. S. Coyne, A. Formia, S. Ngouessono, R. J. Parnell, G.-P. Sounguet, and B. J. Godley. 2008. Satellite tracking highlights difficulties in the design of effective protected areas for critically endangered leatherback turtles *Dermochelys coriacea* during the inter-nesting period. Oryx 42: 296–300.

Witt, M. J., A. C. Broderick, D. J. Johns, C. Martin, R. Penrose, M. S. Hoogmoed, and B. J. Godley. 2007. Prey landscapes help identify potential foraging habitats for leatherback turtles in the NE Atlantic. Marine Ecology Progress Series 337: 231–243.

Wyneken, J. W., and A.G.J. Rhodin. 1999. Sphincter valves in the pulmonary arteries of turtles. Abstract. American Zoologist: 39: 81A (no. 480).

Wyneken, J., and M. Salmon. 1992. Frenzy and postfrenzy swimming activity in loggerhead, green, and leatherback hatchling sea turtles. Copeia 1992: 478–484.

4

Anatomy of the Leatherback Turtle

JEANETTE WYNEKEN

One cannot look at a leatherback turtle (*Dermochelys coriacea*) without instantly knowing that it is structurally very different from other turtles. Many unique characteristics of this living relic species are adaptations that (1) enable capture and digestion of high volumes of salty, low-calorie prey (gelatinous zooplankton); (2) are thermoregulatory and thermoprotective mechanisms facilitating exploitation of cold water feeding grounds and tropical breeding grounds; and (3) provide resistance to dive injuries through collapsible and compressible structures. This chapter discusses aspects of the anatomy not detailed elsewhere. The reader should also see three excellent sources for broader coverage of leatherback anatomy specifically, and sea turtle anatomy generally: Deraniyagala (1939), Schumacher (1973), and Wyneken (2001). Much of this discussion arises from new research (updated from Wyneken 2001), but it also includes information from key historic works and recent literature. This chapter discusses or describes aspects of the anatomy not detailed elsewhere. Several systems are poorly covered in the literature, particularly the neuromuscular and urogenital systems, compared to our understanding of the anatomy of other sea turtle species or other turtles. Consequently, these systems are discussed to the extent possible, or initial basic descriptions are included. In writing this chapter, it became clear that much research still is to be done on the anatomy of leatherback turtles. It is my hope that future researchers will address the gaps identified here.

The leatherback shares similarity with other sea turtles in having a large adult size, a reduced, somewhat streamlined shell, a neck and limbs that cannot retract, forelimbs modified as flippers, and oar-shaped hind limbs. Its body has a number of unique characteristics that set it apart from other sea turtles. These characteristics include five dorsal ridges running the length of the carapace, two additional ridges forming the margins, and two to five ridges occurring ventrally (fig. 4.1A, 4.1B).

The head is rounded in shape and has distinctive upper jaws with a medial notch and another on each side (figs. 4.1C, 4.2A), so leatherbacks almost appear to have "fangs" (they do not have teeth). The flippers are long, often reaching the hind limbs in adults as well as in hatchlings. Flippers and hind feet lack claws (figs. 4.1A, 4.1B, 4.3).

Fig. 4.1. A. Nesting female leatherback and a hatchling from the same beach in St. Croix, US Virgin Islands, show typical adult and hatchling black and white patterns, respectively. A typical nesting female is 154 cm CCL (curved carapace length) while the hatchling is 5.4 cm SCL (straight carapace length). B. Ventrally, neonate leatherbacks have longitudinal stripes that become reduced to interrupted spots in adults. C. The upper mandible characteristically has a single, sharp, shallow midline notch and a pair of deep bilateral notches. D. Callosities along the adult plastron are the remnants of scales and osteoderms.

Hatchlings have skeletons formed of bone, as well as significant proportions as cartilage. Juveniles and adults retain portions of the skeleton as cartilage and do not form many bony structures in comparison with cheloniid sea turtles. Proportions of the leatherback body tend to be more juvenile in form; for example, the manus and pes remain proportionately large throughout all life stages. Ossification patterns are "truncated" so that some bony skeletal elements are distinctly small, such as the individual ribs and plastron bones; limb long bones retain significant cartilaginous ends that are ossified in other species. Consequentially, the leatherback is often described as neotenic (Rhodin et al. 1981; Kordikova 2002; Wyneken 2003; Snover and Rhodin 2008).

Leatherbacks are distinctly colored throughout life, and are typically black with white speckling. Hatchlings, juveniles, and adults differ from one another in the details of their external morphology, which suggests differences in their ecology. Hatchlings have black-scaled bodies with white scales adorning the ridges dorsally as white stripes. Ventrally, the body has black and white stripes with occasional black spots (fig. 4.1B). Flippers have a white edge, and the ventral neck is white to gray. The white jaws have black or gray edges. Most hatchlings have some white over the dorsal (anterior) eyelid (fig. 4.1C), and the head skin is scaled. Juvenile leatherbacks shed their scales as they grow so that scale margins become indistinct. Black areas of the flippers, head, and carapace may turn dark gray or remain black with a few white spots dorsally, but distinct light and dark stripes are retained ventrally. Leatherbacks lack distinctive head scales as adults, develop a patch of mostly

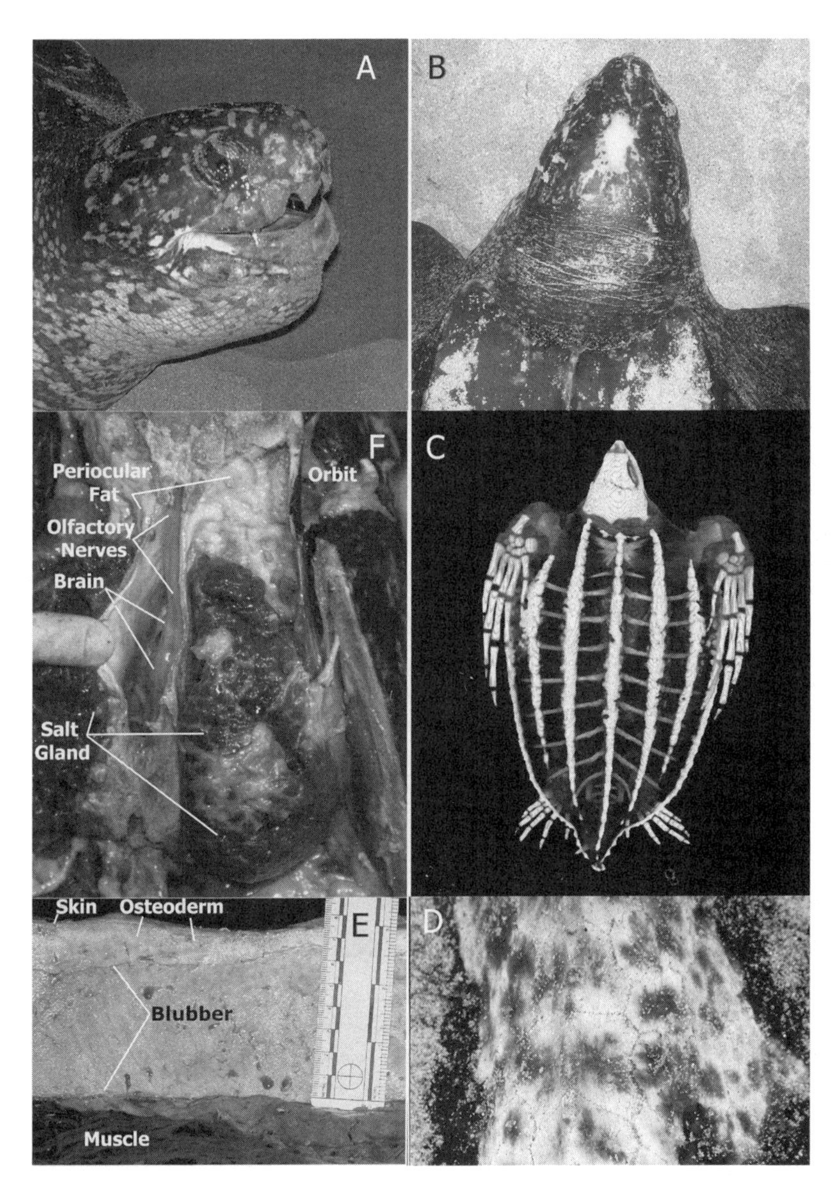

Fig. 4.2. Head, neck, and shell structures. A. Lateral view of adult female head showing jaw form, expanded throat during ventilation, black and white color pattern, and anterior-posterior position of the eyelids. Viscous tears drain from the ventral eye onto the upper jaw. B. Dorsal view of an adult head showing the "pink spot." C. Dorsal CT scan of a 17 cm SCL juvenile leatherback showing the developing osteoderms along the dorsal and lateral ridges, the ribs, and nuchal and sacral bones. This animal had a fracture along the dorsal skull, posterior to the orbits. D. Close-up of fully formed osteoderms on adult, with skin (dark) partially covered by lighter sand to either side. E. Cut section of plastron showing 6.5 cm thick blubber layer and skin approximately 1 cm thick, with embedded osteoderms. F. Coronal cut through skull showing the long tubular brain of an adult male leatherback. The brain is disproportionately small compared to the salt gland.

pigment-free skin on the dorsal head (the "pink spot," fig. 4.2B), and have a minimal keratin covering on the jaws except at the cutting edges, which have slightly thicker knifelike keratinous coverings. While most adult leatherbacks have distinct white spots and patches, particularly ventrally (figs. 4.1D, 4.2A), some melanistic adults are seen with few light spots. Often adults have a few ridges ventrally, and remnant "callosities" can be found along the sites of former ventral ridges (fig. 4.1D). The skin of the shell is smooth and feels waxy.

Skeleton

The bony skeleton is somewhat reduced so that the carapace primarily is formed of vertebrae and ribs; the plastron is a large loop of small bones, and the appendicular bones have significantly greater proportions of cartilage at their ends than is found in other sea turtles (figs. 4.2C, 4.3). The skull is composed of the braincase and overlying dermal bones, jaws, and the hyoid apparatus. Many skull bones articulate via cartilaginous joints that have restricted movement. The axial skeleton includes vertebrae, nuchal bones, ribs, and plastron bones. The appendicular skeleton includes flippers (forelimbs) and paddle-like hind limbs, and supporting structures for both. These structures include the paired tri-radiate pectoral girdles, each formed by the scapulae and procoracoids (sometimes termed coracoids), which support the forelimbs; while the pelvic girdle, formed by the paired bones and cartilages (pubis, ilium, and

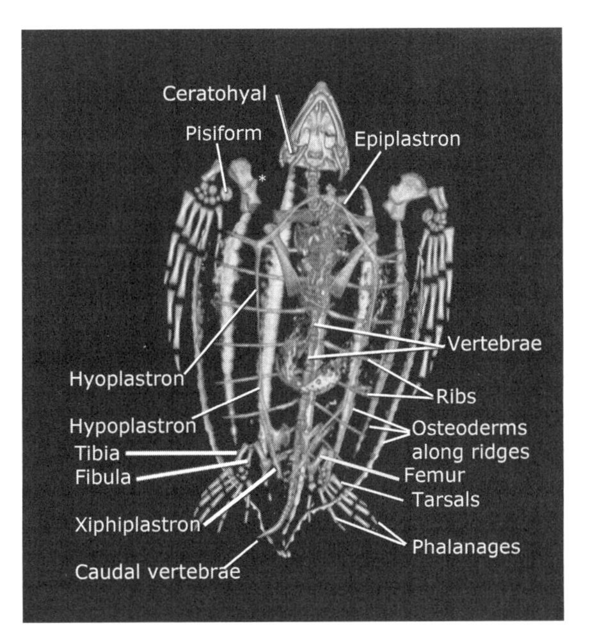

Fig. 4.3. Ventral view CT scan of a 17 cm SCL juvenile leatherback showing the skeleton in situ. The much reduced plastron appears as a "loop" of thin bones (Hypo-, Hyo-. and Epiplastron). Some dense material in the very long esophagus makes the proximal part of the esophagus visible in this scan.

ischium), support the hind limbs (figs. 4.3, 4.4). Many dermal ossicles (osteoderms) form as a thin layer of bone between the skin and blubber layer of the carapace and plastron (fig. 4.2D, 4.2E).

Skull and Hyoid Apparatus

The leatherback skull is unlike that of any other extant sea turtle. It is wide and rounded anteriorly with large orbits. It differs from cheloniid skulls; there are no parietal notches in the posterior skull roof. Many skull bones articulate via synchrondroses that provide restricted flexibility. As in other vertebrates, the skull is organized into an inner bony and cartilaginous braincase (the chondrocranium) that houses the brain, and an outer bony superstructure of dermal bones (dermatocranium) that encase the jaw muscles, salt glands, and sensory organs (plate 1). The hyoid elements (hyoid body and cornua; plates 2, 3), along with the developing (upper and lower) jaws, comprise the splanchnocranium portion of the skull. Dermal bones encase most of the jaws and roof of the mouth. The braincase, located along the midline is internal to the dermatocranial skull roof, snout, and jaws (fig. 4.2F). Skull bones are particularly thin where they overlie the pineal organ (plate 4) and are covered by skin that lacks melanin (fig. 4.2B). The bones are the same in all sea turtle species, however their specific form and some articulations differ slightly.

Leatherbacks lack a secondary palate. Margins of the jaws are sharp and upper jaws possess notches adjacent to the pointed cusps of the anterior maxillary bones. The anterior midline of the lower jaw forms a dorsally directed point at the symphysis. The lower jaw is formed of dermal bones, paired remnants of Meckel's cartilage that are medial to the dentary bones, including an articular cartilage that may not ossify.

Throat and tongue muscles attach to the hyoid body and its paired cartilage and bone skeletal elements are termed horns or cornua (plates 2, 3). During inhalation (ventilation) and buccal pumping, the hyoid apparatus rotates ventrally and anteriorly, depressing the floor of

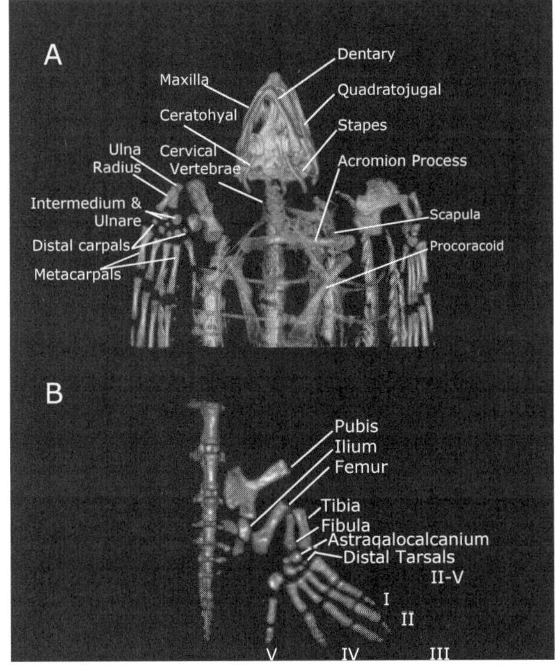

Fig. 4.4. CT scans of the ventral views of the cranial, cervical, pectoral, pelvic, and caudal skeleton. The images are of the skeleton in situ. Cartilages are not detected with this imaging mode; bone alone is shown. A. Details of the proximal flipper skeleton and the supporting pectoral girdles are identified, as are the major bones of the splanchnocranium, mandibular dermal bones, and cervical skeleton. The dermal ossicles of the carapacial ridges are seen as longitudinal structures dorsal to the ribs and vertebrae. B. The pelvic skeleton shows the separate bones of one side of the pelvis (ventrally) and the major bony elements of the hind limb. Cartilages compose substantial parts of these structures and do not show up in this CT scan, so even though the bones appear to not articulate, they are in fact connected by cartilaginous processes and joints.

the mouth and expanding the throat (fig. 4.2A). There are several naming systems for the cornua that reflect a lack of appreciation of the extent of the cartilaginous skeleton of the leatherback. Romer (1956) notes just two pairs of hyoid arch horns while others count three pairs (Schumacher 1973; Fraher et al. 2010).

Axial and Integumentary Skeleton

The leatherback shell is composed of axial skeletal bones and cartilages, blubber (dense connective tissue and fat), and skin. Carapace and plastron are composite structures organized as epithecally derived osteoderms (Vickaryous and Sire 2009) that overlay a thick blubber layer (fig. 4.2C, 4.2D); the ribs, nuchal bone, vertebrae, and plastron bones are found under the blubber layer.

The carapace's axial skeleton is composed solely of an expanded nuchal, ribs, and vertebrae. Ribs remain distinct and do not form the expanded pleural bones that are found in cheloniid turtles. Bilaterally arranged pairs of ribs align with the junction of two vertebral bodies along the trunk. Unlike in cheloniids, there is no formation of peripheral or suprapygal bones (figs. 4.2C, 4.3). Ventrally, the plastron is composed of a loop of reduced plastron bones (epiplastra, hyoplastra, and xiphiplastra). No entoplastron is present (fig. 4.3).

Leatherbacks have seven mobile cervical vertebrae. The atlas is a reduced pair of bones joined by fibrous connective tissue and closely associated with the much larger axis. The next five cervical vertebrae are stout, procoelous in form, and have robust articular surfaces that resist twisting. The eighth vertebra is associated with the large butterfly-shaped nuchal bone. Ten trunk vertebrae are followed by 3 to 4 sacral vertebrae and 12 or more caudal vertebrae. Caudal vertebrae of females are somewhat shorter than those of males. Tails of mature males and females are not hugely different in length as in cheloniid sea turtles.

Limbs and Limb Girdles

Flippers and pectoral girdles comprise the anterior appendicular skeleton. Pectoral girdles are paired; each is composed of two bones. These include the dorsoventrally positioned scapula with a ventromedial positioned acromion process, and the anterolateral-to-posteromedial positioned procoracoid bones (also termed reptilian coracoids). Each procoracoid terminates in a large crescent-shaped coracoid cartilage. The scapula, its acromion process, and procoracoid together form a tri-radiate structure that serves as a major framework from which many flipper muscles originate (figs. 4.3, 4.4). Cartilaginous ends of the scapula and procoracoid on each side form a shoulder joint, the glenoid fossa, which is cartilaginous in leatherbacks.

The leatherback forelimb skeleton is comprised of the humerus, radius, ulna, carpals, metacarpals, and phalanges. Forelimb long bones have extensive cartilages at the ends. The humerus shaft is composed primarily of cancellous bone surrounded by a very thin layer of cortical lamellar bone. The large humerus has a somewhat primitive form with a flattened profile and extended medial process that is formed mostly of cartilage (Romer 1956; Rhodin et al. 1981; Rhodin 1985; Wyneken 2001). The lateral process (=pectorodeltoid process) is distal to the head along the ventral surface, approximately halfway down the shaft (Walker 1973). Two humeral processes are insertion sites for most major swimming muscles, particularly those used during power stroking. Humeral head and distal articulations to the radius and ulna are largely cartilaginous and highly vascular (fig. 4.5A).

A short, mostly straight radius, along with a shorter, slightly curved ulna with 9–10 wide and flat carpal elements and 5 digits (metacarpals and phalanges) support the flipper blade internally. The radius articulates with a small carpal, the radiale. The ulna articulates with two other carpals, the ulnare and intermedium. A centrale is distal to the intermedium and articulates with the second, third, and fourth distal carpals. The large flat pisiform and distal carpals demark the distal carpus (figs. 4.3, 4.4A). The metacarpals articulate with the distal carpals and with the phalanges by synovial joints that are encased in thick fibrous connective tissue. Although it is not part of the skeleton, a thick layer of fibrous connective tissue stiffens and stabilizes the entire leading edge of the flipper blade in adults and juveniles (Wyneken 2013; figs. 4.4A, 4.5E).

The pelvis is composed of three pairs of skeletal elements: the pubis, ischium, and ilium are formed of bone and large cartilage components. Pubis and ischium pairs form the ventrally positioned part of the pelvis and parts of the hip joint (acetabulum). Two ilia extend dorsally from the acetabulum and articulate dorsomedially with the sacral vertebrae (fig. 4.4B). Cartilages of all three bones form the acetabulum. Pelvic bones are connected by synchrondroses throughout life.

Each femur articulates with the deeply cupped acetabulum via a cartilaginous femoral head that is offset from the relatively straight bony shaft of the femur. The proximal femur has two bony and cartilaginous processes, the major and minor trochanters, located distal to the head. These trochanters are insertion sites

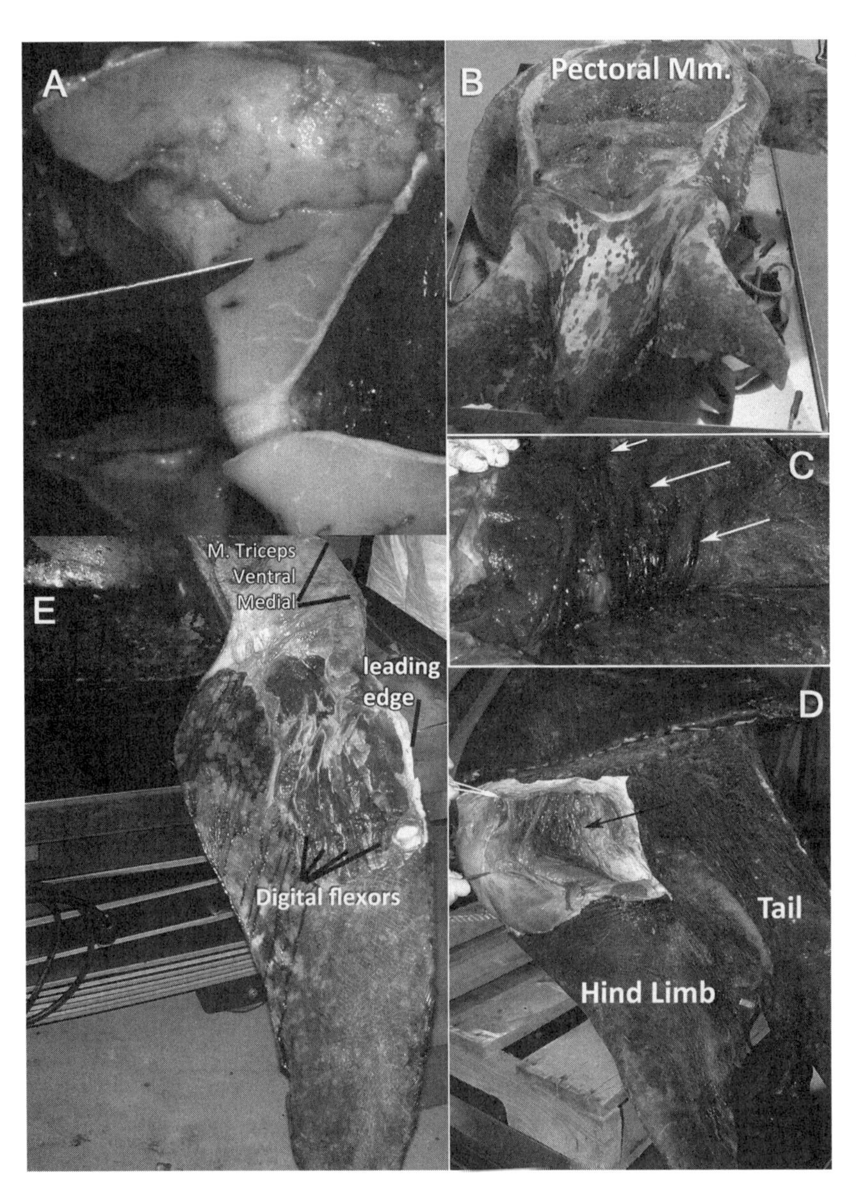

Fig. 4.5. A. Ends of long bones and the girdles are formed of significant parts of hyaline cartilage. Parts of the limbs and girdle skeleton are highly vascular (blood vessels are seen in this cut humeral process and glenoid fossa, adjacent to the knife). B. The plastron was removed from this stranded adult male leatherback revealing the thick fat layer of the plastron and the massive ventral pelvic and pectoral muscle ("Mm.") masses. Anterior is to the top of the image. C. The pectoral muscles are separated here in this oblique ventral view showing the pectoral rete system of arteries and veins. D. Dissection of the hind limb rete shows the extensive network of blood vessels along the lateral thigh. E. Ventral view of the left front flipper that is partially skinned to show the flexible connective tissue enveloping the triceps muscle complex, the thick and stiffened leading edge of the muscles of the palmer surfaces of the flipper, and the extensive muscles retained in the palmar carpus.

for most thigh retractor and adductor muscles, respectively. The distal femur terminates in a thick cartilage that articulates with tibia and fibula bones comprising the shank (crus). The tibia tends to be slightly more stout than the fibula and has a triangular articular surface at its proximal end. Bony ends of the fibula are convex, but in life are covered by articular cartilages at both ends; the shaft is concave toward the tibia (Romer 1956). The short tarsus (ankle) consists of the calcaneum and astragalus, which often form as one bone (the astragalocalcaneum) and four smaller distal tarsals; the latter are mostly cartilaginous with a bony center. They articulate, sequentially, with metatarsals I–IV. Metatarsal V is "hook-shaped" and articulates with the outer (anatomically medial) edge of the distal tarsal IV.

The broad, flat distal hind foot has five sets of widely spaced metatarsals and digits formed of long phalanges that are mostly round in cross section. Digit I is on the tibial side of the hind limb and is supported by one large tapering metatarsal and two phalanges. Digit II is below the fibula and has a short, flat "hook-shaped" metatarsal. Digits III and IV each have one metatarsal that resembles a phalange, except the proximal articular surface is somewhat irregular in shape. Digits II, III and IV have three phalanges each. Digit V has two phalanges (a third mostly cartilaginous distal phalanx sometimes is also found in digit V; fig. 4.4B).

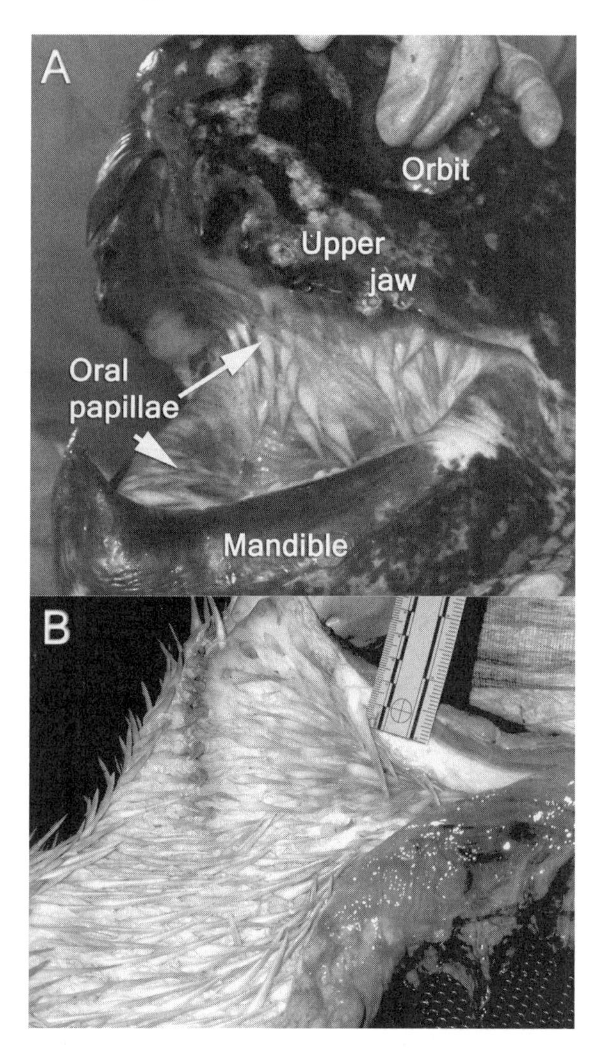

Fig. 4.6. Oral and esophageal papillae. The mouth (A) is lined by flexible oral papillae along the buccal surfaces of the upper and lower jaws and the roof of the mouth. The tongue is relatively small, not protrusible, and lacks these conical papillae except where it joins the esophagus. B. The papillae lining the esophagus point toward the stomach and can be 6–9 cm long in parts of the proximal esophagus. The proximal esophagus has a 1–2 cm thick layer of yellow fat (shown in the cut edge of this dissection); this fatty layer may provide insulation from cold-water prey as it is transported toward the stomach.

Muscles

The muscular system is the least described. The largest muscle masses are those associated with the pectoral limbs. Pectoral major muscles are thick and contain a large network of blood vessels that appear to be a rete system within each side (fig. 4.5B, 4.5C); the pectoral retia are more centrally located that axial and inguinal networks (fig. 4.5D).

Walker (1973) describes the locomotor muscles of turtles and includes discussions of some leatherback limb muscles. That work is careful and comparative with other turtles. My dissections indicated that carpal and metacarpal muscles remain robust in adult leatherbacks (fig. 4.5E), although the digits and carpus flex little, if at all.

Schumacher (1973) describes muscles of the head and larynx along with their innervation (plates 1 to 8). This work is largely correct; however, there are some clear labeling mistakes in the original work, which have been corrected in this volume. Overall, the muscular system of leatherbacks has clearly not been studied enough, because many descriptions are lacking.

Digestive System

The mouth and alimentary canal of leatherbacks have many distinctive features, and the mouth structure is unique within turtles. Keratinous oral papillae form along the buccal side of the maxilla and mandible, adjacent to the tongue, as well as across much of the palate; they also line the muscular angles of the jaws and the most posterior portion of the tongue (fig. 4.6A). Papillae tend to angle toward the interior of the mouth and the esophagus. Oral papillae appear to be continuous with the larger papillae of the esophagus (fig. 4.6B).

Esophagus and Stomach

Perhaps one of the most distinctive structures of leatherbacks is the esophagus, which is extremely long, extending for 20% of the length of the entire gastrointestinal (GI) system in hatchlings and about 10% of GI length in adults. The esophagus passes into the body cavity caudal to the hyoid skeleton and extends posteriorly to the level of the pelvis before curving laterally to the turtle's left, and then extending cranially to the level of the axilla, where it curves medially before entering the anterodorsal aspect of the stomach (fig. 4.7A).

The esophagus is muscular and its proximal portion is lined with a thick layer of yellow fat (fig. 4.6B) that may insulate the body when cold food is swallowed (Davenport, Fraher, Fitzgerald, McLaughlin, Doyle, Harman, and Cuffe 2009). At all ages the esophagus is lined with large sharply pointed keratinous papillae (fig. 4.6B). These sharp papillae clearly trap any prey that is swallowed and prevent its loss when the muscular esophagus contracts, squeezing out excess seawater. Papillae likely macerate gelatinous prey as they are swallowed and squeezed. The sizes of the papillae decrease near the stomach, and they end abruptly at the gastroesophageal junction.

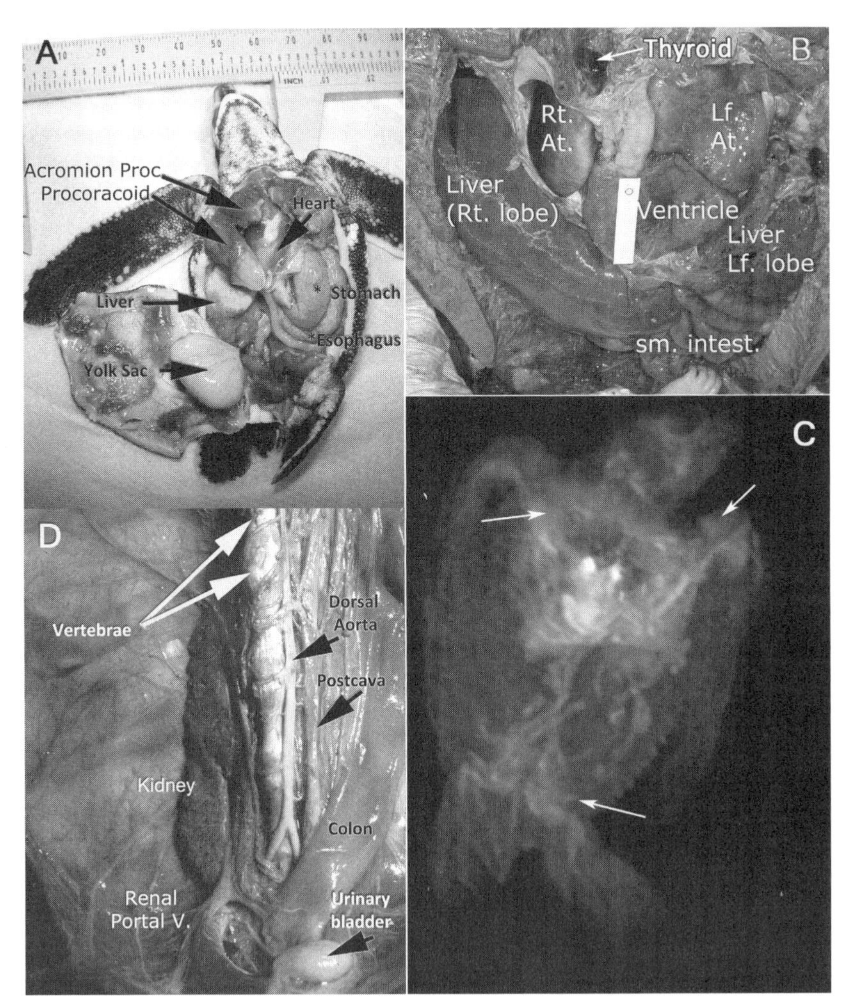

Fig. 4.7. The visceral and circulatory organization of leatherbacks. A. Hatchling ventral view showing the heart framed by the procoracoid processes, pale liver (normal in a hatchling), gastrointestinal system, as well as the large internalized yolk sac. B. Heart and liver of an adult leatherback. White ruler along ventricle is 16 cm long. (Rt. At.: right atrium; Lf. At.: left atrium; sm. intest: small intestine.) C. MRI scan of the circulatory system of a leatherback neonate. The head is at the top of the image, right side of body is on the right side of the image. Paired bright white areas are the atria of the heart. Paired dorsal aortae as well as hepatic portal vein form the major abdominal veins caudal to the heart. Arrows identify retia: axillary, pectoral, and inguinal. Renal portal veins are partially visible proximal to the hind limbs and medial to the right inguinal arrow. D. Left kidney and renal circulation of a juvenile leatherback (approximately 16 cm SCL). Aorta was injected with latex.

The stomach is large, thin walled, highly distensible, and C shaped. Walls are weakly muscular and digestion appears to be largely chemical. There are no histological studies detailing characteristics of the leatherback's glandular gastric epithelium. No regional muscular specializations are apparent, except at the pylorus. The stomach joins the duodenum (first region of small intestine) via a muscular pyloric sphincter. The C-shaped stomach attaches to the left lung via a short gastropulmonary mesentery. The stomach is located such that the heart may be "wrapped" by the stomach when the latter is full (fig. 4.7A).

Intestine

Leatherback intestines from the pylorus to the cloaca include the small intestine with muscular walls and narrow lumen size, and a large intestine characterized by greater lumen diameter and thin walls proximally that transition to a slightly thicker-walled rectum (Den Hartog and Van Nierop 1984; Wyneken 2001). The small intestine is regionally specialized proximally as a duodenum and distally as an ileum. Mammalian subdivisions of small intestines, duodenum, jejunum, and ileum, may not be the same in leatherbacks because they are based upon mammalian structural specializations in the intestinal mucosa. However, the three-part terminology for the mammalian small intestine is often applied as a convenient set of regional identifiers, even though they likely do not reflect functional specializations of leatherback small intestines.

The small intestine is muscular, extends from the stomach and is suspended in the body cavity by a dorsally attached, circular, vascular mesentery that also contains patches of lymphoid tissue. The duodenum's internal mucosal surface is highly textured with crypts and villi that increase surface areas. The duodenum produces mucous, lipases, proteases, and peptidases, as well as cellular products that adjust the pH of the digesta from the stomach (Kardong 2012). Digestive enzymes from the pancreas and liver enter the duodenum together in the proximal duodenum. The more distal duodenum transitions structurally into a middle region of small intestine that lacks a highly textured mucosal surface. Some references refer to the middle region as the jejunum; however, this designation may be inappropriate because the structure is not fully like that of the mammalian jejunum and the functions probably differ as well. The last portion of the small intestine ends in an ileocaecal valve that demarks the entrance to the large intestine. The large intestine or colon is much greater in luminal diameter that the ileum, yet does not form a distinct caecum. A distal loop of small intestine typically crosses the ventral surface of the coiled intestines diagonally from posterior to anterior.

The large intestine is thin walled and relatively short compared to the small intestine. This ileocaecalcolonic region is a frequent site for sequestering foreign bodies by fibrosis (B. Stacy, personal communication). The large intestine is fairly uniform in diameter from the caecum to the rectum, which opens into the cloaca.

Glands

There are two major categories of glands: exocrine and endocrine. Those with ducts are exocrine. Exocrine glands include the lachrymal and Harderian glands associated with the eyes, and the liver and pancreas associated with digestion. Endocrine structures release their products (hormones and other metabolites) into extracellular fluid. The circulation picks up those products and distributes them. Leatherback endocrine glands include the epiphysis (pineal body), pituitary, thyroid, ultimobranchial bodies, parathyroids, thymus, adrenals, and gonads (testes or ovaries). Some organs have a secondary endocrine function. These include the pancreas, kidney, liver, and probably the skin.

Liver

The liver is the largest visceral organ in reptiles generally, and this is true in leatherbacks. It may fill nearly one-third of the body cavity and its shape is variable, so it generally conforms to the shape of the coelomic cavity (fig. 4.7A, 4.7B). The liver usually consists of two lobes. The right lobe is larger and may extend distally, nearly to the inguinal fossa in some individuals. It houses the gallbladder and contains hepatic ducts, the cystic duct, and the common bile duct. The liver is suspended by several mesenteries. The right lobe is suspended from the ventral surface of the right lung via the hepatopulomonary ligament and attaches to the duodenum via the hepatoduodenal ligament. The liver attaches medially to an abbreviated transverse septum along the dorsal pericardium and ventromedially via the coronary ligament. The left lobe is attached to the stomach's ventromedial surface via the gastrohepatic ligament (Wyneken 2001).

The liver is a digestive gland, producing bile, and it is an immunological organ as well as a filter. Breakdown of carbohydrates and proteins from the gastrointestinal circulation and lipids from the lymph vessels occurs in the liver. They are converted to phospholipids and cholesterols, as well as bile salts, which aid in digestion and absorption. Production of bile also is a mechanism for excretion of cholesterols and breakdown of hemoglobin. Bile is transported by the common bile duct to the duodenum and is also transported to the gallbladder via the short cystic duct.

The liver acts as a clearance organ by binding metals such as mercury and other toxicants. It is also both a storage organ rich in iron and lipids and a synthetic organ. For example, vitellogenin, a major precursor component of yolk, is produced there.

Nutrient-rich blood enters the liver via the hepatic portal vein from intestines, abdominal veins from the posterior body, and oxygen-rich blood arrives at the liver via hepatic arteries that usually arise from the gastric artery of the left aorta. Lipids from the digestive system are transported to the liver via the lymphatic system. Blood leaves the liver via the hepatic veins to the sinus venosus, which leads to the right atrium.

Pancreas

The pancreas is both an exocrine and endocrine organ located along the duodenum (Wyneken 2001). Pancreatic cells produce three major kinds of exocrine products: proteolytic enzymes, amylases, and lipases (Kardong 2012). These products enter the duodenum via the pancreatic duct that opens into the ampulla of Vater. The pancreas is also an endocrine organ in other turtles and is presumed to have the same role in leatherbacks; pancreatic islet cells produce insulin and glucagon, which both regulate blood glucose.

Thymus

These paired glands are located distal to the brachiocephalic trunk along the subclavian and axillary arteries (Wyneken 2001). Thymus glands are large in hatchlings and may be robust or thin and diffuse in adults. In other turtles, there is a seasonal cycle of shrinkage and then recovery of the cortical cells (Leceta et al. 1989). The thymus in turtles is inferred to be an immunologic structure, but its function has not been systematically examined.

Thyroid

The thyroid is a single structure in leatherbacks and is located just cranial to the heart at the branching point of the brachiocephalic arteries (fig. 4.7B). There have been no systematic studies of the thyroid gland or of its changes with season or age in leatherbacks. This gland plays a major role in iodine metabolism in all animals and is a regulator of growth, overall metabolism, and reproduction.

Adrenal Glands

The adrenal glands of leatherbacks are located at the anterior and medial aspect of each kidney. In hatchlings, the adrenal tissue is somewhat compact and bright yellow in color. The gland usually becomes more diffuse and occupies part of the cranial pole of each kidney in most adult leatherbacks. In those individuals with compact adrenal glands, the structures are elongate and cylindrical, located along the cranial and medial aspects of each kidney.

Orbital Glands

Paired lachrymal glands are greatly hypertrophied in leatherbacks. Lachrymal (salt) glands are dense, highly lobed, pink or red organs that are dorsal and posterior to each eye (fig. 4.2F; plate 5). They are positioned deep within the skull, close to the parietal and quadratojugal bones. These are the largest organs in the head and exceed the size of the eyeball by at least four times in volume, and exceed the size of the brain by approximately eighteen times in volume. Salt glands play a fundamental role in maintaining hydration by removing excess salt from the body (Nicolson and Lutz 1989; Lutz 1997).

Harderian glands are much smaller, less than half the size of the eye. These are lubricatory glands located anterior and slightly medial to each eye.

Parathyroid Glands and Ultimobranchial Bodies

Parathyroid glands (four to six pairs) secrete parathyroid hormone, which elevates blood calcium levels. There is usually one pair of ultimobranchial bodies (Kardong 2012). These glands secrete calcitonin, which lowers blood calcium; they function as the antagonists of the parathyroid glands. While small and difficult to find, they are physiologically quite important. Both are small pale glands located along the branches of the aorta originating from the brachiocephalic trunk. Sometimes these glands are displaced cranially along the thyroidal, carotid, or hyoidean arteries to locations near the hyoid structures or their derivatives.

Urogenital System

The urogenital system includes the kidneys, ureters, urinary bladder, gonads (ovaries or testes), mesonephric duct (which becomes the Wolffian duct and part of the epididymis), and paramesonephric duct (this last is called the Müllarian duct and forms in both sexes but regress in males). Paramesonephric ducts become the oviducts and uterus of females (Miller and Limpus 2003).

Kidneys

Kidneys are located posterior to the lungs and lie between the coelomic lining and the carapace. In hatchling leatherbacks, kidneys are elongated and formed of two parts, a remnant mesonephric portion and larger metanephric portion (figs. 4.7D, 4.8E). In adults, the kidney is a metanephros alone. Ureters are identifiable along the caudal half of the kidney in hatchlings; they are somewhat obscured by overlying connective tissue in adults. Ureters drain into the proximal cloaca through the urogenital papillae.

Cloaca

Left and right ureters, left and right gonadal ducts, rectum, and urinary bladder each lead into the proximal cloaca (fig. 4.8A, 4.8B). Opening of the urinary bladder is proximal and ventral, where the "floor" of the cloaca begins. The penis or clitoris is located in the floor of the cloaca, extending close to the vent. The clitoris is always contained within the floor of the cloaca. The penis is internal unless engorged with blood (fig. 4.8A). Its base is a shaft of U-shaped fibrous connective tissue surrounded by spade-shaped erectile tissue (fig. 4.8C). When the erectile tissue is filled, the spermatic sulcus forms a distinct channel through which sperm pass during mating (Miller and Limpus 2003).

Urinary Bladder

The urinary bladder is located medially along the dorsal surface of the pubis and is ventral to the rectum (fig.

4.7D). The bladder opens into the floor of the cloaca and is not connected to the kidneys (fig. 4.8B). Urine drains from the urogenital papillae in the dorsal cloaca. It may enter the bladder or drain out the cloaca though the vent.

Gonads

Paired gonads (ovaries or testes) are attached to the coelomic wall, caudal to the lung and ventral to the kidneys. Ovaries and testes of hatchlings are difficult to distinguish from one another due their delayed morphological differentiation.

Testes are fusiform to bean shaped, depending on age and reproductive status. Each is attached to the coelomic membrane overlying its kidney (fig. 4.8D, 4.8E). In juveniles, testes are fusiform, white, and flat. There is little development of the vas deferens. In adults, testes are yellow or pink, fusiform, and surfaces are either smooth or, during the breeding season, irregular due to the expanded vas deferens. The vas deferens is sometimes described as a proximal epididymis and distal vas deferens; it is a highly coiled tube that is well developed in adult males. It is caudal and slightly lateral to its testis (fig. 4.8D). In turtles, the vas deferens stores and then transports sperm to the cloaca. The paramesonephric ducts are incomplete or absent in males.

Each ovary is attached to the coelomic surface by a mesovarium that attaches along the gonad length either by one edge or asymmetrically so that one ovarian edge is wider than the other. Ovaries of juveniles tend to be fusiform and white (fig. 4.8E). Ovaries of mature and maturing turtles have a number of round yellow follicles that appear as small (~2 mm to 2 cm) diameter spheres embedded throughout the length of the elongate ovary. The paramesonephric duct in juvenile females extends from the cloaca to the level of the lung and has a complete lumen (fig. 4.8E).

Turtle oviducts (presumably leatherbacks are similar to other turtles) can be functionally and histologically divided into several regions. Oviductal regions are not functionally identical to those in birds (Aitken and Solomon 1976; Girling 2002). The infundibulum is located most cranially and has a funnel-shaped opening called the ostium. Ova pass from the infundibulum to the uterine tube (a glandular, albumin-forming region) and then through the short (probably a glandular) isthmus to the uterus (shell-forming region). Shelled eggs pass through the muscular vagina to the cloaca during oviposition. Paired oviducts are located laterally to the ovaries and are not attached to them. In mature turtles, the uterus appears "pleated," while it is a thin tube in immature turtles.

Respiratory System

The lungs are attached to the peritoneal wall of the carapace and vertebral bodies; they extend from the nuchal bone posteriorly for slightly less than the carapace length (Lutcavage and Lutz 1997). Their ventral surface is covered by smooth visceral peritoneum (=coelomic membrane) and has short mesenteries to the liver and stomach (fig. 4.9A), while the dorsal surface is secondarily retroperitoneal due to the ventral positioning of the pericardial cavity during development (Hyman 1922). The leatherback has multiple adaptations thought to be associated with deep or prolonged diving and heat conservation. These include dense, highly elastic lungs, a large-bore trachea (fig. 4.9C), and possible regional flow through of air (Schachner et al. 2013). The vasculature to and from the lungs intimately links cardiac structure and function to that of the pulmonary system. Structural adaptations in the pulmonary arteries likely contribute to the presence and reversal of central shunts so that blood from the ventricle goes to the lungs and body, or primarily to the body.

The trachea of leatherbacks is formed of cartilaginous rings stacked closely and joined by elastic and fibrous connective tissue (Davenport et al. 2009; Fraher et al. 2010; fig. 4.9C). The flexible trachea is collapsible under pressures associated with dives (Murphy et al. 2012). The adult trachea is lined with a dense vascular plexus and lymphatics (Davenport et al. 2009). The trachea bifurcates into two bronchi internally (the carina) in advance of the external bifurcation. A cartilaginous bronchus enters each lung, extending nearly to the caudal end (Wyneken 2001). Bronchi are perforated by many cartilage-supported airways within the lung. Lungs are spongy and highly elastic. Gas exchange structures are composed of complex arrangements of boxlike acini and faveoli (after Perry 1998; Wyneken 2001). Together these structures and their material properties are thought to be adaptations for breath-hold diving and life in cooler waters.

Cardiovascular System

Heart, arteries, veins, and lymphatic vessels make up the cardiovascular system. The heart is located just caudal to the medial attachments of the acromial processes to the plastron, and medial to the procoracoid processes (fig. 4.7A). It sits in a thick pericardial sac and is bathed in clear pericardial fluid. The heart includes a large thin-walled chamber dorsally, the sinus venous, which receives low oxygen venous blood from the body (fig. 4.9B). There are two separate atria. The right atrium

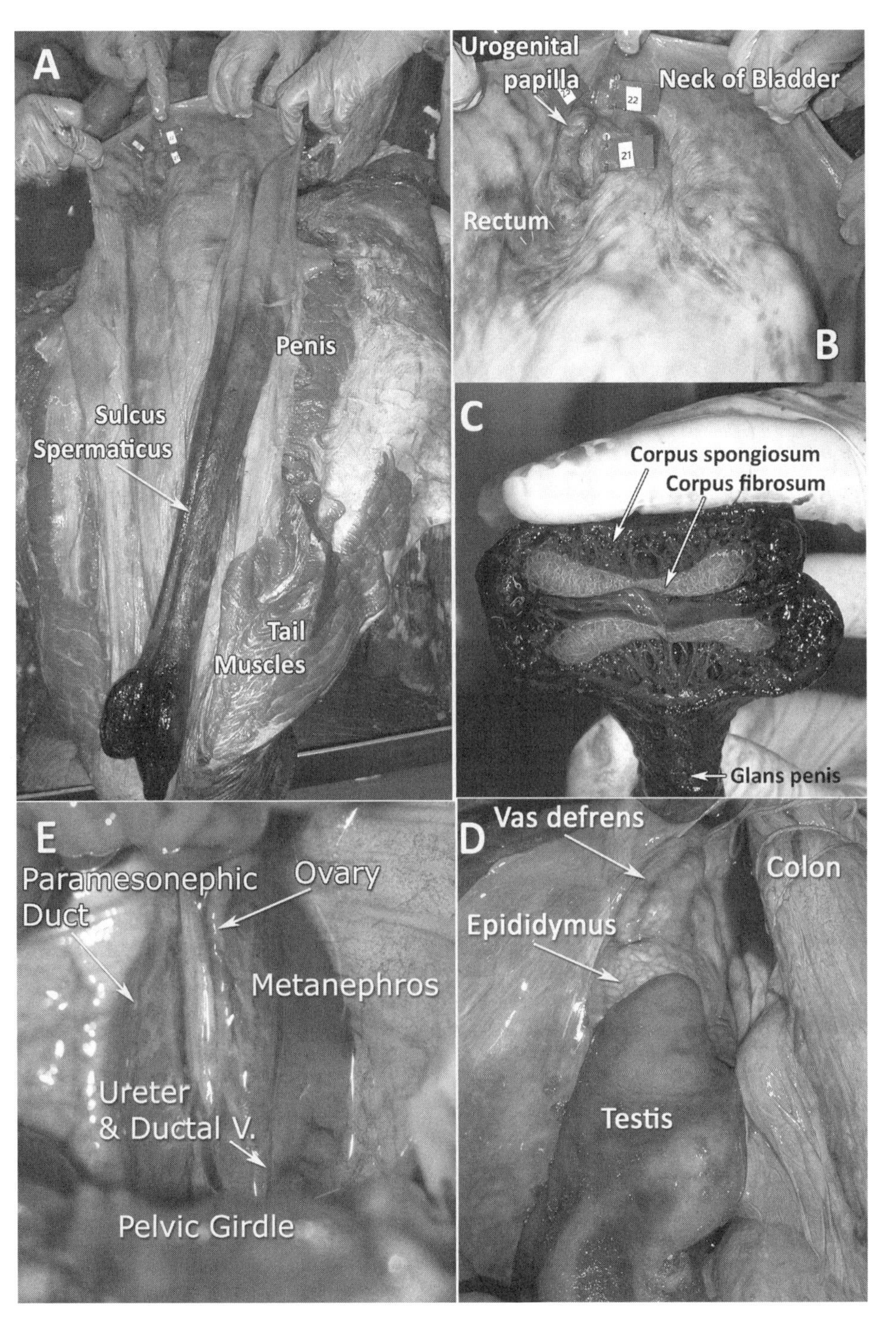

Fig. 4.8. Cloaca and urogenital systems of leatherbacks. A. Dissection of the cloaca of an adult male leatherback, showing the proximal drainage structures from the kidney and gonads, colon, and urinary bladder. The mostly retracted penis is the linear organ in the cloacal floor. The dorsal cloacal wall is cut and reflected to the turtle's right (left side in image). B. Close-up of the proximal cloaca, showing the details of the drainage structures and the base of the penis. C. Cross section of the distal penis showing its fibrous and spongy structure. D. Posterior and ventral view of a testis and its ducts in a stranded adult male in breeding condition. Hypertrophied testis (partially cropped off at the bottom of the image), distended epididymis, and vas deferens are characteristic of a male during breeding season. Left gonadal structures overlay the left kidney, which is deep to the coelomic membrane. E. Hatchling female gonad, duct, and kidney structures. The paramesonephric duct is often spotted with black pigment in leatherbacks, as are parts of the kidney surface. Remnant mesonephros is lateral to the metanephros and ureter.

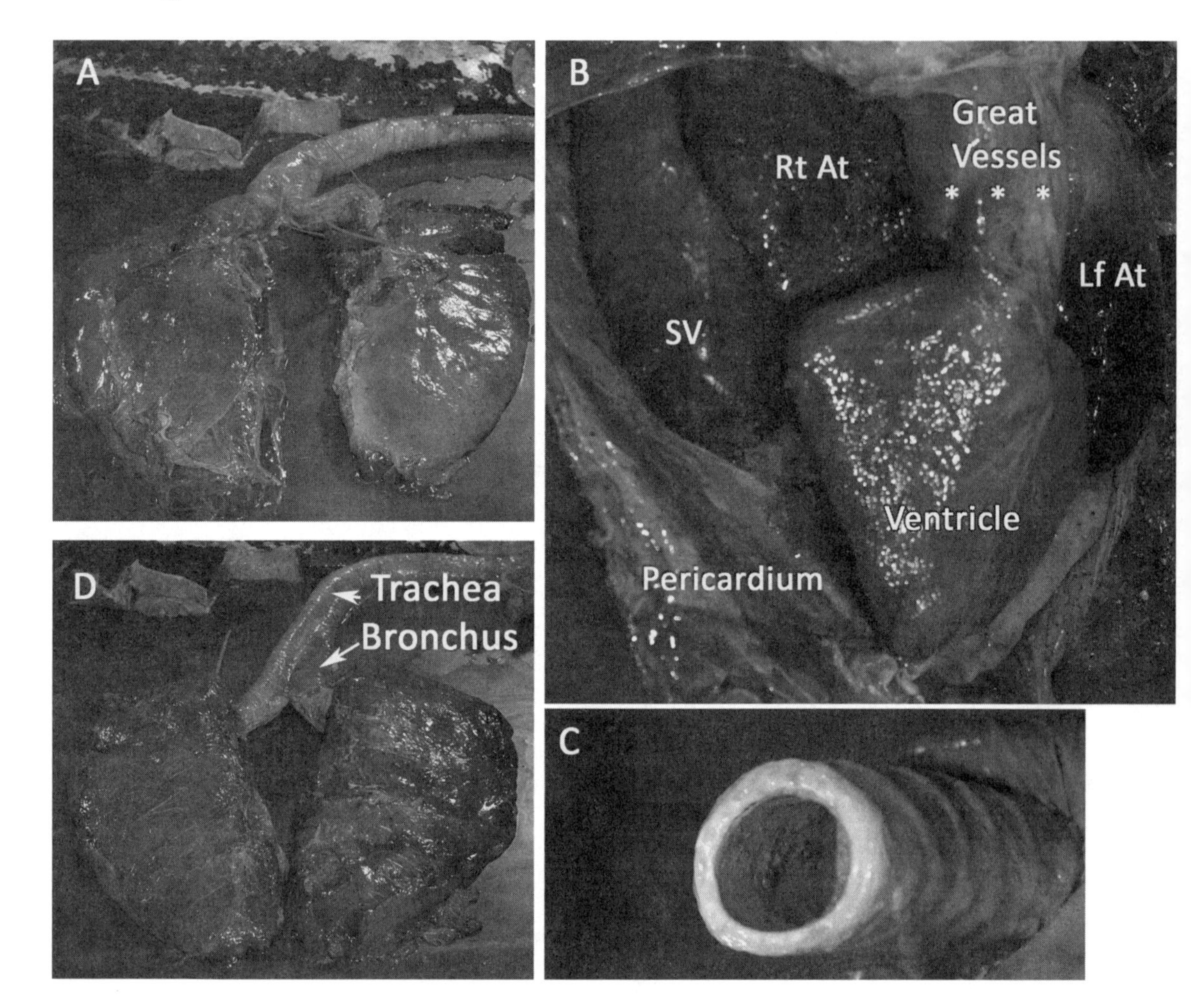

Fig. 4.9. Cardiopulmonary structures. A. Ventral view of paired lungs, bronchi, and trachea. Mesenteries that supported the liver (on right lung) and stomach (on left lung) were cut close to the lung surface. These lungs were partially inflated. B. Ventral view of heart of a dead adult male leatherback, with the ventricle contracted. Apex of ventricle is anchored to the tough pericardium by the fibrous gubernaculum cordis (not labeled). The large distended right atrium is common in dead turtles. The sinus venosus is distended so that it is displaced ventrally and can be seen through the pericardium. This positioning likely is a postmortem change. C. Cut trachea showing the cartilaginous rings proximal to the carina. D. Dorsal view of paired lungs shown in A. Visceral peritoneum is partially retained in this lung preparation along with the segmental arteries and nerves from the carapace.

is often larger than the left (fig. 4.9B). Blood from the sinus venosus flows into the right atrium.

Oxygenated arterial blood from the lungs flows into the left atrium. Both atria separately open into the single ventricle that is partially subdivided into three compartments (O'Donoghue 1918). The ventricle is triangular in adults but tends to be rounded and wider than it is long, particularly in young animals (figs. 4.7A, 4.7B, 4.9B). The caudal apex of the ventricle is attached to the pericardium via a fibrous connective tissue tendon, the gubernaculum cordis, which anchors the ventricle as it contracts.

Walls of the atria are formed of thin trabecular cardiac muscle and relatively thin, compact cardiac muscle. The ventricle is composed of a thin layer of compact cardiac muscle and thick, complex trabecular muscle that gives the ventricular compartments a spongy appearance.

Arterial System

The arterial system is grossly similar to that of other sea turtles except in the elaboration of extensive networks of blood vessels that are associated with countercurrent heat exchange (Greer et al. 1973). These systems, known as retia (singular is rete), are best described for the axillary and inguinal regions (figs. 4.5D, 4.7C). There may also be similar adaptations to conserve or release body heat associated with the throat, and possibly the plastron (Penick et al. 1998).

There are three great vessels (pulmonary trunk, left aorta, and right aorta) that arise from the ventral and anterior part of the ventricle (fig. 4.9B). Their bases are supported by hyaline cartilage. The right aorta branches almost immediately, as the brachiocephalic trunk, which gives rise to paired, large subclavian arteries and common carotid arteries. Thyroid arteries from the brachiocephalic trunk supply the single, gelatinous thyroid gland, which is surrounded by connective tissue and fat.

Paired common carotid arteries arise from either the brachiocephalic trunk directly or from the subclavian arteries (lateral to the thyroid arteries). Common carotids supply blood to the head. They bifurcate as they enter the skull and form the small internal carotid arteries and large external carotid arteries (plate 5).

Each subclavian artery continues laterally toward its flipper and becomes the axillary artery near the medial aspect of the scapulocoracoid junction. The axillary artery branches multiple times giving rise to (1) the anterior scapular artery to the scapular musculature; (2) the marginocostal artery, which extends posteriorly along the lateral carapace, and (3) multiple pectoral branches to the ventral pectoral muscles (these, with the pectoral veins, form a pectoral rete). The axillary artery continues distally, branching to form multiple arteries that travel with axial veins in the proximal shoulder muscles as a rete. The distal axillary branches join (anastomose) at the brachial artery, which gives rise to the radial artery and ulnar artery. They form a pair of vascular arches (vascular circumflex) in the carpus of the ventral flipper. Small digital arteries arise from this arch and lie between the flipper's digits.

The left aorta, the middle of the three great vessels, curves dorsolaterally and passes the more cranial aspect of the stomach before giving rise to three major arteries: the gastric, coeliac, and the superior mesenteric arteries (Wyneken 2001). The gastric artery usually arises more cranially and bifurcates into branches to the greater (lateral) and lesser (medial aspect) curvatures of the stomach. The large coeliac artery branches, forming (1) the anterior pancreaticoduodenal artery to much of the pancreas, duodenum, and stomach; and (2) the posterior pancreaticoduodenal artery to the distal pancreas, duodenum, liver, and gallbladder. The superior mesenteric artery arises next and supplies the many branches in the intestinal mesenteries to the small intestines. Caudal to the origin of the superior mesenteric artery, the left aorta continues posteriorly where it joins the right aorta (typically) to form a single dorsal aorta. The anterior-posterior location where the two aortae join is variable; generally it is in the middle third of the body's length.

The dorsal aorta continues posteriorly, supplying paired costal arteries that enter the carapace and form multiple branches in the blubber. Next, gonadal arteries arise from the dorsal aorta and enter the mesenteries as multiple branches supplying the ovaries or testes. At least one pair of adrenal arteries, and three or more pairs of renal arteries, arise from the dorsal aorta and supply each kidney (fig. 4.7D). Adrenal and renal arteries may arise as left-right pairs, but staggered in location. A pair of epigastric arteries branch off the dorsal aorta, near the kidney, and travel to the left and right; each joins a marginocostal artery.

Posterior to the renal arteries, the dorsal aorta gives rise to external and internal iliac arteries, sometimes via the common iliac. The external iliac branches multiple times to supply the hind limb rete system (fig. 4.5D). Several large femoral vessels continue into the hind limb and form the femoral and sciatic arteries. The internal iliac artery branches to form the relatively large haemorrhoidal artery supplying the large intestine. It also provides branches to the bladder and gonadal ducts. The remaining single dorsal aorta then extends to the tail as the vertebral artery.

The other great vessel, the pulmonary trunk, divides shortly after leaving the heart, forming right and left pulmonary arteries to the lungs. Pulmonary arteries have thick muscular sphincters as they approach the lung, extending into the vessel as they enter the lung. These sphincters are thought to function in initiating intracardiac shunting. The pulmonary arteries travel with each bronchus, giving off multiple branches throughout the lung (Wyneken 2001).

Venous System

Blood from the body enters the sinus venosus via major veins through the left and right superior vena cavae from the head, neck, and flippers; left hepatic vein; and the inferior vena cava (=right hepatic vein). Each superior vena cava receives branches from the internal and external jugular veins, subclavian veins, and azygos veins. The azygos vein receives blood from a series of vessels located deep within the pectoral muscles.

The external jugular vein is large but deeply positioned in leatherbacks. (It is often termed the dorsal cervical sinus, a commonly used venipuncture [blood collection] site in hard-shelled sea turtles and hatchlings, but is difficult to access in leatherbacks due to their extensive fat layers around the neck.) The left and right external jugular veins are paired and drain structures of the head to the common cardinal veins, which drain into the sinus venous. Transverse cervical veins join external jugulars and provide drainage for the dorsal cervical muscles, cervical vertebrae, and the spinal meninges.

Venous supply from each flipper and shoulder to the paired subclavian veins is difficult to trace because each flipper is drained by several variable rete systems. As a result, the description of the vasculature starts centrally, combining the results of my anatomic vascular preparations with descriptions from the literature (Greer et al. 1973). Each subclavian vein receives a thyroscapular vein with branches from the thyroid gland and the scapular musculature. The subclavian vein also drains the pectoral rete system directly or via one or more axillary veins. The transverse scapular vein appears to arise from a small rete system, which includes one or more the cephalic veins from the dorsal, posterior, and ventral flipper. After receiving the thyroscapular branch, the subclavian vein extends laterally and forms the multiple axillary veins in the axilla.

Many small axillary veins form the venous component of the axillary rete system. The axillary veins can be traced to the brachial vein in the upper forelimb. The brachial vein drains blood from the flipper via the internal brachial vein to the posterior (postaxial) flipper and the dorsal brachial vein to the anterodorsal flipper. There is a vascular network of veins distal to the carpus that receives drainage from the digital veins that travel dorsal and ventral to each digit.

Veins draining the hind limbs and tail form the renal portal system that drain to the kidneys (fig. 4.7D). Because of the extensive hind limb rete system, it appears that the anatomical potential for blood to bypass the renal portal system may be greater in leatherbacks than in other turtles.

Nervous System

In leatherbacks, the tubular brain passes from the nasal cavity to the spinal cord along a path of dorsoventral curves as it proceeds from anterior to posterior. The sense organs (eyes and ears) are also discussed below.

Brain

The brain of the leatherback is similar to that of other sea turtles except in the relative length of the epiphysis and pineal. The pineal is extended into a pineal recess in the dorsal skull. In most vertebrates, this neuroendocrine organ is sensitive to long-term changes in light levels, such as day length. In the leatherback, it is housed in a cartilaginous space adjacent to the thinnest part of the skull roof under the pink spot visible near the center of the head.

The tubular brain (figs. 4.2F, 4.10; plate 3) has very long, paired olfactory tracts (cranial nerve I) leading to the olfactory bulbs, which are adjacent to the cerebral hemispheres. The pituitary gland is located ventrally, along with the optic chiasma and the optic (cranial nerve II) nerves. The epiphysis and pineal (pineal complex) arise dorsally and extend into a cartilaginous recess in the dorsal skull that is associated with the pink spot on the dorsal head (figs. 4.2B, 4.10; plate 3). These structures plus cranial nerve I are parts of the forebrain. The midbrain structures include the paired optic lobes that are dorsal and posterior to the cerebrum. The paired optic (cranial nerve II), oculomotor (cranial nerve III), and trochlear nerves (cranial nerve IV) arise lateral and ventral to the optic lobes (fig. 4.10). Posterior to the midbrain is the hindbrain; this includes the singular cerebellum and medulla (plate 3). Several prominent pairs of cranial nerves arise in this area (fig. 4.10): trigeminal (V), abducens (VI), facial (VII), statoaccoustic (VIII), glossopharyngeal (IX), and vagus (X). The medulla is relatively long in leatherbacks. The trigeminal, facial, and glossopharyngeal nerves innervate most of the jaws, tongue, and throat muscles (plates 7, 8) in leatherbacks as in other sea turtles (Schumacher 1973; Werneburg 2011). The last two sets of cranial nerves, the spinal accessory (XI) and hypoglossal (XII), arise laterally just anterior to the spinal cord.

Sense Organs

The eye is the best studied of the sensory systems in leatherback turtles (Brudenall et al. 2008). The eyeball is protected by paired lids (palpebrae) and a highly pleated conjunctiva. The sclera contains a ring of scleral ossicles peripherally and the back of the eyeball is supported by hyaline cartilage. A layer of loose connective tissue and thick fat surrounds it. The extrinsic ocular muscle anatomy is described, in detail, by Schumacher (1973) (plates 6, 7). Interiorly, the leatherback retina includes both rods and cones and there is a specialized area temporalis, which provides increased acuity ahead and below the turtle (Oliver et al. 2000).

The ear structure has received little attention recently. There is no external ear. The ear is composed of an unspecialized tympanum (plate 1) formed of skin overlying a fat pad in the temporal region of the face. There is a single ear bone, the stapes (=columella) that extends posteriorly and medially where its footpad articulates with the saccule. Connection of the stapes to

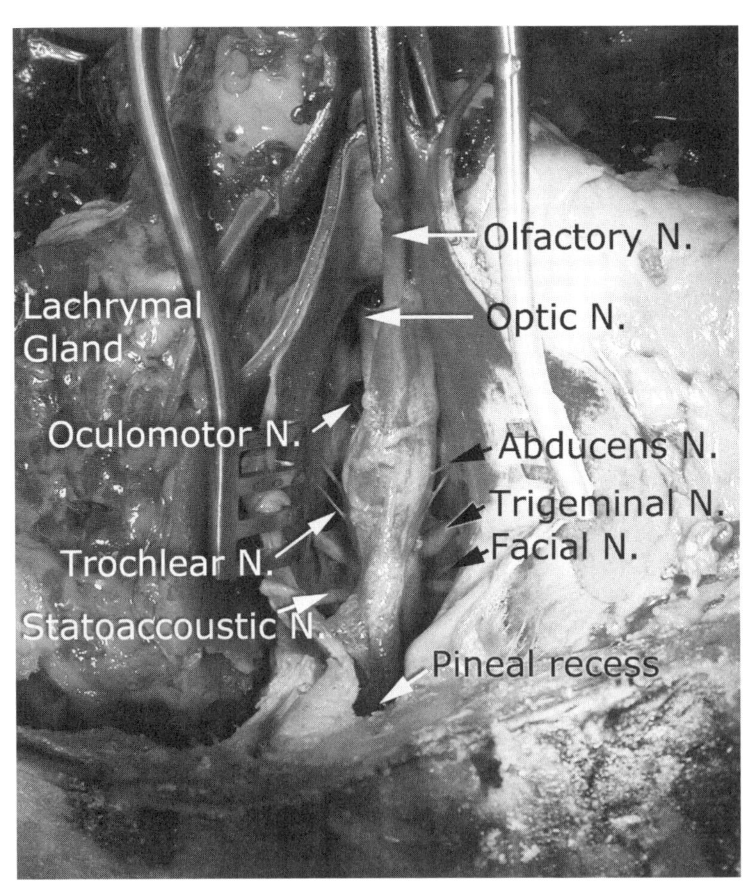

Fig. 4.10. Dorsal view of the forebrain and midbrain of an adult leatherback. Skull roof was cut away exposing brain from the posterior extent of the olfactory tracts, to just anterior to the dorsal extension of the pineal complex. Dura mater and leptomenix cover the brain. Chondrocranial portion of the neurocranium is cartilaginous and tough but flexible so that walls were spread to take this image. Several cranial nerves can be traced to the locations where they exit the braincase. Large left salt gland (lachrymal gland) can be seen to the left of the brain.

the oval window is via stapedosaccular strands (Wever and Vernon 1956; Wever 1978; Lenhardt et al. 1985).

Conclusions

While much is known about the anatomy of the leatherback, it became clear in writing this chapter that many systems have received little or no attention. Undoubtedly, some omissions are due to the daunting size of adults and the often poor condition of this imperiled species' carcasses when they become available. Nevertheless, the mouth and upper GI tract, heart, eyes, and bony skeleton receive much attention while the cartilaginous skeleton and most other organs remain described only grossly. Neuromuscular anatomy remains largely undescribed. Similarly, most viscera, including the reproductive systems, are described only grossly. Ontogenetic changes and seasonal changes in morphology are rarely discussed (Davenport et al. 2009; Davenport et al. 2011; Davenport et al. 2014). It is my hope that gaps in our understanding of morphology and its relationships of structure and function will intrigue readers and inspire study so that those long-standing data gaps disappear.

LITERATURE CITED

Aitken, R.N.C., and S. E. Solomon. 1976. Observations on the ultrastructure of the oviduct of the Costa Rican green turtle (*Chelonia mydas* L.). Journal of Experimental Marine Biology and Ecology 21: 75–90.

Brudenall, D. K., I. R. Schwab, and K. A. Fritsches. 2008. Ocular morphology of the leatherback sea turtle (*Dermochelys coriacea*). Veterinary Ophthalmology 11: 99–110.

Davenport, J., J. Fraher, E. Fitzgerald, P. McLaughlin, T. Doyle, L. Harman, and T. Cuffe. 2009. Fat head: an analysis of head and neck insulation in the leatherback turtle, *Dermochelys coriacea*. Journal of Experimental Biology 212: 2753–2759.

Davenport, J., J. Fraher, E. Fitzgerald, P. McLaughlin, T. Doyle, L. Harman, T. Cuffe, and P. Dockery. 2009. Ontogenetic changes in tracheal structure facilitate deep dives and cold water foraging in adult leatherback sea turtles. Journal of Experimental Biology 212: 3440–3447.

Davenport, J., T. T. Jones, T. M. Work, and G. H. Balazs. 2014. Unique characteristics of the trachea of the juvenile leatherback turtle facilitate feeding, diving and endothermy. Journal of Experimental Marine Biology and Ecology 450: 40–46.

Davenport, J., V. Plot, J. V. Georges, T. K. Doyle, and M. C. James. 2011. Pleated turtle escapes the box-shape changes in *Dermochelys coriacea*. Journal of Experimental Biology 214: 3474–3479.

Den Hartog, J. C., and M. M. Van Nierop. 1984. A study on the gut contents of six leathery turtles *Dermochelys coriacea* (Linnaeus) (Reptilia: Testudines: Dermochelyidae) from British waters and from the Netherlands. Zoologische Verhandelingen 209: 1–36.

Deraniyagala, P.E.P. 1939. Tetrapod reptiles of Ceylon, Vol. 1. Testudinates and crocodilians. Dulau, London, UK.

Fraher, J., J. Davenport, E. Fitzgerald, P. McLaughlin, T. Doyle, L. Harman, and T. Cuffe. 2010. Opening and closing mechanisms of the leatherback sea turtle larynx: a crucial role for the tongue. Journal of Experimental Biology 213: 4137–4145.

Girling, J. E. 2002. The reptilian oviduct: a review of structure and function and directions for future research. Journal of Experimental Zoology 293: 141–170.

Greer, A. E., J. D. Lazell, and R. M. Wright. 1973. Anatomical evidence for a countercurrent heat exchanger in the leatherback turtle (*Dermochelys coriacea*). Nature 244: 181.

Hyman, L. H. 1922. A laboratory manual for comparative vertebrate anatomy. University of Chicago Press. Chicago, IL, USA.

Kardong, K. V. 2012. Vertebrates: Comparative anatomy, function, evolution. 6th ed. McGraw-Hill, New York, NY, USA.

Kordikova, E. G. 2002. Heterochrony in the evolution of the shell of Chelonia, Part 1. Terminology, Cheloniidae, Dermochelyidae, Trionychidae, Cyclanorbidae and Carettochelyidae. Neues Jahrbuch für Geologie und Paläontologie 226: 343–417.

Leceta, J., E. Garrido, M. Torroba, and A. G. Zapata. 1989. Ultrastructural changes in the thymus of the turtle *Mauremys caspica* in relation to the seasonal cycle. Cell and Tissue Research 256: 213–219.

Lenhardt, M. L., R. C. Klinger, and J. A. Musick. 1985. Marine turtle middle-ear anatomy. Journal of Auditory Research 25: 66–72.

Lutcavage, M. E., and P. L. Lutz. 1997. Diving physiology. In: P. L. Lutz and J. A. Musick (eds.), The biology of sea turtles. CRC Press, Boca Raton, FL, USA, pp. 277–296.

Lutz, P. L. 1997. Salt, water and pH balance in the sea turtle. In: P. L. Lutz and J. A. Musick (eds.), The biology of sea turtles. CRC Press, Boca Raton, FL, USA, pp. 343–362.

Miller J. D., and C. J. Limpus. 2003. *Ontogeny of marine turtle gonads*. In: P. L. Lutz, J. A. Musick, and J. Wyneken (eds.), The biology of sea turtles, Vol. 2. CRC Press, Boca Raton, FL, USA, pp. 199–224,

Murphy, C., D. Kelliher, and J. Davenport. 2012. Shape and material characteristics of the trachea in the leatherback sea turtle promote progressive collapse and reinflation during dives. Journal of Experimental Biology 215: 3064–3071.

Nicolson, S. W., and P. L. Lutz. 1989. Salt gland function in the green sea turtle *Chelonia mydas*. Journal of Experimental Biology 144: 171–184.

O'Donoghue, C. H. 1918. The heart of the leathery turtle, *Dermochelys (Sphargis) coriacea* with a note on the septum ventriculorum in the Reptilia. Journal of Anatomy 52: 467–480.

Oliver, L. J., M. Salmon, J. Wyneken, R. Heuter, and T. Cronin. 2000. Retinal anatomy of hatchling sea turtles: anatomical specializations and behavioral correlates. Marine and Freshwater Behavior and Physiology 33: 233–248.

Penick, D. N., J. R. Spotila, M. P. O'Connor, A. C. Steyermark, R. H. George, C. J. Salice, and F.V. Paladino. 1998. Thermal independence of muscle tissue metabolism in the leatherback turtle, *Dermochelys coriacea*. Comparative Biochemistry and Physiology A 120: 399–403.

Perry, S. F. 1998. Lungs: comparative anatomy, functional morphology and evolution. In: C. Gans and T. Gaunt (eds.), Biology of the Reptilia, Vol. 19. Society for the Study of Amphibians and Reptiles, Ithaca, NY, USA, pp.1–92.

Rhodin, A.G.J. 1985. Comparative chondro-osseous development and growth of marine turtles. Copeia 1985: 752–771.

Rhodin, A.G.J., J. A. Ogden, and G. J. Conlogue. 1981. Chondro-osseous morphology of *Dermochelys coriacea*, a marine reptile with mammalian skeletal features. Nature 290: 244–246.

Romer, A. S. 1956. Osteology of the reptiles. University of Chicago Press, Chicago, IL, USA.

Schachner, E. R., J. R. Hutchinson, and C. G. Farmer. 2013. Pulmonary anatomy in the Nile crocodile and the evolution of unidirectional airflow in Archosauria. PeerJ 1: e60. doi: 10.7717/peerj.60.

Schumacher G.-H. 1973. Die kopf-und halsregion der lederschildkrote *Dermochelys coriacea* (Linnaeus 1766). Abhandlungen der Akademie der Wissenschaften der DDR. Akademie-Verlag, Berlin, Germany.

Snover, M. L., and A.G.J. Rhodin. 2008. Comparative ontogenetic and phylogenetic aspects of chelonian chondro-osseous growth and skeletochronology. In: J. Wyneken, M. H. Godfrey, and V. Bels (eds.), Biology of turtles. CRC Press, Boca Raton, FL, pp. 17–43.

Vickaryous, M. K., and J.-Y. Sire. 2009. The integumentary skeleton of tetrapods: origin, evolution, and development. Journal of Anatomy 214: 441–464.

Walker, W. F. 1973. The locomotor apparatus of Testudines. In: C. Gans and T. S. Parsons (eds.), Biology of the Reptilia, Vol. 4. Academic Press, London, UK, pp. 1–100.

Werneburg, I. 2011. The cranial musculature of turtles. Palaeontologia Electronica 14: 15A. http://palaeo-electronica.org/2011_2/254/index.html.

Wever, E. G. 1978. The reptile ear. Princeton University Press, Princeton, NJ, USA.

Wever, E. G., and J. A. Vernon. 1956. Sound transmission in the turtle's ear. Proceedings of the National Academy of Science USA 42: 292–299.

Wyneken, J. 2001. Guide to the anatomy of sea turtles. NMFS Technical Publication. NOAA Technical Memorandum NMFS-SEFSC-470. NOAA, Miami, FL, USA.

Wyneken, J. 2003. The external morphology, musculoskeletal system, and neuro-anatomy of sea turtles. In: P. L. Lutz, J. A. Musick, and J. Wyneken (eds.), The biology of sea turtles, Vol. 2. CRC Press, Boca Raton, FL, USA, pp. 39–77.

Wyneken, J. 2013. The skeleton; an *in vivo* view of structure. In: J. Wyneken, K. J. Lohmann, and J. A. Musick (eds.), The biology of sea turtles, Vol. 3. CRC Press / Taylor and Frances Group, Boca Raton, Florida, USA, pp 79–95.

Plates 1–8: Schumacher (1973) provides a comprehensive description of the anatomy of the leatherback head. Several of the detailed color drawings are reproduced here. The original treatise is written in German, as were some labels; others are given by the Latin names for each structure. Where appropriate, those labels were translated to their contemporary English form for this book. In some cases, the original labels were incorrect or incomplete and have been corrected here.

Dissections were conducted in layers, exposing more superficial structures first and then progressively removing them to show the deeper structures and their relationships. Color plates are organized here from the superficial to progressively deeper structures. The brain was removed and its cranial nerves numbered in the brain drawings; however, cranial nerves and their branches are named and numbered as they were drawn in the head dissections.

All plates are from Schumacher G.-H. 1973. Die kopf-und halsregion der lederschildkrote *Dermochelys coriacea* (Linnaeus 1766). Abhandlungen der Akademie der Wissenschaften der DDR. Akademie-Verlag, Berlin, Germany. The original plate designations follow the current designations in parentheses.

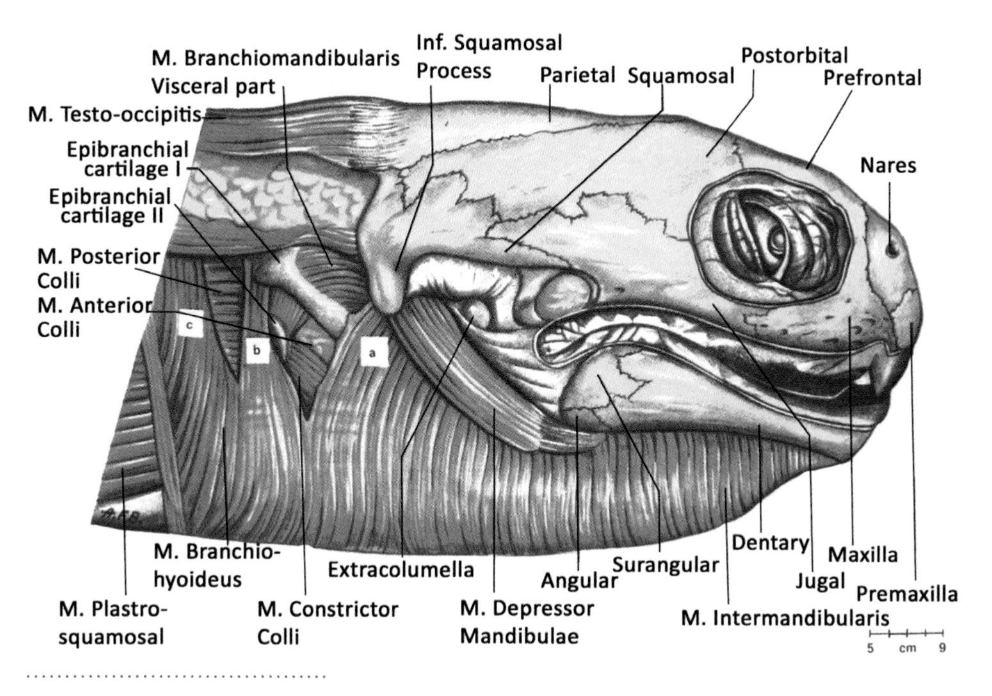

Plate 1. (Tabula XVIII.) Superficial lateral neck and throat muscles.

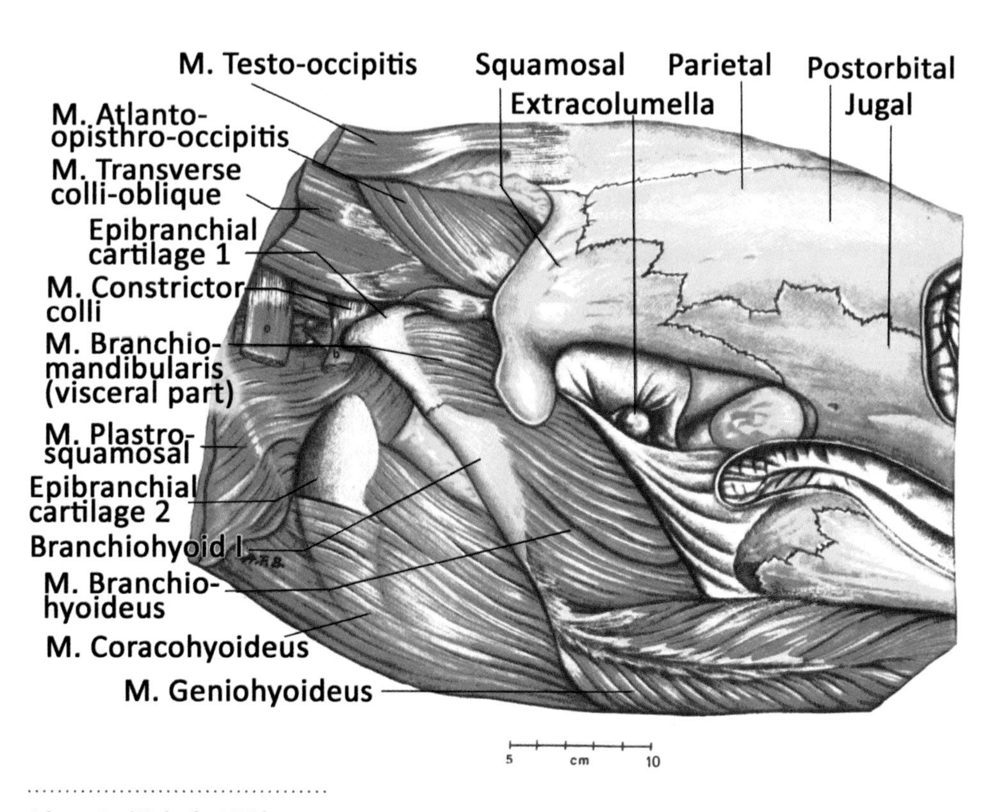

Plate 2. (Tabula XX.) Detail of lateral neck and hyoid arch muscles (superficial muscles removed).

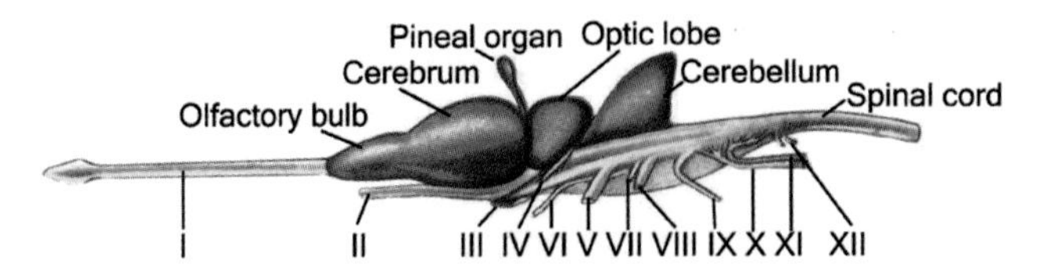

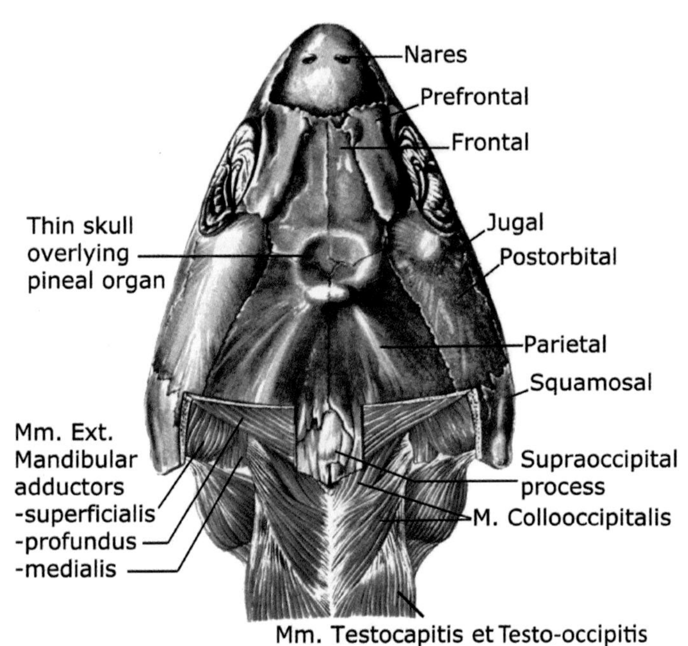

Plate 3. Top (Tabula XVII., Fig. 1) Lateral view of brain. ***Left (Tabula IV.)*** Dorsal dermatocranial skull bones, pineal recess, and dorsal neck muscles.

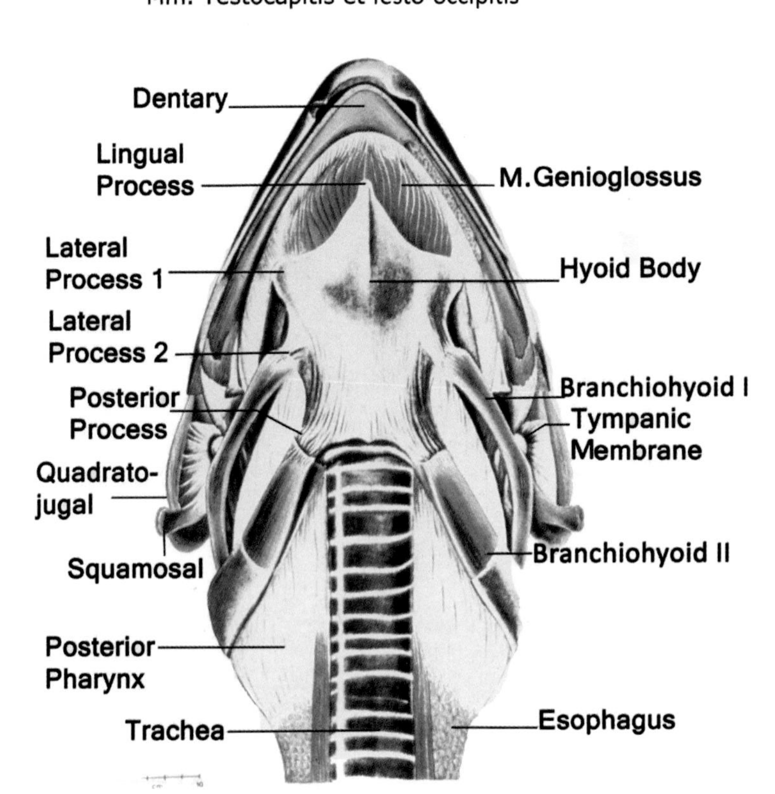

Plate 4. (Tabula XXV.) Ventral hyoid apparatus, in situ (most muscles removed).

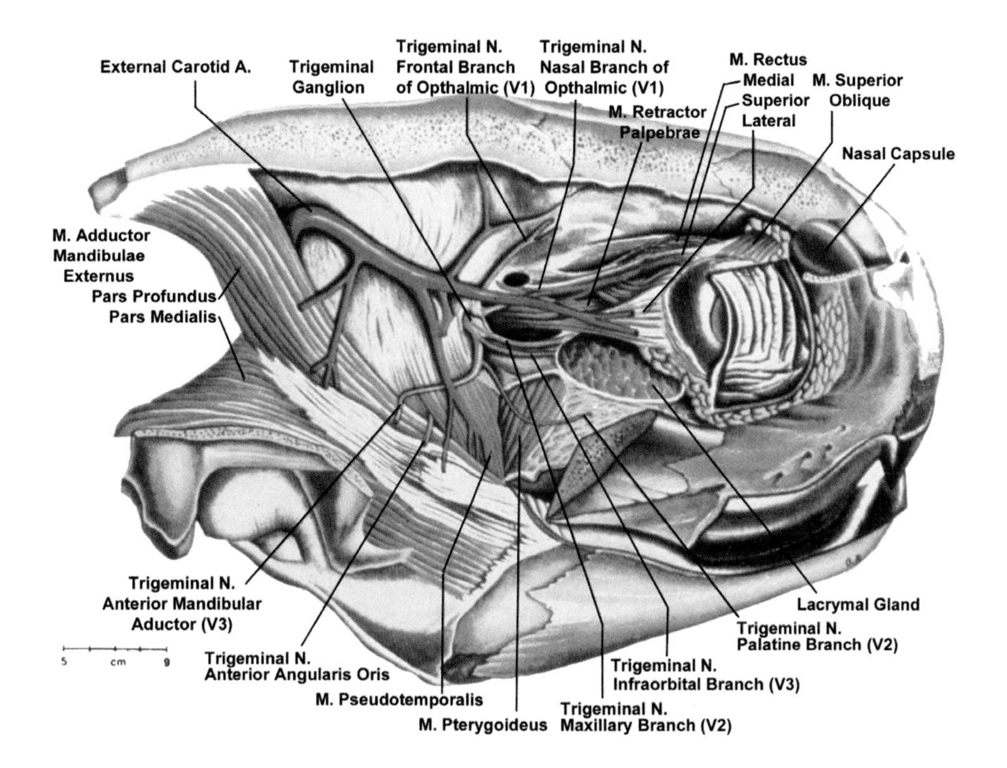

Plate 5. (Tabula X.) Deep jaw muscles, trigeminal nerve, and cut extrinsic eye muscles.

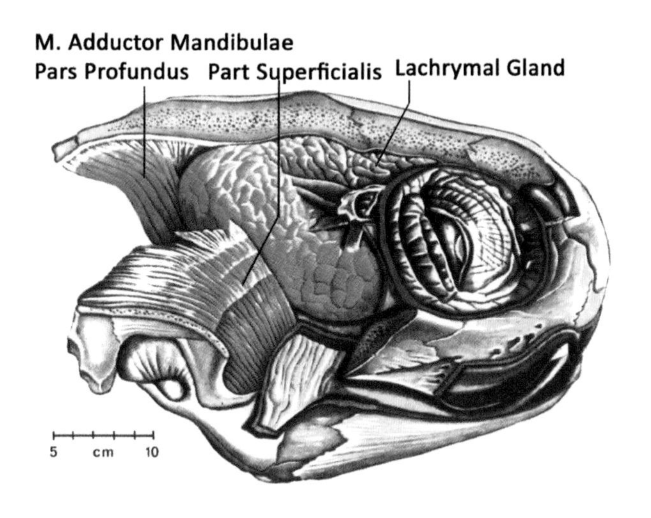

Plate 6. (Tabula V.) Lachrymal gland and eye in situ and excised.

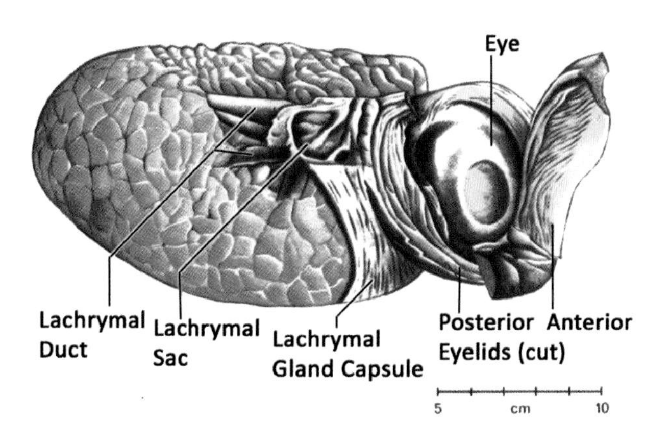

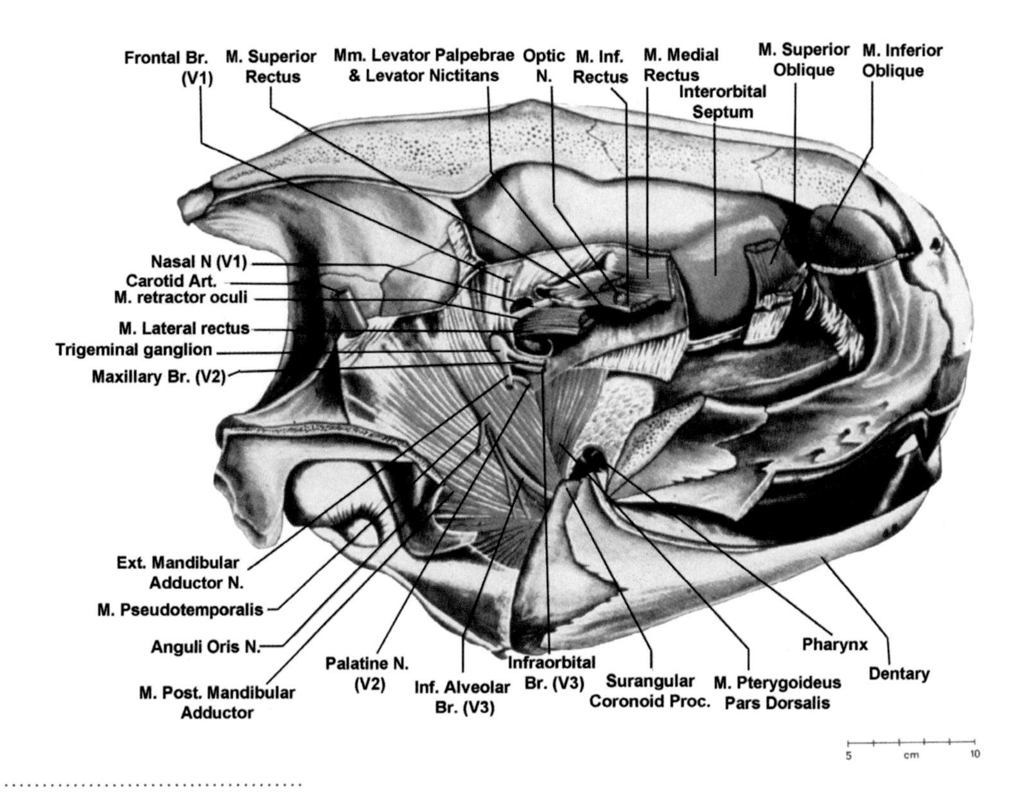

Plate 7. (Tabula XIII.) Lateral head arterial circulation and intermediate jaw muscles.

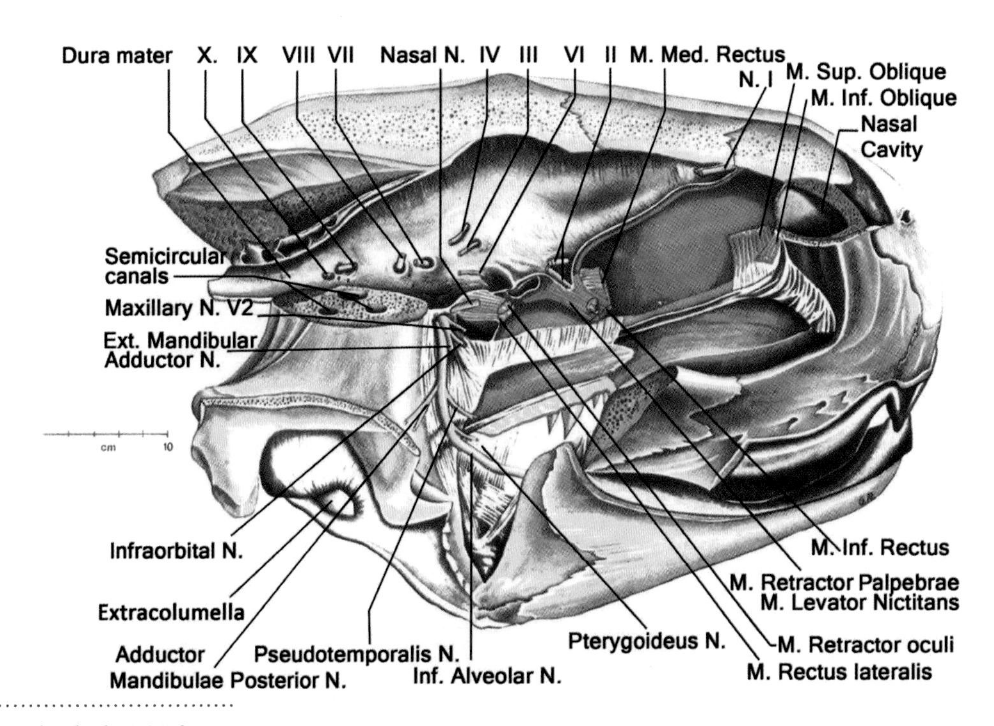

Plate 8. (Tabula XVI.) Braincase and cranial nerve roots.

Part II LIFE HISTORY AND REPRODUCTION

5

Reproductive Biology of the Leatherback Turtle

DAVID C. ROSTAL

The leatherback turtle (*Dermochelys coriacea*) is the largest extant turtle species and belongs to the monophyletic family Dermochelyidae. Leatherback turtles display many unique characteristics relative to other sea turtle species, both with respect to their gross anatomy (Carr 1952; Wyneken 2003), diving behavior (Eckert and Eckert 1989), thermal biology (Frair et al. 1972; Greer et al. 1973; Mrosovsky 1980), and nesting ecology (Tucker and Frazer 1991; Tucker and Frazer 1994; Rostal et al. 1996; Rostal et al. 2001). The nesting ecology of *D. coriacea* is unusual in that females nest up to twelve times in a single nesting season (each time is one nesting event), display the shortest inter-nesting interval of any sea turtle (9 to 10 days), and produce a small number of eggs per nest relative to their large body size (Hirth 1980; Tucker and Frazer 1991; Reina et al. 2002). They are also the only species that produces numerous albumin filled egg-like structures with each clutch. These are termed shelled albumin gobs, also called SAGs (Wallace et al. 2004; Wallace et al. 2006). Various functions for SAGs have been proposed, such as thermal buffering (Frazier and Salas 1984), satiation of nest predators (Hirth 1980), and increased gas exchange in the nest (Pritchard and Trebbau 1984); however, their function still remains a mystery. The SAGs may be "production over-runs" of leatherback oviducts producing copious albumin for eggs (Wallace et al. 2006; chapter 12).

The reproductive physiology of several cheloniid sea turtle species is known; these include the (1) green turtle, *Chelonia mydas* (Owens 1976; Licht et al. 1979; Wibbels et al. 1992); (2) olive ridley, *Lepidochelys olivacea* (Licht et al. 1982; Rostal 2007); (3) loggerhead, *Caretta caretta* (Wibbels et al. 1990; Wibbels et al. 1992); and (4) Kemp's ridley, *Lepidochelys kempii* (Rostal et al. 1997; Rostal et al. 1998; Rostal 2005, 2007). Seasonal and nesting endocrine cycles for these species display similar patterns; this commonality is most likely a result of their monophyletic origin. The reproductive functions of luteinizing hormone (LH), follicle stimulating hormone (FSH), testosterone, estradiol, progesterone, and plasma calcium are relatively well understood for most members of the family Cheloniidae.

Most information on the reproductive endocrinology of sea turtles comes from studies on nesting females. Plasma estradiol, testosterone, and progesterone have been measured in wild female *C. mydas* (Licht et al.

1980; Wibbels et al. 1990; Wibbels et al. 1992), *L. olivacea* (Licht et al. 1982), *C. caretta* (Wibbels et al. 1990; Wibbels et al. 1992; Whittier et al. 1997), and *L. kempii* (Rostal et al. 1997). In addition, LH and FSH have been measured in female *C. mydas* (Licht et al. 1979; Licht et al. 1980), *L. olivacea* (Licht et al. 1982), and *C. caretta* (Wibbels et al. 1992). Hormonal data during the entire reproductive cycle for both males and females are known for captive *C. mydas* (Licht et al. 1979), captive L. *kempii* (Rostal et al. 1998), and wild *C. caretta* (Wibbels et al. 1990). Circulating corticosterone has been measured in wild *C. mydas* (Jessop et al. 1999; Jessop 2001), *L. olivacea* (Valverde et al. 1999), *C. caretta* (Gregory et al. 1996; Whittier et al. 1997), and hawksbill turtles, *Eretmochelys imbricata* (Jessop 2001). The reproductive biology of the leatherback turtle is far less known. In this chapter I present information on the reproductive biology of leatherback turtles with an emphasis on their physiology.

Courtship and Mating

Based primarily on captive studies, courtship and mating in sea turtles occur prior to nesting (Rostal et al. 1997). Once nesting begins a significant decline in mating activity is expected. In captive *L. kempii*, mating occurs one month prior to nesting. After a female mates several times, she becomes nonreceptive to male advances and avoids males or uses her hind flippers to obstruct copulation if a male mounts her (Rostal 2005, 2007). At the beginning of the nesting season of many sea turtles, mating attempts are often observed off the nesting beach; however, at this point in time it is highly probable that these are simply mounting events and not true copulations. As the season progresses, the number of males in the area declines rapidly. In *D. coriacea* at Parque Nacional Marino Las Baulas (PNMB), on the Pacific coast of Costa Rica, male turtles attempt to mate with females shortly before and during the early portion of the nesting season. Cameras on the backs of females recorded mating attempts in which females avoided males and copulation did not appear to be successful (Reina et al. 2005) At this time, females have fully developed ovaries with large preovulatory follicles (Rostal et al. 1996). Good accounts of courtship behavior in wild leatherback turtles have been difficult to obtain due to the deeper water usually associated with their nesting beaches and their long migrations.

Males

Reproductive System

Hamann et al. (2003) reviewed the male sea turtle reproductive system plus spermatogenesis. Similar to other sea turtles, the male leatherback has a paired reproductive tract with the testes located intraperitoneally and medial to the kidney. The epididymis is coiled and is adjacent to the testis. It is presumed that fully mature sperm are stored in the epididymis until mating, as in other sea turtles (Blanvillain et al. 2008; Blanvillain et al. 2010). Male leatherback turtles display secondary sex characteristics such as an elongated tail with enclosed penis (chapter 4), also similar to other sea turtles. However, unlike other sea turtles, the male leatherback lacks claws on the front and hind flippers (Wyneken 2003). There is no decornified region on the midline of the plastron as occurs in other male sea turtles, which is hypothesized to be a secondary sex characteristic resulting from increased androgen levels (Wibbels et al. 1991; Owens 1997). However, the plastron of a male leatherback is flexible and compressed in a concave shape even at shallow water depths during the breeding season (Reina et al. 2005) and that may aid in mounting a female. Mating is presumed to be seasonal due to the long migrations made by leatherbacks, observations of mating shortly before the beginning of the nesting season and observations of mating attempts that continue into the early nesting season (Reina et al. 2005), as seen in other sea turtles (Owens 1997).

Females

Reproductive System

Dermochelys coriacea is a seasonal nester. Tucker and Frazer (1991) reported that female *D. coriacea* nest up to 10 times in a single nesting season. Females at PNMB nest up to 14 times (Reina et al. 2002; Santidrián Tomillo et al. 2012). Although some leatherback turtles nest during most months at PNMB, the main nesting season begins in September or October (Steyermark et al. 1996). At the time of arrival at the beach, females display mature ovaries containing hundreds of large, preovulatory vitellogenic follicles (Rostal et al. 1996). As each female repeatedly nests during the season, follicles are ovulated from the ovaries and a gradual decrease in the number of follicles and size of the ovary occurs. These declines correlate with a decline in circulating testosterone and estradiol levels (Rostal et al. 1996; Rostal et al. 2001).

Ovarian Cycle

Rostal et al. (1996) used ultrasonography at PNMB to classify the ovarian state or reproductive condition of the female according to the following criteria (fig. 5.1):

Fig. 5.1. Image of noninvasive ultrasound scanning procedure on a nesting female leatherback turtle (*Dermochelys coriacea*) at Parque Nacional Marino Las Baulas, Costa Rica. The ultrasound equipment was placed behind the female and the 3.5 MHz convex linear transducer was positioned in the inguinal region. From Rostal et al. (1996), with permission from Allen Press Publishing Services.

Mature Ovary (Early Preovulatory): Presence of multiple, tightly grouped, large preovulatory vitellogenic follicles measuring more than 3.0 cm in diameter. No atretic follicles observed (fig. 5.2).

Intermediate Ovary (Late Preovulatory): Presence of multiple, loosely grouped, large preovulatory vitellogenic follicles measuring more than 3.0 cm in diameter. Atretic follicles observed in some females. Coelomic space also observed at the apex of the ovary (fig. 5.2).

Depleted Ovary (Postovulatory): Fewer than ten large, preovulatory vitellogenic follicles observed per ovary (fig. 5.2). Multiple atretic follicles are observed in most females.

Nesting females begin arriving at PNMB in September and continue to nest up to March of the following year. Using ultrasonography, we determined the reproductive condition of a subset of nesting females through the major portion of the season (Rostal et al. 1996; fig. 5.3). The majority of females scanned in November displayed mature ovaries (82%). In December, there were fewer females displaying mature ovaries (40%) and an increase in females displaying intermediate ovaries (60%). By January, the number of females displaying either mature or intermediate ovaries (23% each) continued to decrease while females displaying depleted ovaries at the end of their nesting cycle increased (54%).

Preovulatory vitellogenic follicles occur throughout the season. Mean follicular diameter does not appear to vary during the nesting season and it ranges per female from 2.95 to 3.55 cm with an overall mean diameter of 3.33 ± 0.02 cm (SE, $n = 30$; Rostal et al. 1996; fig. 5.2). During the nesting sequence, enlarging previtellogenic follicles do not occur in females, supporting the conclusion that vitellogenesis is complete prior to the arrival of the female at the nesting beach (Rostal et al. 1996; Rostal et al. 2001). Smelker et al. (2014) reported that low levels of vitellogenin were in circulation during the nesting season in *C. caretta*, but noted that these levels declined along with testosterone over the course of the season. In addition, the levels of vitellogenin measured during the nesting season are below the levels in females that are actively enlarging ovarian follicles several months prior to the nesting season when estradiol levels are significantly elevated. We expect a similar pattern may be occurring in *D. coriacea*, but with the rapid inter-nesting cycle of 9–10 days, and energy use shifted to oviductal egg formation, this circulating vitellogenin may function in maintaining preovulatory follicles for later ovulation.

Perrault et al. (2014) confirmed that nesting *D. coriacea* in the Caribbean are fasting throughout the nesting season as proposed for other sea turtles, and therefore the female is maintaining herself on fat reserves and not expending energy on growing ovarian follicles. Vitellogenesis is an energy demanding process. Rostal et al. (1996) detailed the reproductive characteristics of leatherback turtles. There were atretic follicles in females with intermediate or depleted ovaries (Rostal et al. 1996). There was no correlation between follicle size and carapace length; however, it is interesting to note that the smallest follicles occurred in the smallest female observed (CCL [curved carapace length] = 125 cm). Both eggs and SAGs occurred in the oviduct during egg laying (Rostal et al. 1996). Eggs ranged from 5.0 to 5.6 cm in diameter with mean diameter of 5.31 ± 0.06 cm (SE, $n = 16$). Yolk diameters ranged from 3.2 to 3.8 cm with a mean diameter of 3.53 ± 0.04 (SE, $n = 20$).

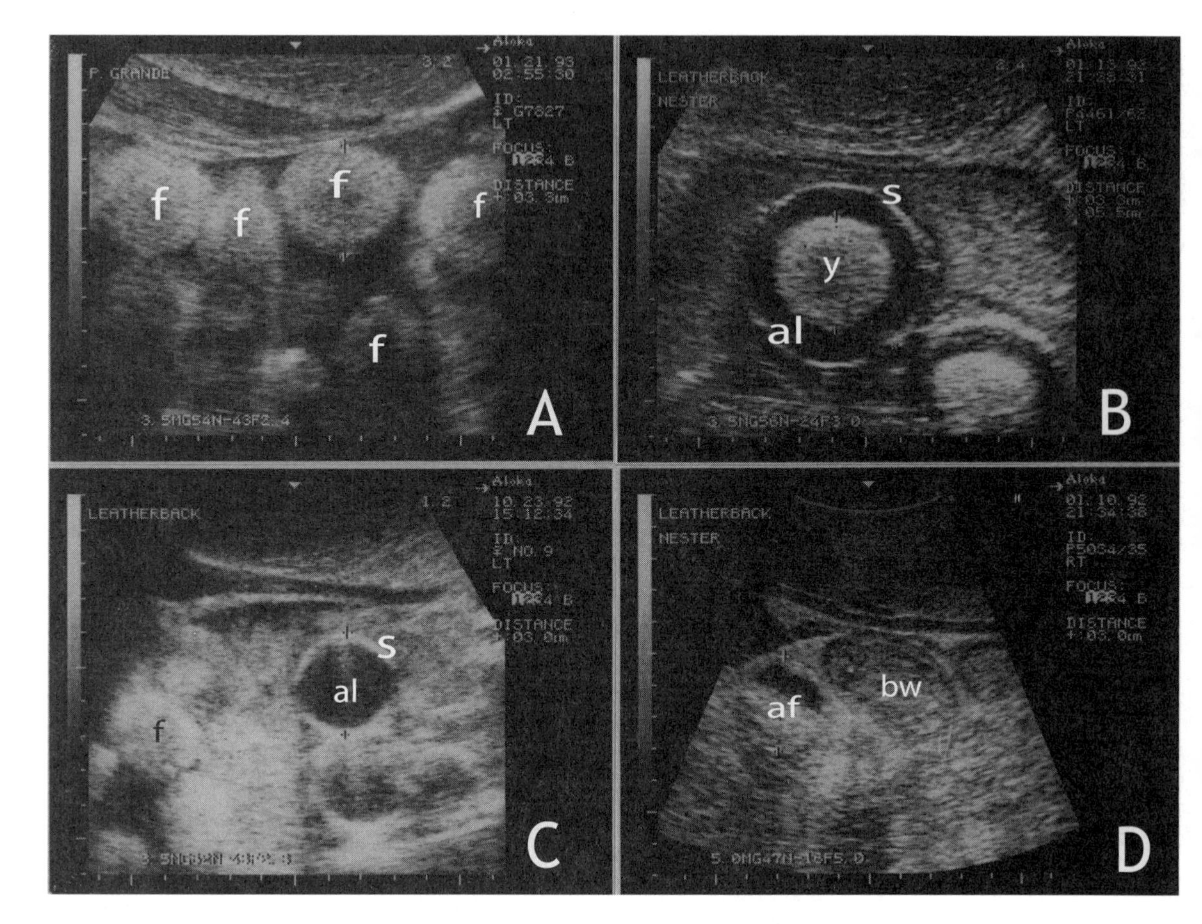

Fig. 5.2. Ultrasonography of leatherback turtle (*Dermochelys coriacea*) reproductive structures. A. Ultrasound image of large preovulatory vitellogenic follicles (f: 3.3 cm diameter) in a female with mature ovaries. B. Ultrasound image of a shelled oviductal egg at the time of nesting showing a clearly defined yolk (y: 3.3 cm diameter) and calcified shell (s: 5.5 cm diameter). The anechoic albumin layer (al) is also clearly visible. C. Ultrasound image of an oviductal yolkless egg-like structure (3.0 cm diameter) showing a dark anechoic center (al) and calcified shell (s). D. Ultrasound image of an atretic follicle (af: 3.0 cm diameter) in a female with depleted ovaries. Note the absence of vitellogenic follicles in the image. Intestinal bowel (bw) is clearly visible at this stage in the nesting cycle. Modified from Rostal et al. (1996), with permission from Allen Press Publishing Services.

Mean yolk diameter was similar to that of preovulatory vitellogenic follicles (3.33 ± 0.02 cm, SE, $n = 30$). The SAGs were the last structures produced in the oviduct and were not interspersed between eggs while in the oviduct. The SAGs were, however, interspersed among eggs in the nest. This may be the result of one oviduct releasing its eggs prior to the completion of egg deposition by the other oviduct (Rostal et al. 1996).

Plasma Testosterone, Progesterone, and Estradiol

Over the course of the nesting season, circulating testosterone, estradiol, and progesterone levels show similar patterns to those of other sea turtle species (Rostal et al. 1996; Rostal et al. 2001; Blanvillain et al. 2010; fig. 5.4). Testosterone levels decline as ovarian size decreases in *D. coriacea*, in a similar fashion to that observed in *C. mydas* (Licht et al. 1979), *C. caretta* (Wibbels et al. 1990), and *L. kempii* (Rostal et al. 1997; Rostal et al. 1998; Rostal 2005; Rostal et al. 2007). Estradiol levels in *D. coriacea* also decline slightly during the nesting season (fig. 5.4) and are relatively low, similar to those in *C. mydas* and *C. caretta* during the nesting season (Licht et al. 1979; Wibbels et al. 1990). Physiological and behavioral functions of testosterone and estradiol are only partially understood in sea turtles. Testosterone may play a role in stimulating migration, mating, and nesting behavior in sea turtles (Licht et al. 1979; Owens 1997; Wibbels et al. 1990; Rostal 2005). Circulating testosterone lev-

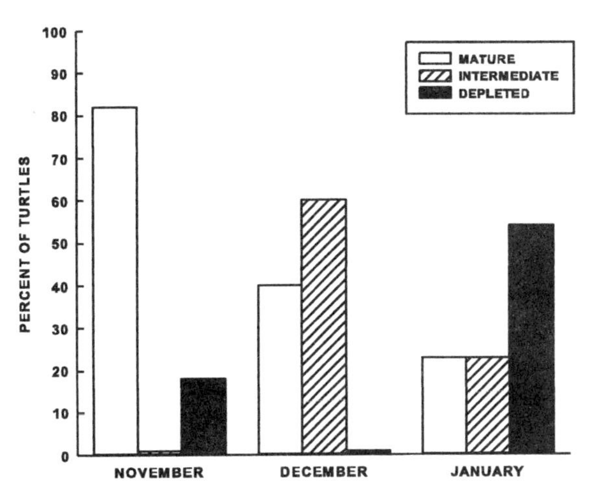

Fig. 5.3. Relationship of reproductive condition to nesting season period. Nesting *D. coriacea* reproductive condition was determined using ultrasonography, and percentage of females with mature, intermediate, and depleted ovaries per month was established. From Rostal et al. (1996), with permission from Allen Press Publishing Services.

els peak during the mating period prior to nesting in captive female *C. mydas* (Licht et al. 1979) and *L. kempii* (Rostal et al. 1998; Rostal 2005, 2007), while they peak in wild female *C. caretta* with the onset of migration (Wibbels et al. 1990). The difference may be a result of captivity. Estradiol stimulates vitellogenesis in reptiles (Ho 1987), including sea turtles (Heck et al. 1997). Licht et al. (1979) reported elevated estradiol levels in captive *C. mydas* only during the prebreeding period when follicular growth is occurring in the ovary. Progesterone levels do not vary greatly in *D. coriacea* during the nesting season (Rostal et al. 2001; fig. 5.4), but this could be due to sampling females within 30 minutes after nesting. Progesterone is associated with ovulation and increases sharply 24–48 hours post-nesting in other sea turtles (Licht et al. 1979; Licht et al. 1982; Wibbels et al. 1992).

Circulating levels of testosterone and estradiol are higher overall in nesting *D. coriacea* compared to those in other sea turtles (Licht et al. 1979; Wibbels et al. 1990; Rostal et al. 1998). However, some of the highest gonadal steroid levels measured in turtles occur in relatively small species (e.g., desert tortoise, *Gopherus agassizii*, Rostal et al. 1994; stinkpot turtle, *Sternotherus odoratus*, McPherson et al. 1982). Factors that influence overall circulating steroid levels in different taxa are unknown.

Circulating gonadal steroid levels vary over the course of the nesting season (fig. 5.4). Rostal et al. (1996) elucidated endocrine patterns by comparing hormonal levels with the reproductive condition of the female as determined by ultrasonography. Plasma testosterone levels decline significantly over the course of the nesting cycle from elevated levels in females at the beginning of their nesting cycle when their ovaries are mature, to intermediate levels midway through their nesting cycle, to low levels in females displaying depleted ovaries at the end of their nesting cycle. Plasma estradiol levels decline in a similar fashion over the course of the nesting cycle; they are highest in females displaying mature ovaries, intermediate in females that are midway through their nesting cycle, and lowest in females displaying depleted ovaries at the end of their nesting cycle. Plasma progesterone levels are highly variable between individuals and are not correlated with reproductive condition (Rostal et al. 1996). This lack of pattern could be due to sampling of females during nesting or within 30 minutes post-nesting. Progesterone increases sharply 24–48 hours post-nesting in association with an increased secretion of LH prior to ovulation of subsequent clutches in other sea turtles (*L. olivacea*, Licht et al. 1982; *C. mydas* and *C. caretta*, Wibbels et al. 1992). Unlike testosterone and estradiol, which appear to be secreted at a constant rate, progesterone secretion is more pulsatile and associated with ovulation. The production of progesterone by the corpora lutea following ovulation is relatively unstudied and may be near background levels at the time of sampling.

Vitellogenesis

Vitellogenesis appears complete prior to the arrival of the female at the nesting beach (Rostal et al. 1996; Rostal et al. 2001). There is no indication of multiple vitellogenic periods during the nesting season in *D. coriacea*. Multiple size classes of vitellogenic follicles are not observed using ultrasonography and plasma calcium levels do not vary relative to ovarian condition. In contrast, multiple size classes of follicles are observed using ultrasonography in multiclutching *G. agassizii* during vitellogenesis, as well as during the nesting season (Rostal et al. 1994). This response in *G. agassizii* is related to the availability of resources in the spring following emergence from hibernation (Rostal 2014). Nesting *G. agassizii* feed throughout the nesting season while nesting *D. coriacea* fast throughout the nesting season (Perrault et al. 2014). This limitation of resources in *D. coriacea* along with the much higher investment in reproduction supports the advantage of completing vitellogenesis prior to arrival at the nesting beach.

Increases in total calcium levels are correlated with vitellogenesis (Ho 1987) and ovarian follicular growth in

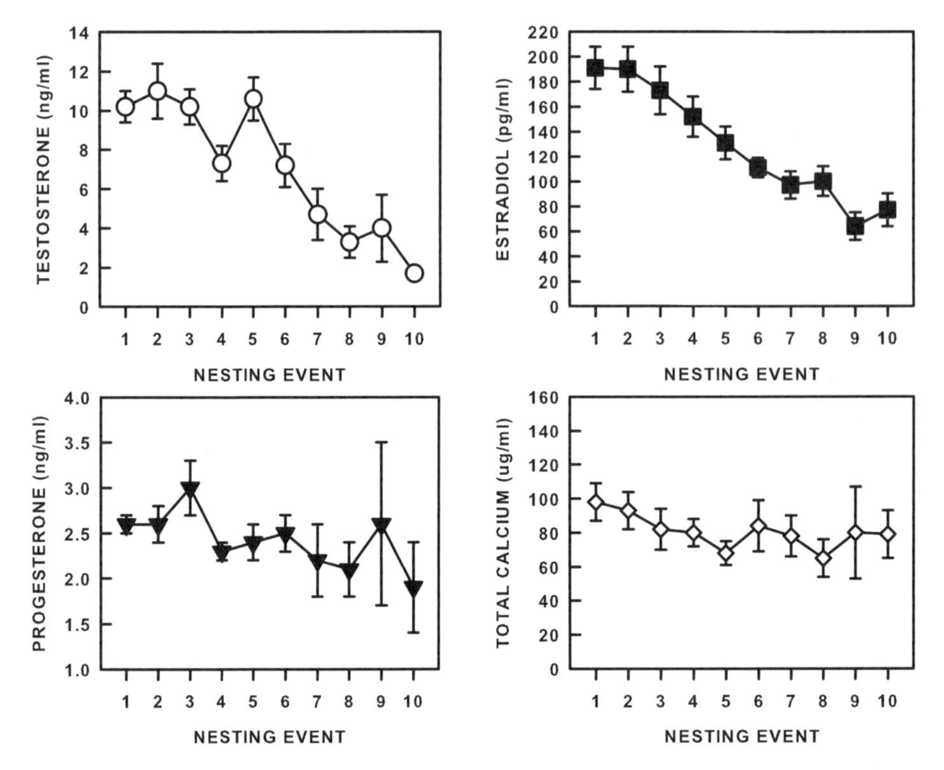

Fig. 5.4. Mean plasma testosterone, estradiol, progesterone, and calcium levels plotted relative to recorded nesting events for leatherback turtles (*Dermochelys coriacea*) at Parque Nacional Marino Las Baulas, Costa Rica, during the 1996–97 and 1997–98 seasons. Blood samples were collected during nesting from the interdigital vein on the front flipper. Values are means ± SE. From Rostal et al. (2001), with permission from Elsevier.

a variety of reptiles. These include the cobra, *Naja naja* (Lance 1976); painted turtle, *Chrysemys picta* (Callard et al. 1978); American alligator, *Alligator mississippiensis* (Lance et al. 1983; Lance 1987); *L. kempii* (Rostal et al 1998); *G. agassizii* (Rostal et al. 1994); and tuatara, *Sphenodon punctatus* (Cree et al. 1991). In captive *L. kempii*, plasma calcium levels rise twofold from the nonvitellogenic to vitellogenic state in females (mean = 201.8 ± 12.7 ug/ml in December). Calcium levels, however, rapidly decline following mating and the first nesting in *L. kempii* and remain low throughout the nesting season, with a mean of 107.2 ± 10 ug/ml in May (Rostal et al. 1998; Rostal 2005). In nesting *D. coriacea*, plasma calcium levels remained relatively constant throughout the nesting season (Rostal et al. 2001; fig. 5.4). Basal calcium levels plus the observations of only one size of vitellogenic follicle observed in nesting *D. coriacea* support the conclusion that vitellogenesis is complete prior to the arrival of the nesting female at the nesting beach (Rostal et al. 1996; Rostal et al. 2001). Unfortunately, no samples from non-nesting females have been available for analysis. Since females caught feeding would most likely be highly stressed due to the capture procedure, blood chemistries may be highly altered from glucocorticoid and steroid release by the adrenal cortex as observed in other species (Lance and Rostal 2002). Circulating calcium levels in nesting *D. coriacea* are similar to those observed in wild nesting *L. kempii* (Rostal et al. 1998; Rostal 2005) and post-nesting *G. agassizii* (Rostal et al. 1994; Rostal 2014).

Ovulation

Following courtship and mating, the first ovulation of the nesting season occurs. Because turtles may be far from the nesting beach when this occurs it has been difficult to document. In captive *L. kempii*, ovulation occurred shortly after successful mating when the turtles were no longer behaviorally receptive to males (Rostal 2005). This was between three to four weeks prior to the first nesting event of the season. It is known that subsequent ovulations following the first nesting event occur within 48 hours following the completion of nesting and that specific hormonal events occur during this period.

Progesterone is primarily associated with ovulation in sea turtles. Progesterone levels increase sharply

24–48 hours following nesting in *L. olivacea* (Licht et al. 1982), *C. mydas*, and *C. caretta* (Licht et al. 1979; Wibbels et al. 1992). As the nesting season progresses, a steady decline in basal levels of progesterone levels occurred as the ovary went through repeated ovulations in *D. coriacea* (Rostal et al. 2001). While the trend is not significant in the leatherback, similar statistically significant patterns in declining levels of progesterone occur in other multi-clutching species (*C. caretta*, Wibbels et al. 1992; *L. kempii*, Rostal et al. 1997; Rostal et al. 1998; Rostal 2005; *G. agassizii*, Lance and Rostal 2002), suggesting that a similar process occurs in *D. coriacea*.

Luetinizing hormone (LH) increases sharply after nesting, concurrent with a surge in progesterone and ovulation of the ovum from the follicle into the upper oviduct, where it will be fertilized (Licht et al. 1979; Licht et al. 1982; Wibbels et al. 1992). Unlike the sequential decline observed in progesterone over the course of the nesting season, LH is only elevated during the brief period of ovulation and appears to have a very limited half-life in the blood, unlike progesterone. No data are available for LH levels in *D. coriacea*. Once the ovum has passed from the follicle in the ovary it enters the oviduct. As the developing egg travels down the oviduct, a new series of events occurs as the follicle is fertilized, the extra-embryonic membranes are formed, the albumin layers are laid down, and the shell is formed.

Oviductal Egg Development

Oviductal egg formation and development are not well studied in sea turtles. Ultrasound has been used successfully to track egg development in both captive and wild sea turtles (fig. 5.2; Rostal et al. 1990; Rostal et al. 1996; Rostal et al. 1997; Rostal et al. 1998; Blanco et al. 2012). Ovarian follicles (fig. 5.2A) are ovulated in response to LH and progesterone surges. Once ovulated, the ovum is first fertilized in the upper albumin gland region of the oviduct. As the fertilized ovum travels through the albumin gland region of the oviduct it is surrounded by a non-echoic albumin layer. Once the albumin layer deposition is completed, the ovum plus albumin enters into the shell gland region of the oviduct where the shell membrane and the eggshell form (fig. 5.2B; day six to nine). The SAGs appear as smaller shelled structures but lack a discernible yolk mass (fig. 5.2C). Following the completion of the nesting season, the remaining follicles that have not been ovulated undergo atresia (fig. 5.2D).

Atresia is the process by which energy originally stored as yolk platelets in the oocyte during vitellogenesis and follicular growth is reabsorbed from the oocyte and potentially reused by the fasting turtle (Perrault et al. 2014). During this process, the follicle forms a unique "cat eye" appearance, as observed with ultrasound, where there is a non-echoic line that forms in the echoic yolk (fig. 5.2D). This "cat eye" appearance is unique to all sea turtles (*L. kempii*, Rostal et al. 1990; Rostal et al. 1997; *L. olivacea*, Rostal 2007; *C. caretta*, D. C. Rostal, unpublished data; and *C. mydas*, D. C. Rostal, unpublished data), but has not been observed in freshwater or terrestrial turtles (*Gopherus agassizii*, Rostal et al. 1994, Rostal 2014; *Geochelone nigra*, Robeck et al. 1990; *Macrochelys temminckii*, D. C. Rostal, unpublished data; *Gopherus polyphemus*, Rostal 2014; *Trachemys scripta scripta*, D. C. Rostal, unpublished data; or *Alligator mississippiensis*, Lance et al. 2009). The band of non-echoic material widens until the follicle appears to loose spherical integrity and becomes irregular in shape. Then the follicle reduces in size over time as it eventually becomes a corpora albicans, leaving behind a small scar.

Oviposition

Clutch size is correlated with female size in most sea turtle species (Hirth 1980; Van Buskirk and Crowder 1994), but Hirth (1980) notes that mean nesting female size is not correlated with mean hatchling size for *C. mydas* populations. The relationship of follicle size to female size has not been studied in green turtles. In leatherback turtles there is no correlation between follicle size and female size. Though a slight effect of female size on egg size may occur, larger female sea turtles tend to produce larger clutches rather than larger eggs.

A minimum amount of yolk is required to nurture a viable size hatchling with an adequate yolk reserve. Congdon (1989) notes that larger hatchlings do not directly equate to increased fitness. Therefore, evolutionarily speaking, larger females should not necessarily produce larger follicles or larger hatchlings. While leatherback turtles may be capable of producing larger eggs compared to other sea turtles, selection appears to have favored larger numbers of relatively smaller eggs as well as an increased number of clutches. Interestingly, the flatback turtle, *Natator depressus*, produces smaller clutches (less than 50 eggs) and larger hatchlings relative to other sea turtles. Hirth (1980) suggested that the larger hatchling size should offset the lower number of hatchlings produced and have similar survival value. Factors influencing hatchling size and survival need further research.

Mean clutch size, including SAGs, does not vary significantly during the nesting season in leatherbacks

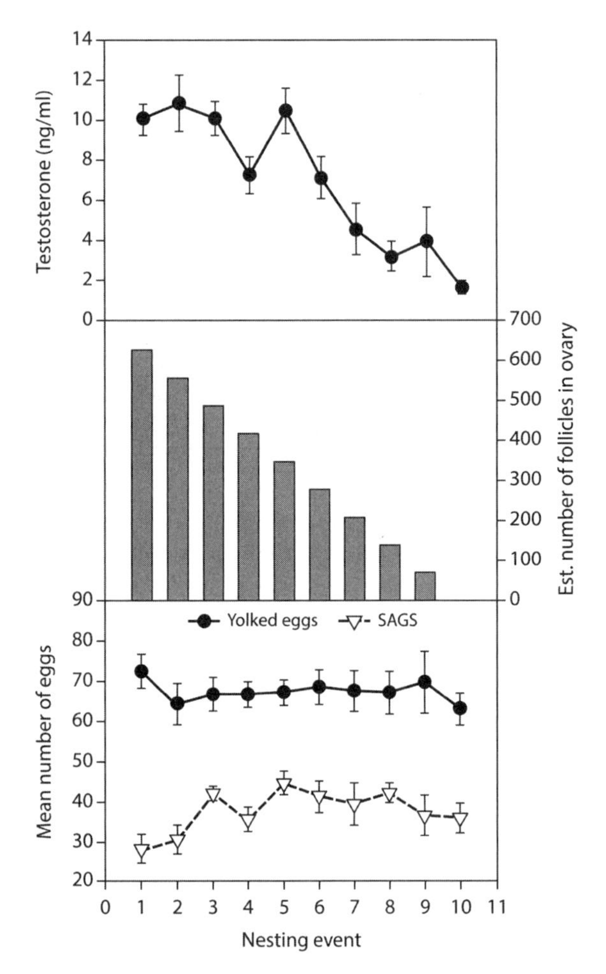

Fig. 5.5. Mean testosterone levels (*top*), estimated number of ovarian follicles per nesting event (*middle*), and mean number of eggs with yolks and SAGs per clutch (*bottom*) plotted relative to known nesting event for leatherback turtles nesting at Parque Nacional Marino Las Baulas, Costa Rica, during the 1996-97 and 1997-98 seasons. Sample size per nesting event ranged from a high of $n = 22$ (first nesting event) to a low of $n = 4$ (ninth nesting event). A total number of 123 nests were sampled. Values are means ± SE.

(Rostal et al. 2001; fig. 5.5). Females ovulate equal numbers of follicles per nesting event. Since leatherbacks ovulate a relatively small proportion of the available follicles at each nesting event, it is not unexpected that the decline in circulating testosterone and estradiol is more gradual than observed in other sea turtle species that may ovulate up to 50% of their available follicles at a single nesting event, as in *L. olivacea* (Licht et al. 1982; fig. 5.5) and *L. kempii* (Rostal et al. 1997). Testosterone levels in wild *L. kempii* decline by as much as 50% from the first nesting event to the second nesting event.

The function of SAGs still remains a mystery. A selective advantage for SAGs remains to be demonstrated. In the eastern Pacific leatherback population nesting at PNMB the mean number of eggs and mean number of SAGs is relatively constant (Rostal et al. 2001; fig. 5.5). It is possible that these SAGs may not have a function. The size and number of SAGs is more variable in contrast to the uniform size and number of eggs. Studies of nest success have not demonstrated an advantageous function for SAGs to date (Eckert 1987; Hirth and Ogren 1987; Eckert and Eckert 1990; Leslie et al. 1996; Steyermark et al. 1996). Further studies on these unique structures are needed to determine energy investment for their production and their potential cost to reproduction.

Nesting Patterns

At PNMB leatherbacks nest up to 14 times in a season, laying 700 plus eggs (Santidrián Tomillo et al. 2012). The mean number of clutches per season for a turtle is 9.5. Nesting occurs from September to April with the majority of females nesting in November, December, and January (Steyermark et al. 1996). Rostal et al. (1996) correlated steroid levels with reproductive condition determined by ultrasonography; however, the exact nesting sequence for a given female is not known.

When the female first arrives at the nesting beach, she displays mature ovaries containing hundreds of large, pre-ovulatory vitellogenic follicles at 3.33 cm diameter (Rostal et al. 1996). Vitellogenesis is complete prior to her arrival at the nesting beach (Rostal et al. 2001). Like other sea turtles, as the female repeatedly nests during the season, follicles are ovulated from both right and left ovaries and a gradual decrease in number of follicles and size of ovary occurs (Rostal et al. 1996).

Concurrent with this decline in the number of follicles and size of the ovary, a decline in circulating testosterone and estradiol occurs (figs. 5.4, 5.5). During the nesting season, circulating plasma testosterone, estradiol, and progesterone levels display patterns similar to those of other sea turtle species. Testosterone levels decline as ovarian size decreases—more specifically, as the number of follicles per ovary decrease—in a similar fashion to that observed in *C. mydas* (Licht et al. 1979), *C. caretta* (Wibbels et al. 1990), and *L. kempii* (Rostal et al. 1997; Rostal et al. 1998). Estradiol levels also decline significantly during the nesting season and are relatively low, similar to those observed in *C. mydas* or *C. caretta* (Wibbels et al. 1990). Progesterone levels do not vary greatly during the nesting season, but this could be due to our sampling of females during

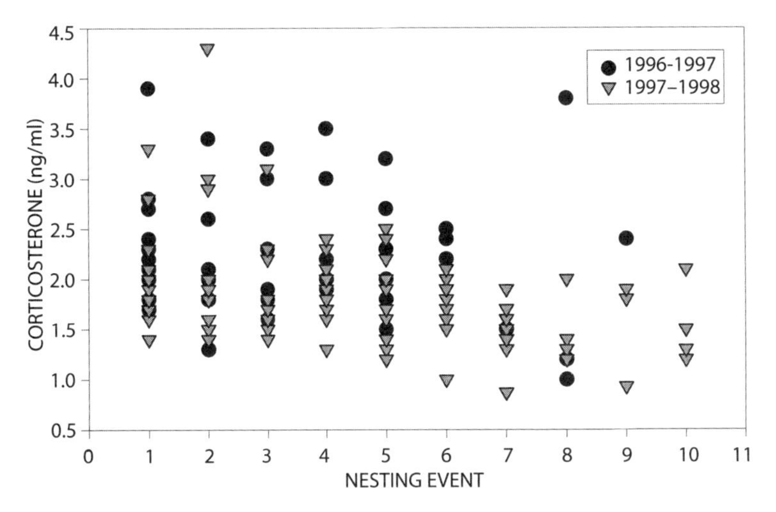

Fig. 5.6. Corticosterone levels for nesting leatherback turtles (*Dermochelys coriacea*) at Parque Nacional Marino Las Baulas, Costa Rica, in the 1996 and 1997 nesting seasons. Samples were collected within 30 min of nesting. Levels were indicative of nonstressed animals.

nesting or within 30 minutes post-nesting. In other sea turtles, progesterone increases sharply 24–48 hours post-nesting in association with an increased secretion of LH prior to ovulation of subsequent clutches (*L. olivacea*, Licht et al. 1982; *C. mydas* and *C. caretta*, Wibbels et al. 1992). Circulating levels of testosterone and estradiol are higher overall in leatherback turtles compared to those reported for other sea turtles (Licht et al. 1979; Wibbels et al. 1990; Rostal et al. 1997; Rostal et al. 1998). Factors that influence overall circulating steroid levels in different taxa are unknown but may be due to the number of ovarian follicles and the size of the ovarian follicles present in the ovary at the time of blood sampling.

Corticosterone levels are low and relatively constant over the course of the nesting season in *D. coriacea* (Rostal et al. 2001). This pattern is similar to the observations of corticosterone levels in nesting *C. mydas* (Jessop et al. 1999; Jessop et al. 2000; Jessop 2001), *E. imbricata* (Jessop 2001), and L. *olivacea* (Valverde et al. 1999) during the breeding and nesting cycle. Acute stress stimulates the adrenocortical stress response in male and nonreproductive female *C. mydas* (Jessop et al. 1999; Jessop et al. 2000; Jessop 2001), *E. imbricata* (Jessop 2001), and *L. olivacea* (Valverde et al. 1999). Nesting *C. caretta* become more responsive to acute stress as the nesting season progresses and there is a rapid surge in corticosterone within 30 minutes of a stress event (Drake 2001). These observations support the hypothesis that elevated gonadal steroid levels inhibit the stress response during the early part of the nesting season in sea turtles (Drake 2001). The relatively constant corticosterone levels in *D. coriacea* show a slight decline as the nesting season progresses (fig. 5.6). Thus, the suppression of the adrenocortical stress response in leatherbacks and other sea turtles may be an important component of the trade-off between survival and reproduction because nesting and egg laying are stressful activities in these animals.

Suppression of the "normal" adrenocortical stress response would be advantageous to a reproductive female, because she would be more likely to remain near the nesting beach to complete the nesting process. There is no correlation between corticosterone levels and testosterone levels (Rostal et al. 2001). These observations are contrary to those of Whittier et al. (1997) who observed a correlation between testosterone and corticosterone in nesting *C. caretta* in Australia. However, mean corticosterone levels reported in the Whittier study were minimal and ranged from nondetectible to 5 ng/ml, similar to those observed in *D. coriacea* (Rostal et al. 2001).

Generalized Reproductive Model and Future Research Needs

Owens (1997) proposed a general model for sea turtle reproduction. However, much of our understanding of reproduction for *D. coriacea* is based on data collected from nesting females. Aspects of male and female reproductive physiology while at sea can only be speculated on based on general similarities with other sea turtles. Since much of what we know from nesting *D. coriacea* is comparable to other sea turtles, I propose that we can draw reliable comparisons and generate a general reproductive model (fig. 5.7).

The reproductive physiology of the leatherback turtle is similar to that of other sea turtles with respect to overall seasonal patterns. Testosterone, progesterone, and estradiol appear to have similar circulating

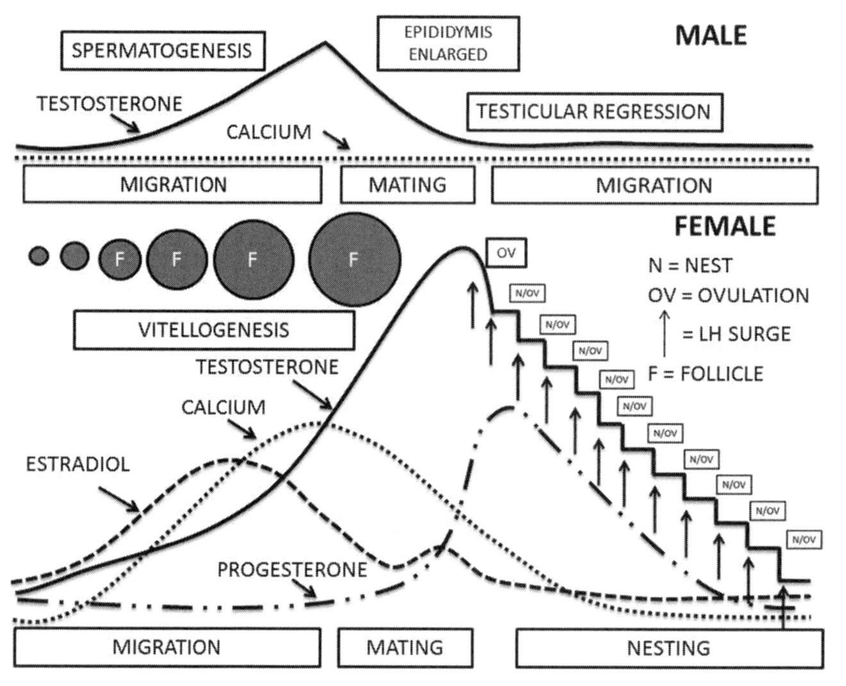

Fig. 5.7. A generalized seasonal model for leatherback reproduction (*Dermochelys coriacea*) based on data from Parque Nacional Marino Las Baulas, Costa Rica. Non-nesting events are based on our knowledge of other sea turtle species.

patterns to those observed in other sea turtles; however, further research is needed to clarify their physiological functions. While this chapter has concentrated on nesting females, there is still a need for data on leatherback turtles at sea, and particularly on male leatherback turtles. Most of the nesting information is only available from the eastern Pacific population at PNMB. More research is needed on the Atlantic and western Pacific populations to see if they are similar.

LITERATURE CITED

Blanco, G. S., S. J. Morreale, E. Vélez, R. Piedra, W. M. Montes, F. V. Paladino, and J. R. Spotila. 2012. Reproductive output and ultrasonography of an endangered population of East Pacific green turtles. Journal of Wildlife Management 76: 841–846

Blanvillain, G., D. Owens, and G. Kuchling. 2010. Hormones and reproductive cycles in turtles. In: D. O. Norris and K. H. Lopez (eds.), Hormones and reproduction of vertebrates, Vol. 3. Elsevier, San Diego, CA, USA, pp. 277–303.

Blanvillain, G., A. P. Pease, A. L. Segars, D. C. Rostal, A. J. Richards, and D. W. Owens. 2008. Comparing methods for the assessment of reproductive activity in adult male loggerhead sea turtles *Caretta caretta* at Cape Canaveral, Florida. Endangered Species Research 6: 75–85.

Callard, I. P., V. A. Lance, A. R. Salhanick, and D. Barad. 1978. The annual ovarian cycle of *Chrysemys picta*: correlated changes in plasma steroids and parameters of vitellogenesis. General and Comparative Endocrinology 35: 245–257.

Carr, A. 1952. Handbook of turtles. Cornell University Press, Ithaca, NY, USA.

Congdon, J. D. 1989. Proximate and evolutionary constraints on energy relations of reptiles. Physiological Zoology 62: 356–373.

Cree, A., L. J. Guillette Jr., M. A. Brown, G. K. Chambers, J. F. Cockrem, and J. D. Newton.1991. Slow estradiol-induced vitellogenesis in the tuatara, *Sphenodon punctatus*. Physiological Zoology 65: 1234–1251.

Drake, K. K. 2001. Reproductive parameters of nesting *Caretta caretta* on Georgia's Barrier Islands. MS thesis, Georgia Southern University, Statesboro, GA, USA.

Eckert, K. L. 1987. Environmental unpredictability of leatherback sea turtle (*Dermochelys coriacea*) nest loss. Herpetologica 43: 315–323.

Eckert, K. L., and S. A. Eckert. 1990. Embryo mortality and hatch success in *in situ* and translocated leatherback sea turtle, *Dermochelys coriacea*, eggs. Biological Conservation 53: 37–46.

Eckert, S. A., and K. L. Eckert. 1989. Diving and foraging behavior of leatherback sea turtles (*Dermochelys coriacea*). Canadian Journal of Zoology 67: 2834–2840.

Frair, W., R. G. Ackman, and N. Mrosovsky. 1972. Body temperature of *Dermochelys coriacea*: warm turtle from cold water. Science 177: 791–793.

Frazier, J., and S. Salas. 1984. The status of marine turtles in the Egyptian Red Sea. Biological Conservation 30: 41–67.

Greer, A. E., J. D. Lazell, and R. M. Wright. 1973. Anatomical evidence for a countercurrent heat exchanger in the leatherback turtle (*Dermochelys coriacea*). Nature 244: 181.

Gregory, L. S., T. S. Gross, A. B. Bolten, K. A. Bjorndal, and L. J. Guillette Jr. 1996. Plasma corticosterone concentrations associated with acute captivity stress in wild loggerhead sea turtles (*Caretta caretta*). General and Comparative Endocrinology 104: 312–320.

Hamann, M., C. J. Limpus, and D. W. Owens. 2003. Reproductive cycles of males and females. In: P. L. Lutz, J. A. Musick, and J. Wyneken (eds.), The biology of sea turtles. CRC Press, Boca Raton, FL, USA, pp. 135–161.

Heck, J., D. S. MacKenzie, D. C. Rostal, K. Medlar, and D. W. Owens. 1997. Estrogen induction of plasma vitellogenin in the Kemp's ridley sea turtle (*Lepidochelys kempii*). General and Comparative Endocrinology 107: 280–288.

Hirth, H. F. 1980. Some aspects of the nesting behavior and reproductive biology of sea turtles. American Zoology 20: 507–523.

Hirth, H. F., and L. H. Ogren. 1987. Some aspects of the ecology of the leatherback turtle, *Dermochelys coriacea*, at Laguna Jalova, Costa Rica. NOAA Technical Report NMFS 56. NOAA, Springfield, VA, USA.

Ho, S. 1987. Endocrinology of vitellogenesis. In: D. Norris and R. Jones (eds.), Hormones and reproduction in fishes, amphibians, and reptiles. Plenum, New York, USA, pp. 145–169.

Jessop, T. S. 2001. Modulation of the adrenocortical stress response in marine turtles (Cheloniidae): evidence for a hormonal tactic maximizing maternal reproductive investment. Journal of Zoology 254: 57–65.

Jessop, T. S., M. Hamann, M. A. Read, and C. J. Limpus. 2000. Evidence of a hormonal tactic maximizing green turtle reproduction in response to a passive ecological stressor. General and Comparative Endocrinology 118: 407–417.

Jessop, T. S., C. J. Limpus, and J. M. Whittier. 1999. Plasma steroid interactions during high-density green turtle nesting and associated disturbance. General and Comparative Endocrinology 115: 90–100.

Lance, V. 1976. Studies on the annual reproductive cycle of the female cobra, *Naja naja*. Seasonal variation in plasma inorganic ions. Comparative Biochemistry and Physiology A 53: 285–289.

Lance, V. 1987. Hormonal control of reproduction in crocodilians. In: G.J.W. Webb, S. C. Manolis, and P. J. Whitehead (eds.), Wildlife management. Crocodiles and alligators. Surrey Beatty and Sons, Chipping Norton, NSW, Australia, pp. 409–415.

Lance, V., T. Joanen, and L. McNease. 1983. Selenium, vitamin E, and trace elements in the plasma of wild and farm-reared alligators during the reproductive cycle. Canadian Journal of Zoology 61: 1744–1751.

Lance, V. A., and D. C. Rostal. 2002. The annual reproductive cycle of the male and female desert tortoise: physiology and endocrinology. Chelonian Conservation and Biology 4: 302–312.

Lance, V. A., D. C. Rostal, R. M. Elsey, and P. L. Trosclair III. 2009. Ultrasonography of reproductive structures and hormonal correlates of follicular development in female American alligators, *Alligator mississippiensis*, in southwest Louisiana. General and Comparative Endocrinology 162: 251–256.

Leslie, A. J., D. N. Penick, J. R. Spotila, and F. V. Paladino. 1996. Leatherback, *Dermochelys coriacea*, nesting and nest success at Tortuguero, Costa Rica. Chelonian Conservation Biology 2: 159–168.

Licht, P., D. W. Owens, K. Cliffton, and C. Penaflores. 1982. Changes in LH and progesterone associated with the nesting cycle and ovulation in the olive ridley sea turtle, *Lepidochelys olivacea*. General and Comparative Endocrinology 48: 247–253.

Licht, P., W. Rainey, and K. Cliffton. 1980. Serum gonadotropin and steroids associated with breeding activities in the green sea turtle *Chelonia mydas*, II. Mating and nesting in natural populations. General and Comparative Endocrinology 40: 116–122.

Licht, P., J. Wood, D. W. Owens, and F. E. Wood. 1979. Serum gonadotropins and steroids associated with breeding activities in the green sea turtle *Chelonia mydas*, I. Captive animals. General and Comparative Endocrinology 39: 274–289.

McPherson, R. J., L. R. Boots, R. MacGregor III, and K. R. Marion. 1982. Plasma steroids associated with seasonal reproductive changes in a multiclutched freshwater turtle, *Sternotherus odoratus*. General and Comparative Endocrinology 48: 440–451.

Mrosovsky, N. 1980. Thermal biology of sea turtles. American Zoologist 20: 531–547.

Owens, D. W. 1976. Endocrine control of reproduction and growth in the green turtle, *Chelonia mydas*. PhD diss., University of Arizona, Tucson, AZ, USA.

Owens, D. W. 1997. Hormones in the life history of sea turtles. In: P. Lutz and J. A. Musick (eds.), The biology of sea turtles, CRC Press. Boca Raton, FL, USA, pp. 315–341.

Perrault, J. R., J. Wyneken, A. Page-Karjian, A. Merrill, and D. L. Miller. 2014. Seasonal trends in nesting leatherback turtle (*Dermochelys coriacea*) serum proteins further verify capital breeding hypothesis. Conservation Physiology 2: 1–15.

Pritchard, P.C.H., and P. Trebbau. 1984. The turtles of Venezuela. Society for the Study of Amphibians and Reptiles, Oxford, OH, USA.

Reina, R. D., K. J. Abernathy, G. J. Marshall, and J. R. Spotila. 2005 Respiratory frequency, dive behavior and social interactions of leatherback turtles, *Dermochelys coriacea*, during the inter-nesting interval. Journal of Experimental Marine Biology and Ecology 316: 1–16.

Reina, R. D., P. A. Mayor, J. R. Spotila, and F. V. Paladino. 2002. Nesting ecology of the leatherback turtle, *Dermochelys coriacea* at Parque Nacional Marino Las Baulas, Costa Rica 1988–89 to 1999–2000. Copeia 2002: 653–664.

Robeck, T. R., D. C. Rostal, P. M. Burchfield, D. W. Owens, and D. C. Kraemer. 1990. Ultrasound imaging of reproductive organs and eggs in Galapagos tortoises, *Geochelone elephantopus*. Zoo Biology 9: 349–359.

Rostal, D. C. 2005. Seasonal reproductive biology of the Kemp's ridley sea turtle (*Lepidochelys kempii*): comparison of captive and wild populations. Chelonian Conservation and Biology 4: 788–800.

Rostal, D. C. 2007. Reproductive physiology of the ridley sea turtle. In: P. Plotkin (ed.), Biology and conservation of ridley sea turtles. Johns Hopkins University Press, Baltimore, MD, USA, pp. 151–165.

Rostal, D. C. 2014. Reproductive physiology of North American tortoises. In: D. C. Rostal, E. McCoy, and H. Mushinsky (eds.), Biology and conservation of North American tortoises. John Hopkins University Press, Baltimore, MD, USA.

Rostal, D. C., J. S. Grumbles, R. A. Byles, R. Marquez, and D. W. Owens. 1997. Nesting physiology of wild Kemp's ridley turtles, *Lepidochelys kempii*, at Rancho Nuevo, Tamaulipas, Mexico. Chelonian Conservation and Biology 2: 538–547.

Rostal, D. C., J. S. Grumbles, K. S. Palmer, V. A. Lance, J. R. Spotila, and F. V. Paladino. 2001. Changes in gonadal and adrenal steroid levels in the leatherback sea turtle (*Dermochelys coriacea*) during the nesting cycle. General and Comparative Endocrinology 122: 139–147.

Rostal, D. C., V. A. Lance, J. S. Grumbles, and A. C. Alberts. 1994. Seasonal reproductive cycle of the desert tortoise (*Gopherus agassizii*) in the eastern Mojave desert. Herpetological Monographs 8: 72–82.

Rostal, D. C., D. W. Owens, J. S. Grumbles, D. S. MacKenzie, and M. S. Amoss. 1998. Seasonal reproductive cycle of the Kemp's ridley sea turtle (*Lepidochelys kempii*). General and Comparative Endocrinology 109: 232–243.

Rostal, D. C., F. V. Paladino, R. M. Patterson, and J. R. Spotila. 1996. Reproductive physiology of nesting leatherback turtles (*Dermochelys coriacea*) at Las Baulas de Guanacaste National Park, Costa Rica. Chelonian Conservation and Biology 2: 230–236.

Rostal, D. C., T. R. Robeck, D. W. Owens, and D. C. Kraemer. 1990. Ultrasound imaging of ovaries and eggs in Kemp's ridley sea turtles (*Lepidochelys kempii*). Journal of Zoo and Wildlife Medicine 21: 27–35.

Santidrián Tomillo, P., V. S. Saba, G. S. Blanco, C. A. Stock, F. V. Paladino, and J. R. Spotila. 2012. Climate driven egg and hatchling mortality threatens survival of eastern Pacific leatherback turtles. PloS One 7(5): e 37602. doi: 10.1371/journal.pone.0037602.

Smelker, K., L. Smith, M. Arendt, J. Schwenter, D. Rostal, K. Selcer, and R. Valverde. 2014. Plasma vitellogenin in free-ranging loggerhead sea turtles (*Caretta caretta*) of the northwest Atlantic Ocean. Journal of Marine Biology 267: 1–10. doi.org/10.1155/2014/748.

Steyermark, A. C., K. Williams, J. R. Spotila, F. V. Paladino, D. Rostal, S. Morreale, M. T. Koberg, and R. Arauz. 1996. Nesting leatherback turtles at Las Baulas National Park, Costa Rica. Chelonian Conservation and Biology 2: 173–183.

Tucker, A. D., and N. B. Frazer. 1991. Reproductive variation in leatherback turtles, *Dermochelys coriacea*, at Culebra National Wildlife Refuge, Puerto Rico. Herpetologica 47: 115–124.

Tucker, A. D., and N. B. Frazer. 1994. Seasonal variation in clutch size of the turtle, *Dermochelys coriacea*. Journal of Herpetology 28: 102–109.

Valverde, R. A., D. W. Owens, D. S. Mackenzie, and M. S. Amoss. 1999. Basal and stress-induced corticosterone levels in olive ridley sea turtles (*Lepidochelys olivacea*) in relation to their mass nesting behavior. Journal of Experimental Zoology 284; 652–662.

Van Buskirk, J., and L. B. Crowder. 1994. Life-history variation in marine turtles. Copeia 1994: 66–81.

Wallace, B. P., P. R. Sotherland, P. Santidrián Tomillo, S. S. Bouchard, R. D. Reina, J. R. Spotila, and F. V. Paladino. 2006. Egg components, egg size, and hatchling size in leatherback turtles. Comparative Biochemistry and Physiology A 145: 524–532.

Wallace, B. P., P. R. Sotherland, J. R. Spotila, R. D. Reina, B. F. Franks, and F. V. Paladino. 2004. Biotic and abiotic factors affect the nest environment of embryonic leatherback turtles, *Dermochelys coriacea*. Physiological and Biochemical Zoology 77: 423–432.

Whittier, J. M., F. Corrie, and C. Limpus. 1997. Plasma steroid profiles in nesting loggerhead turtle (*Caretta caretta*) in Queensland, Australia: relationship to nesting episode and season. General and Comparative Endocrinology 106: 39–47.

Wibbels, T., D. W. Owens, P. Licht, C. J. Limpus, P. C. Reed, and M. S. Amoss Jr. 1992. Serum gonadotropins and gonadal steroids associated with ovulation and egg production in sea turtles. General and Comparative Endocrinology 87: 71–78.

Wibbels, T., D. W. Owens, C. J. Limpus, P. C. Reed, and M. S. Amoss Jr. 1990. Seasonal changes in serum gonadal steroids associated with migration, mating, and nesting in the loggerhead sea turtle (*Caretta caretta*). General and Comparative Endocrinology 79: 154–164.

Wibbels, T., D. W. Owens, and D. C. Rostal. 1991. Soft plastra of adult male sea turtles: an apparent secondary sexual characteristic. Herpetological Review 22: 47–49.

Wyneken, J. 2003. The external morphology, musculoskeletal system, and neuro-anatomy of sea turtles. In: P. L. Lutz, J. A. Musick, and J. Wyneken (eds.), The biology of sea turtles, Vol. II. CRC Press, Boca Raton, FL, USA, pp. 39–77.

6

Nesting Ecology and Reproductive Investment of the Leatherback Turtle

KAREN L. ECKERT,
BRYAN P. WALLACE,
JAMES R. SPOTILA, AND
BARBARA A. BELL

Although the large majority of the lives of leatherback turtles (*Dermochelys coriacea*) and other sea turtle species are spent at sea, terrestrial reproduction is a requirement imposed by their reptilian evolutionary lineage. This highly stereotyped behavior is probably the most widely and carefully observed of any that sea turtles perform, resulting in observations and data from many nesting beaches around the world. However, despite the obvious universality of the nesting process itself, leatherback reproductive behavior and investment patterns are shaped by environmental factors and show geographic variation. In this chapter, we provide a review of published literature on leatherback nesting, including reproductive output parameters. In particular, we cover leatherback courtship and mating, nesting behavior and site selection, density dependence, breeding phenology, seasonal and long-term reproductive output patterns; and we highlight areas in need of further research.

Courtship and Mating

Leatherback courtship and/or mating are only rarely described in the literature, but in each case the observation has occurred proximal to a nesting ground (Rathbun et al. 1985; Carr and Carr 1986; Tucker 1988; Godfrey and Barreto 1998). Lazell (1980) suggested that virtually all males might migrate to and from the nesting beach every year, inseminating females prior to their first oviposition and then leaving the breeding grounds before females complete the laying season. Adult males migrate from temperate Atlantic foraging grounds to residence areas adjacent to known nesting beaches, arrive around nesting colonies before the nesting season, and remain until the peak of the season (James et al. 2005; Dodge et al. 2014). Based on a study of the age composition of the pantropical barnacle (*Conchoderma* sp.) attached to leatherbacks nesting in the US Virgin Islands, Eckert and Eckert (1988) suggested that females arrive asynchronously at the nesting beach from temperate latitudes outside the biogeographic range of the barnacle; they then begin nesting within a relatively few days of arrival. The authors inferred that females do not arrive far enough in advance of their first nesting to accommodate mating near the site. Therefore, they concluded that mating occurs prior to or during migration, meaning that "courtship" in tropical waters might represent opportunistic behavior on the part of adult males.

In Costa Rica (Pacific coast), Reina et al. (2005) mounted cameras on female leatherbacks and recorded interactions with males early in the nesting season. Those video images provided the first documented glimpses of interactions between males and females at and below the surface of the water. In those recordings, males come to the front of a female or approach from the rear and attempt to mount her. They sometimes bite females on the head, anterior carapace, and front flippers. Males also strike females with their front flippers. In one instance a male attempts to mount a female and grasps her with his front flipper, but copulation could not be confirmed. Females frequently exhibit apparent avoidance behavior when approached by males, descending to the seafloor where they then either remain motionless or turn to face the approaching male. Males proceed to circle, approach, and repeatedly pass over females lying on the ocean bottom, with the male usually making some contact with his body, flippers, and/or head. In all recorded events, females remain on the bottom until the male is no longer visible in the frame. Whether such interactions result in successful mating or represent harassment and avoidance behavior has yet to be determined.

In other observations (Carr and Carr 1986) leatherback courtship involved a series of lunges by the male at the female as he attempted to mount her. He eventually positioned his plastron just posterior to the center of her carapace and curved his muscular tail beneath her tail. A semi-erect penis was clearly visible, but there was no long embrace like that seen in some green turtle (*Chelonia mydas*) pairs.

Early studies using microsatellites indicate very infrequent, or no, multiple paternity within or among successive clutches of a female (Dutton and Davis 1998; Rieder et al. 1998; Dutton et al. 2000). Similarly, genetic analyses of leatherbacks nesting at Playa Grande, Costa Rica, confirmed that while polygyny and polyandry are present, single paternity is the most prevalent mating strategy (Crim et al. 2002). Thus, the apparent evasion of male leatherback courtship attempts by female leatherbacks at this site (Reina et al. 2005) are consistent with the observation that female sea turtles store sperm during a reproductive season, and thus do not need to mate more than once (Owens 1980; Miller 1997; FitzSimmons 1998). However, in the most thorough analysis of multiple paternity to date (>1,000 hatchlings from 38 clutches and 12 known females), Stewart and Dutton (2011) reported a surprisingly high rate (>58%) of multiple paternity at the Sandy Point nesting colony in the US Virgin Islands. This suggests that its frequency might reflect the relative availability of males, because the Playa Grande colony has declined dramatically in recent decades (Santidrián Tomillo et al. 2007; chapter 10), while the Sandy Point colony has increased exponentially during the same period (Dutton et al. 2005; chapter 9).

Based on all of these data it appears that leatherback mating occurs before females arrive at the nesting site, or at the nesting area prior to commencement of nesting. Some opportunistic mating may occur early in the nesting season.

Nesting Behavior

Nesting takes place primarily in the dry season; this occurs from March to June along Caribbean Costa Rica, and October to February along Pacific Costa Rica. In the North Atlantic and Caribbean colonies it is typically in spring and early summer. Nesting usually occurs at night but some daylight nesting takes place if turtles are still on the beach when the sun comes up after a late tide. Preferred nesting grounds are mainly in the tropics, but span latitudes from 34°S in Western Cape, South Africa, to 38°N in Maryland, United States (summarized by Eckert et al. 2012), and have been described as "deep, clean, high-energy beaches with either a deep water oceanic approach or . . . a shallow water approach with mud banks but without coral or rock formations" (Turtle Expert Working Group 2007).

Hendrickson and Balasingam (1966) described five interrelated factors that characterize beaches on which *Dermochelys* nests: coarse-grained sand; steep, sloping littoral zone; obstacle-free approach; proximity to deep water; and oceanic currents affecting the coast. Strong waves and tides assist the females in their emergence from the sea (Reina et al. 2002), and a steep profile enables the heavy-bodied turtles to reach high, dry sand with less energy spent crawling (Hendrickson and Balasingam 1966; Pritchard 1971; Hendrickson 1980).

While fidelity to a discrete nesting ground is characteristic of all sea turtle species, the degree to which this is apparent in leatherbacks can vary according to nesting populations and is generally exhibited to a somewhat lesser degree than in other sea turtle species (Miller 1997). For example, on continuous coastlines in the western Atlantic region, leatherbacks are known to nest on beaches in Costa Rica and adjoining Panama (Chacón and Eckert 2007), and on beaches in Suriname and adjoining French Guiana (Hilterman and

Goverse 2007). This is similarly true in the United States where Global Positioning System (GPS) tags deployed on nesting females in Florida indicated that individuals deposited their nests up to 139.8 km apart within a single nesting season (Stewart 2007). Leatherbacks also deposit eggs on multiple Caribbean islands during a single nesting season. For example, "at least eight turtles were recorded moving among nesting beaches in Guadeloupe, Martinique, and Dominica" in 2007, and one such turtle "was first marked on the northeast coast of Trinidad in 2004 and returned to the Caribbean in 2007 to nest in Martinique and later Dominica" (Stapleton and Eckert 2007). In contrast, eastern Pacific leatherbacks nesting in Parque Nacional Marino Las Baulas (Costa Rica) tended to nest across seasons on one of three beaches that cover 5–6 km (Santidrián Tomillo et al. 2007). Reasons for this variation are unknown, but may be related to the presence of offshore currents and suitable nesting sites in the region.

Because of environmental requirements of embryos (chapter 12), successful *Dermochelys* clutches typically incubate above the natural high tide line. Nest site selection appears to generally reflect the opposing selection pressures of (1) embryonic mortality resulting from suboptimal environmental conditions within nests, including tidal inundation and erosion below the high tide line (Mrosovsky 1983; Whitmore and Dutton 1985; Eckert 1987); and (2) embryonic mortality and hatchling disorientation resulting from nests placed too far inland (Godfrey and Barreto 1995; Kamel and Mrosovsky 2004). Early analyses concluded that a scatter-nesting approach may minimize nest loss in uncertain environments where erosion and wave wash claim more than 50% of nests per year (Mrosovsky 1983; Eckert 1987).

Two studies have examined inter- and intra-individual variation in leatherback nest site selection. Nordmoe et al. (2004) reported that nest site selection along the ocean-to-vegetation axis at Playa Grande was independent and leatherbacks typically nested in the open beach between the high tide line and the vegetation regardless of the site of their previous nest. Kamel and Mrosovsky (2004) reported that leatherbacks at Ya:lima:po-Awa:la, French Guiana, also typically nested in the open sand (85%). Nordmoe et al. (2004) also identified a pattern of nest site dependence on earlier sites along the coastal axis, perpendicular to the ocean-to-vegetation axis, supporting the notion of spatial proximity. Along this axis, leatherbacks tended to nest close to the site of their previous nest. However, Kamel and Mrosovsky (2004) found that individual female leatherbacks tended to nest at particular distances from the highest spring tide line, but not at particular sections of the beach along the coastal axis; locations of subsequent nests in French Guiana could not be predicted based on knowledge of previous nest choices. Nevertheless, in both studies, a female that laid one nest below the high tide line was more likely to nest on the open beach than below the high tide the next time.

Whether intra-individual nest site selection patterns exist and are heritable has important implications for nest relocation programs employed by conservation projects worldwide. Specifically, if particular females consistently place nests in "doomed" sites, nest relocation programs would be inadvertently selecting for poor nest site selection traits in those individual turtles (Mrosovsky 2006). However, data available to date do not support that conclusion. Both Nordmoe et al. (2004) and Kamel and Mrosovsky (2004) concluded that the lack of strong predictability in nest placement does not allow us to determine if a particular nest is laid by a turtle with an inherited trait for a tendency to nest lower on the beach or is simply the lowest nest made by a turtle with a tendency to nest high on the beach. There are no data to suggest that doomed nests are laid by particularly bad nesters. Kamel and Mrosovsky (2004) concluded that nest relocation need not have any detrimental effects and that relocating nests can be used to increase recruitment to the population. Nordmoe et al. (2004) reached the same conclusion.

Whatever factors attract turtles to nesting beaches, the geographic location of the beach must be considered in relation to the immediate offshore areas and dominant currents, specifically for hatchling dispersal. Hughes (1974) summarized the situation at five nesting areas and concluded that offshore currents at major rookeries, during the main period of hatching, would carry hatchlings into waters of 26.5°C to 31.5°C. On a regional scale, Shillinger et al. (2012) used particle drift modeling to simulate leatherback hatchling dispersal patterns from four nesting beaches in the eastern Pacific. They found that hatchlings (e.g., inanimate particles) are advected away more quickly from coastal areas that are thought to have highest predator densities, if offshore currents are more accessible. The authors speculated that oceanographic conditions driving variation in effective hatchling dispersal could contribute to evolutionary trends in nest site selection by leatherbacks. For more discussion of hatchling dispersal, see chapter 14.

Phases of Nesting Behavior

The nesting process is a stereotypic sequence of behaviors that are similar for sea turtles in general and has been described by Miller (1997). Nesting behavior is constrained by the anatomy of the turtle and the beach environment in which the nest is constructed. Most differences among species are due to differences in their size that affect their speed of movement and depth of their nest.

Pritchard and Trebbau (1984) described the nesting process of leatherback, green, loggerhead (*Caretta caretta*), and olive ridley (*Lepidochelys olivacea*) turtles. Deraniyagala (1936, 1939) provided the first detailed description of nesting for leatherbacks and accompanied it with hand-drawn illustrations of the process. Later descriptions were given by Carr and Ogren (1959), Pritchard (1971), and others. Briefly, the sequence of behaviors involves emergence from the sea onto the nesting beach, travel up the beach and selection of a suitable nest site, excavation of a body pit, excavation of the nest chamber, oviposition (egg laying), filling the nest chamber, covering and concealing the nest site, and returning to the sea. In leatherback turtles the entire sequence, from emergence to return to the sea, takes about 80 to 140 minutes, depending on local conditions (table 6.1)

Table 6.1. Time periods of nesting behavior of leatherback turtles as summarized from Carr and Ogren (1959) for 1 nesting event at Matina on the Caribbean coast of Costa Rica; by Eckert and Eckert (1988) for 113 nesting events at Sandy Point, St. Croix, US Virgin Islands; and by Reina et al. (2002) for 82–173 nesting events at Playa Grande, Costa Rica. Time is mean duration in minutes.

	Matina	St. Croix	Playa Grande
Nesting Stage			
Emerging from sea and ascending the beach	3	8.5	22.0
Nest preparation (body pit)	17	9.4	16.5
Nest excavation	5	22.9	17.4
Oviposition (egg laying)	15	10.8	12.7
Filling and concealing	45	34.0	47.3
Return to sea	8	4.2	22.0
Total	93	112.6	117.8

EMERGENCE, MOVEMENT UP THE BEACH, AND NEST SITE SELECTION

Tide and lunar cycles, wave action, and steepness of the beach profile all play a role in influencing the timing and success of a leatherback's typically nocturnal emergence from the sea (Miller 1997). Timing of leatherback nesting is loosely associated with the timing of high tide at Playa Grande, Costa Rica, but that association does not continue when high tide is near dusk or dawn. Then turtles nest throughout the night. Phase of the moon also has some influence on time of emergence (Reina et al. 2002). However, in other locations leatherbacks nest during the darkest periods of the night. Additional research is needed to quantify the role of these factors in the timing of emergence of nesting leatherbacks.

Gravid females walk on land using a simultaneous gait where all four limbs move at once to move the turtle forward. As the front limbs act as crutches pulling the turtle forward, the hind limbs simultaneously push it forward. In contrast, loggerheads, olive ridleys, and hawksbills (*Eretmochelys imbricata*) use an alternating gait (Wyneken 1997). The resulting leatherback tracks are symmetrical, with diagonal grooves formed on the outer portion of the track by the front flippers that frame smaller depressions formed by the rear flippers as well as marks left by the plastron and tail drag in the center of the track. The track of a leatherback is approximately 150+ cm wide as compared to about 100 cm for a green turtle and about 70 cm for an olive ridley. Because of the developmental requirements of embryos, successful *Dermochelys* clutches typically incubate above the natural high tide line.

EXCAVATION OF A BODY PIT

After reaching the dry upper portion of the nesting beach, a gravid female often touches the sand with her snout and then begins to lower, scoop, and sweep her large front flippers from anterior to posterior, thus removing and throwing large amounts of sand. She also moves her body side to side. As she continues this activity, she removes surface debris and dry sand from the area. This is important because during the later portions of the nesting season very dry sand is likely to collapse into the eventual nest chamber. The resulting crater formed by the repeated removal and throwing of sand is referred to as a "body pit" (Carr 1967; Miller 1997).

Between sand-throwing sequences using the front flippers, the hind flippers move in a side-to-side fashion and smooth mounds of sand thus form behind the turtle during the sand throwing. The rear half of the turtle's carapace becomes covered in sand, resulting in the alternative name of "sand bath" given to this phase (Deraniyagala 1936). The depth of the body pit depends on the thickness of the dry surface sand layer and the size of the turtle. The turtle will continue body pit construction until reaching firmer, wetter, cooler subsurface sand, at which point she will shift to nest (egg) chamber excavation.

EXCAVATION OF A NEST CHAMBER

Leatherbacks construct the nest chamber with their hind flippers. The turtle first scoops and flicks surface sand away from the area immediately behind her to begin the cavity. Then she performs repeated scooping motions with the ventral side of a rear flipper, after which she gently cups, removes, and then places some sand to her side. Meanwhile, the second flipper holds the weight of the rear portion of the turtle over the cavity by bracing against the side of the cavity. When the first flipper is removed, the second flipper flicks forward sand that has fallen on its dorsal side and then is lowered into the cavity to scoop just as the first flipper had. These actions alternate between flippers until they can no longer bring up sand. This is a stereotypic behavior and if one flipper is missing, the turtle goes through the same motion with the stub of the missing flipper carrying out its action while the good flipper waits and then digs. The process then takes twice as long. If both flippers are missing the turtle will still go through the process but will not be able to complete a nest chamber (without human intervention) and will return to the sea.

The shape of the nest cavity is not uniformly cylindrical, but instead the lower portion (egg chamber) generally takes a rounded shape, such that the entire nest is often described as pear or flask shaped (Miller 1997). Billes and Fretey (2001) used thermal curing polyurethane foam to produce castings of the subterranean nest, documenting several variations on the flask shape and reporting a mean depth of 70 cm and a mean volume of the egg chamber (the bottom ~45% of the nest) of 15 liters.

OVIPOSITION

Generally, after the rear flippers can no longer reach to the bottom of the nest chamber to remove sand, oviposition begins. One rear flipper is moved to the midline just over the top of the cloaca and tail and remains there during oviposition. In this process leatherbacks at Playa Grande are "right flippered." That is, they have a spontaneous limb preference and are more likely to cover the egg chamber with their right flipper than with their left flipper (Sieg et al. 2010). Eggs, typically laid in pulses, are coated in a clear viscous liquid. During the later part of oviposition "shelled albumin gobs" (SAGs), often incorrectly referred to as "yolkless eggs" (chapter 12) are laid along with eggs (fig. 6.1). During oviposition, leatherbacks appear to be in a trancelike state and are relatively indifferent to modest external stimuli, including egg collection, tagging, measuring, attachment of instrumentation under low lighting conditions, and so forth.

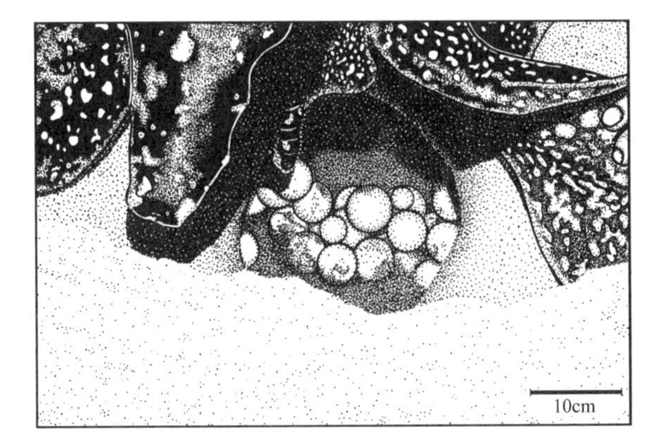

Fig. 6.1. Oviposition by a leatherback turtle. By holding back the right flipper the egg chamber is exposed, revealing both eggs (50 mm diameter) and small, shelled albumin gobs (SAGs) that look like small eggs but only contain albumin. Eggs are deposited singularly or in in groups of up to three at a time. SAGs are deposited toward the end of the run of eggs from a given oviduct. Drawing by Barbara Bell.

FILLING THE NEST

Once oviposition is complete, the female refills the nest chamber by scooping sand into the cavity with her rear flippers in an alternating pattern. As the cavity is filled, the turtle compacts the sand by shifting the weight of the posterior portion of her body onto her hind flippers with each scoop of sand. She uses her tail to probe the sand, and continues the process until the cavity is filled back to the level of the body pit or sand surface.

COVERING AND CONCEALING THE NEST SITE

After refilling the nest chamber with sand, a leatherback begins to cover and conceal the nest by swinging her outstretched rear flippers and tail rapidly from side to side. Then she performs sand-throwing behaviors essentially identical to those exhibited during body pit excavation. This begins when the turtle suddenly jerks both front flippers back, simultaneously showering sand behind her body (fig. 6.2). She continues to use powerful scooping and throwing motions to move large amounts of sand in the general area of the nest chamber. Meanwhile she continues to work the hind flippers as well. She continues these actions as she changes direction, moves forward, turns, and sometimes does entire circles in the sand. The end result is that an area much larger than the actual nest site is greatly disturbed in no discernible pattern, which likely makes detection and locating of the clutch more difficult for potential egg predators.

Fig. 6.2. When the female leatherback finishes covering the nest chamber with her hind flippers she rests with her front flippers a little forward. Suddenly she jerks them back throwing a large amount of sand behind her body. Then she continues in a rhythmic fashion to throw sand and cover up the nesting site. Drawing by Barbara Bell.

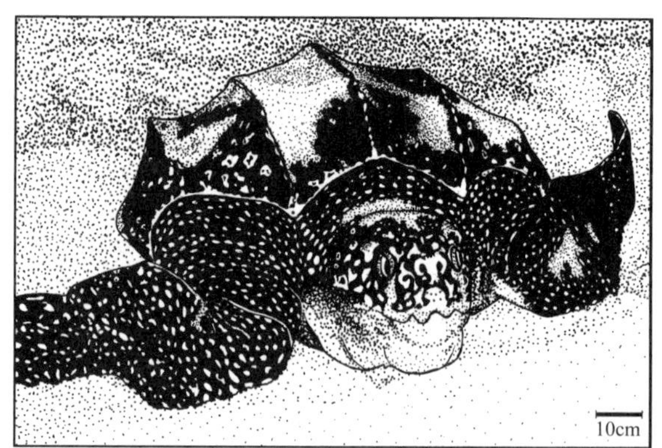

Fig. 6.3. Upon completion of covering the nest, the female leatherback turns toward the ocean and crawls back to the sea. The sand on her back shows that she has completed covering her nest. Drawing by Barbara Bell.

RETURNING TO THE SEA

After a variable period of nest covering, the leatherback turns and begins her return to the sea, utilizing the same simultaneous gait as during her initial emergence from the sea (fig. 6.3). Now her back is covered with sand from the covering process. The return may or may not follow a straight route, and disorientation by coastal lighting is often a significant distraction. Small, quickly executed circles sometimes take place. These circles occurred in 30% of descent crawls at Matura Bay, Trinidad (Bacon 1973), and occurred on mainland Atlantic (Pritchard 1971) and Pacific (Cornelius 1976) coasts, and elsewhere (we have observed them both at St. Croix and Playa Grande). While the function of circling is unclear, it may be related to orientation or finding the sea. That remains to be determined. The duration of the return to sea depends on the distance between the tide line and the nest site, and may range between less than a minute to more than 60 minutes (Reina et al. 2002; table 6.1).

Nesting Success

Nesting success refers to the ratio of successful to unsuccessful nesting attempts, the latter defined as a landing that does not result in oviposition. Nesting success varies both temporally (within and across seasons) and geographically, with reported annual averages ranging between 50% and more than 90% (for a global summary, see Eckert et al. 2012). In general, the ratio reflects the effects of disturbance, as well as beach conditions (e.g., erosion berms and rocks, wave wash, debris, excessive sand dryness) that reduce the successful completion of the overland traverse or the excavation of a suitable body pit or nest chamber.

Leatherbacks were long thought to be more tolerant of disturbances during nesting than other sea turtle species (Deraniyagala 1936; Carr and Ogren 1959). On Culebra Island, Puerto Rico, disturbance during emergence and overland traverse accounted for 28% of unsuccessful nesting attempts (Tucker and Hall 1984), but fewer than 8% of nesting attempts were aborted at Playa Grande during the peak nesting season, partly due to protection from people on the beach (Reina et al. 2002). During the core phases of the nesting process, and particularly during oviposition, leatherbacks are largely unresponsive to ambient stimuli. Nevertheless, aborted nesting attempts do occur due to the presence of humans and predators (e.g., Pritchard 1969, 1971; Bacon 1973). While it is intuitive that turtles do not always survive an encounter with a human intent on causing them harm, it is less well known that nesting success is also reduced by predators worthy of these giants: crocodiles (*Crocodylus porosus*) in the western Pacific (Kinch 2006) and jaguars (*Panthera onca*) in the western Atlantic (Veríssimo et al. 2012).

Density Dependence of Nesting

The highest levels of leatherback nest density in the world occur at Grande Riviere on the north coast of the Caribbean island of Trinidad; in 2007 there were 23,869 body pits on 800 m of beach. At Matura Beach, on Trinidad's east coast, there were 16,911 body pits on 3 km of beach (Nature Seekers, unpublished data in

Eckert et al. 2012). Other large colonies (e.g., Terengganu, Malaysia; Papua, Indonesia; see chapter 10) have declined significantly in recent decades and historical nesting densities may never be known.

In French Guiana, nightly nest density reached a maximum concentration of 256 females per km at Ya:lima:po-Awa:la Beach in 1986 (Fretey and Girondot 1987), and the maximum number of nests was 60,000 in 1992 (Girondot and Fretey 1996). Density-dependent nest destruction occurs on that beach. Using a mathematical model, Girondot et al. (2002) predicted that the number of nests on the beach for maximum hatchling production was 95,000, but Caut et al. (2006) calculated that effective carrying capacity of Ya:lima:po was approximately the same as the maximum nest number on the beach (~60,000 nests). Thus, while density-dependent nest destruction may affect leatherback nest success, this phenomenon is not common worldwide because very few leatherback nesting colonies are sufficiently large to have significant density-dependent effects.

Reproductive Cycles and Seasonality

Leatherback nesting can occur at any time of year at most sites, but almost all nesting activity occurs seasonally and lasts roughly three to six months. The annual timing of nesting seasons varies geographically. Eckert et al. (2012) generalized that nesting occurs between March and August (with a peak in May or June) at nesting beaches in the western Atlantic (except Espírito Santo, Brazil), Sri Lanka, Malaysia, and Jamursba-Medi in Indonesia; at beaches in the eastern Atlantic, eastern and western Pacific, and western Indian Ocean, nesting occurs between October and March (with a peak in December).

The phenology of leatherback nesting seasons roughly approximates a bell shaped curve for distribution of nests laid over time. This is described empirically from comprehensive nesting beach coverage. However, because such a survey effort is challenging due to logistical issues, a variety of statistical methods have been employed to describe nesting phenology of leatherbacks (e.g., Girondot et al. 2006; Gratiot et al. 2006; Girondot et al. 2007; Hilterman and Goverse 2007; Girondot 2010). Acquiring robust estimates of nest numbers is important for deriving indices of abundance, and population sizes and trends.

Reproductive Investment

Leatherbacks exhibit the largest absolute investment (in terms of biomass) in seasonal egg production of any oviparous amniote (Wallace et al. 2007). Leatherback clutch sizes (i.e., number of eggs in a single clutch) can range from 20 to well over 100, with rookery averages ranging from 60 to 100 eggs (Eckert et al. 2012). Correct counts of clutch size do not include SAGs. Clutch sizes of eastern Pacific leatherbacks are noticeably smaller than those of leatherbacks from other regions of the world, while those of leatherbacks in the Indian Ocean (nesting in South Africa and Sri Lanka) appear to be the largest (Eckert et al. 2012).

Table 6.2. Clutch frequency (number of clutches) and inter-nesting interval in leatherback turtles. Data summarized from Miller (1997) and Eckert et al. (2012).

Location	Clutch Frequency (*n*)	Inter-nesting Interval (days)
Western Atlantic		
French Guiana	8.3	9–10
Suriname	4.5	10
Venezuela	4.4	
St. Croix, USVI	5.1	9–10
Culebra, Puerto Rico	6.6	9
Florida	4.2	10
Eastern Atlantic		
Gabon		10
Eastern Pacific		
PNMB, Costa Rica	9.5	9–10
Mexico	5.5	9–10
Western Pacific		
Terengganu, Malaysia	5.5	9–10
Papua, New Guinea	2.2	11–15
Papua, Indonesia		9–10
Australia		9
Indian Ocean		
Tongaland, South Africa	7.3	10
Mozambique	2.1	
Andaman and Nicobar Islands, India	3.8	11

Contrary to early speculations that leatherbacks only nest once during each reproductive season (Dunlap 1955), gravid females are now known to nest roughly six times (range: 1–14 clutches) every 9–10 days (with a range of 7–14 days) during the nesting season (table 6.2). Beaches with high beach coverage by researchers (i.e., Playa Grande), where most individuals are observed every time they nest, have higher clutch frequencies than beaches with lower coverage. Low clutch frequencies reported for some beaches may reflect lower beach coverage rather than a physiological difference in the leatherbacks nesting on that beach. Due to physiological constraints associated with egg production, the minimum period between consecutive nesting events (i.e., inter-nesting period) is six days

(Miller 1997). Considering this minimum inter-nesting period, nesting beach monitoring efforts can assume that any inter-nesting period of greater than 14 days probably includes another nesting event that went undetected (Frazer and Richardson 1985; Reina et al. 2002; Rivalan et al. 2006).

Remigration intervals (i.e., the nonbreeding duration between consecutive nesting seasons) exhibited by most populations of leatherbacks are 2–3 years (range 1–6 years), but are 3–4 years (range 1–11 years) in some populations (Eckert et al. 2012). Inter-individual and interpopulation differences in remigration intervals can reveal differences in foraging habitat quality and/or variability in resource acquisition and assimilation abilities of leatherbacks. This phenomenon was described in a series of studies that demonstrated delayed remigration in eastern Pacific leatherbacks relative to other leatherback populations; this delay was caused by highly variable, less available resources in areas where eastern Pacific leatherbacks feed compared to areas where other populations feed (see Wallace and Saba 2009 for review). As a consequence of these longer remigration intervals, the eastern Pacific leatherback population is less resilient to high levels of mortality, because it is unable to match the long-term reproductive output of other populations. This suite of factors led to the precipitous decline in this population's abundance in recent decades (Wallace and Saba 2009). For further discussion of these interrelationships see chapters 13 and 15.

Factors Influencing Reproductive Output

Santidrián Tomillo et al. (2009) demonstrated that leatherback clutches laid early in the nesting season at Playa Grande, Costa Rica, developed under more favorable conditions for embryos than clutches laid during the latter months of the nesting season, a pattern that reflected decreasing precipitation and increasing nest temperatures as the season progressed. At the same study site, Rafferty et al. (2011) reported that "remigrant" leatherback females (i.e., turtles that had nested in previous seasons) tended to arrive earlier in the nesting season, produced more clutches, and had higher hatching success than "new" females (i.e., turtles presumed to be neophytes). Robinson et al. (2014) reported that the median nesting date at Playa Grande has occurred later in the season over time because there were fewer "remigrant" turtles in the population and more "new" females. If this trend continues, then the nesting season at Playa Grande will coincide with increasingly adverse conditions for hatchling success, further threatening this population. Thus, reproductive experience appears to play an important role in determining seasonal, and thus longer-term, reproductive success in leatherbacks. Future research should aim to elucidate similar patterns in these parameters at other nesting rookeries to provide more insights into how reproductive investment strategies vary across leatherback populations and how these might influence conservation priorities.

Conclusions

The nesting ecology of leatherback turtles is similar to that of other sea turtles. Within this context leatherback turtles are unique in having the largest reproductive output of any egg-laying amniote. The combination of clutch frequency, clutch size, and hatching success in a given nesting season determines seasonal reproductive output and success. Considering these factors, along with remigration interval, determines the long-term reproductive output and success for individual females. In general, leatherbacks can increase seasonal reproductive output by increasing the number and size of clutches, nesting during periods characterized by the most favorable nest environment conditions to promote higher hatching success, and/or shortening remigration intervals. Thus, examining patterns of all of these factors collectively can reveal inter-individual and interpopulation variation in reproductive investment and success.

LITERATURE CITED

Bacon, P. R. 1973. The orientation circle in the beach ascent crawl of the leatherback turtle, *Dermochelys coriacea,* in Trinidad. Herpetologica 29: 343–348.

Billes, A., and J. Fretey. 2001. Nest morphology in the leatherback turtle. Marine Turtle Newsletter 92: 7–9.

Carr, A. 1967. So excellent a fishe: A natural history of sea turtles. Natural History Press, Garden City, New York, USA.

Carr, A., and L. Ogren. 1959. The ecology and migrations of sea turtles, 3. *Dermochelys* in Costa Rica. American Museum Novitates 1958: 1–29.

Carr, T., and N. Carr. 1986. *Dermochelys coriacea* (leatherback sea turtle) copulation. Herpetological Review 17: 24–25.

Caut, S., V. Hulin, and M. Girondot. 2006. Impact of density-dependent nest destruction on emergence success of Guianan leatherback turtles (*Dermochelys coriacea*). Animal Conservation 9: 189–197.

Chacón, D., and K. L. Eckert. 2007. Leatherback sea turtle nesting at Gandoca Beach in Caribbean Costa Rica: management recommendations from 15 years of conservation. Chelonian Conservation and Biology 6: 101–111.

Cornelius, S. E. 1976. Marine turtle nesting activity at Playa Naranjo, Costa Rica. Brenesia 8: 1–27.

Crim, J. L., L. D. Spotila, J. R. Spotila, M. P. O'Connor, R. D. Reina, C. J. Williams, and F. V. Paladino. 2002. The leatherback turtle, *Dermochelys coriacea,* exhibits both polyandry and polygyny. Molecular Ecology 11: 2097–2106.

Deraniyagala, P.E.P. 1936. The nesting habit of the leathery turtle, *Dermochelys coriacea.* Ceylon Journal of Science B 19: 331–336.

Deraniyagala, P.E.P. 1939. The tetrapod reptiles of Ceylon, Vol. 1. Testudinates and crocodilians. Dulau, London, UK.

Dodge, K. L., B. Galuardi, T. J. Miller, and M. E. Lutcavage. 2014. Leatherback turtle movements, dive behavior, and habitat characteristics in ecoregions of the Northwest Atlantic Ocean. PLoS ONE 9(3): e91726. doi: 10.1371/journal.pone.0091726.

Dunlap, C. E. 1955. Notes on the visceral anatomy of the giant leatherback turtle (*Dermochelys coriacea* Linnaeus). The Bulletin of the Tulane Medical Faculty 14: 55–69.

Dutton, P. H., E. Bixby, and S. K. Davis. 2000. Tendency toward single paternity in leatherbacks detected with microsatellites. Abstract. In: F. A. Abreu-Grobois, R. Briseño-Dueñas, R. Márquez-Millán, and L. Sarti-Martínez (compilers), Proceedings of the Eighteenth International Sea Turtle Symposium. NOAA Technical Memorandum NMFS-SEFSC-436. NOAA, Miami, FL, USA, p. 39.

Dutton, P. H., and S. K. Davis. 1998. Use of molecular genetics to identify individuals and infer mating behavior in leatherbacks. Abstract. In: R. Byles and Y. Fernandez (compilers), Proceedings of the Sixteenth International Symposium on Sea Turtle Conservation and Biology. NOAA Technical Memorandum NMFS-SEFC-412. NOAA, Miami, FL, USA, p. 42.

Dutton, D. L., P. H. Dutton, M. Chaloupka, and R. H. Boulon. 2005. Increase of a Caribbean leatherback turtle *Dermochelys coriacea* nesting population linked to long-term nest protection. Biological Conservation 126: 186–194.

Eckert, K. L. 1987. Environmental unpredictability and leatherback sea turtle (*Dermochelys coriacea*) nest loss. Herpetologica 43: 315–323.

Eckert, K. L., and S. A. Eckert. 1988. Prereproductive movements of leatherback sea turtles (*Dermochelys coriacea*) nesting in the Caribbean. Copeia 1988: 400–406.

Eckert, K. L., B. P. Wallace, J. G. Frazier, S. A. Eckert, and P.C.H. Pritchard. 2012. Synopsis of the biological data on the leatherback sea turtle (*Dermochelys coriacea*). Fish and Wildlife Service, Biological Technical Publication BTP-R4015-2012. US Department of the Interior, Washington, DC, USA.

FitzSimmons, N. N. 1998. Single paternity of clutches and sperm storage in the promiscuous green turtle (*Chelonia mydas*). Molecular Ecology 7: 575–584.

Frazer, N. B., and J. I. Richardson. 1985. Annual variation in clutch size and frequency for loggerhead turtles, *Caretta caretta,* nesting at Little Cumberland Island, Georgia, USA. Herpetologica 41: 246–251.

Fretey, J., and M. Girondot. 1987. Recensement des pontes de tortue luth, *Dermochelys coriacea* (Vandelli, 1761), sur les plages de Yalimapo-Les Hattes a Awara (Guyane Francaise) pendant la saison 1986. Bulletin de la Societe Herpetologique de France 43: 1–8.

Girondot, M. 2010. Estimating density of animals during migratory waves: a new model applied to marine turtles at nesting sites. Endangered Species Research 12: 95–105.

Girondot, M., and J. Fretey. 1996. Leatherback turtles, *Dermochelys coriacea,* nesting in French Guiana, 1978–1995. Chelonian Conservation and Biology 2: 204–208.

Girondot, M., M. H. Godfrey, L. Ponge, and P. Rivalan. 2007. Modeling approaches to quantify leatherback nesting trends in French Guiana and Suriname. Chelonian Conservation and Biology 6: 37–47.

Girondot, M., P. Rivalan, R. Wongsopawiro, J.-P. Briane, V. Hulin, S. Caut, E. Guirlet, and M. H. Godfrey. 2006. Phenology of marine turtle nesting revealed by statistical model of the nesting season. BMC Ecology 6: 11.

Girondot, M., A. D. Tucker, P. Rivalan, M. H. Godfrey, and J. Chevalier. 2002. Density-dependent nest destruction and population fluctuations of Guianan leatherback turtles. Animal Conservation 5: 75–84.

Godfrey, M. H., and R. Barreto. 1995. Beach vegetation and seafinding orientation of turtle hatchlings. Biological Conservation 74: 29–32.

Godfrey, M. H., and R. Barreto. 1998. *Dermochelys coriacea* (leatherback sea turtle) copulation. Herpetological Review 29: 40–41.

Gratiot, N., G. Gratiot, L. Kelle, and B. de Thoisy. 2006. Estimation of the nesting season of marine turtles from incomplete data: statistical adjustment of a sinusoidal function. Animal Conservation 9: 95–102.

Hendrickson, J. R. 1980. The ecological strategies of sea turtles. American Zoologist 20: 597–608.

Hendrickson, J. R., and E. Balasingam. 1966. Nesting beach preferences of Malayan sea turtles. Bulletin of the National Museum of Singapore 33: 69–76.

Hilterman, M. L., and E. Goverse. 2007. Nesting and nest success of the leatherback turtle (*Dermochelys coriacea*) in Suriname, 1999–2005. Chelonian Conservation and Biology 6: 87–100.

Hughes, G. R. 1974. The sea turtles of south-east Africa, I. Status, morphology and distributions. South African Association for Marine Biological Research, Oceanographic Research Institute. Investigational Report No. 35. Oceanographic Research Institute, Durban, South Africa.

James, M. C., S. A. Eckert, and R. A. Myers. 2005. Migratory and reproductive movements of male leatherback turtles (*Dermochelys coriacea*). Marine Biology 147: 845–853.

Kamel, S. J., and N. Mrosovsky. 2004. Nest site selection in leatherbacks, *Dermochelys coriacea*: individual patterns and their consequences. Animal Behaviour 68: 357–366.

Kinch, J. P. 2006. A socio-economic assessment of the Huon Coast leatherback turtle nesting beach projects (Labu Tale, Busama, Lababia and Paiawa), Morobe Province, Papua New Guinea. Western Pacific Regional Fisheries Management Council, Honolulu, Hawaii, USA.

Lazell, J. D., Jr. 1980. New England waters: critical habitat for marine turtles. Copeia 1980: 290–295.

Miller, J. D. 1997. Reproduction in sea turtles, In: P. L. Lutz and J. A. Musick (eds.), The biology of sea turtles. CRC Press, Boca Raton, Florida, USA, pp. 51–82.

Mrosovsky, N. 1983. Ecology and nest-site selection of leatherback turtles, *Dermochelys coriacea*. Biological Conservation 26: 47–56.

Mrosovsky, N. 2006. Distorting gene pools by conservation: assessing the case of doomed turtle eggs. Environmental Management 38: 523–531.

Nordmoe, E. D., A. E. Sieg, P. R. Sotherland, J. R. Spotila, F. V. Paladino, and R. D. Reina. 2004. Nest site fidelity of leatherback turtles at Playa Grande, Costa Rica. Animal Behaviour 68: 387–94.

Owens, D. W. 1980. The comparative reproductive physiology of sea turtles. American Zoologist 20: 549–563.

Pritchard, P.C.H. 1969. Sea turtles of the Guianas. Bulletin of the Florida State Museum, Biological Sciences 13: 85–140.

Pritchard, P.C.H. 1971. The leatherback or leathery turtle, *Dermochelys coriacea*. International Union for the Conservation of Nature (IUCN) Monograph 1. IUCN, Morges, Switzerland.

Pritchard, P.C.H., and P. Trebbau. 1984. The turtles of Venezuela. Society for the Study of Amphibians and Reptiles, Ithaca, New York, USA.

Rafferty, A. R., P. Santidrián Tomillo, J. R. Spotila, F. V. Paladino, and R. D. Reina. 2011. Embryonic death is linked to maternal identity in the leatherback turtle (*Dermochelys coriacea*). PLoS ONE 6(6): e21038. doi: 10.1371/journal.pone.0021038.

Rathbun, G. B., T. Carr, N. Carr, and C. A. Woods. 1985. The distribution of manatees and sea turtles in Puerto Rico, with emphasis on Roosevelt Roads Naval Station. NTIS Publication PB86-1518347AS. National Technical Information Service, Springfield, VA, USA.

Reina, R. D., K. J. Abernathy, G. J. Marshall, and J. R. Spotila. 2005. Respiratory frequency, dive behavior and social interactions of leatherback turtles, *Dermochelys coriacea* during the inter-nesting interval. Journal of Experimental Marine Biology and Ecology 316: 1–16.

Reina, R. D., P. A. Mayor, J. R. Spotila, R. Piedra, and F. V. Paladino. 2002. Nesting ecology of the leatherback turtle, *Dermochelys coriacea*, at Parque Nacional Marino Las Baulas, Costa Rica: 1988–1989 to 1999–2000. Copeia 2002: 653–664.

Rieder, J. P., P. G. Parker, J. R. Spotila, and M. E. Irwin. 1998. The mating system of the leatherback turtle: a molecular approach. Bulletin of the Ecological Society of America 77 (Suppl. 3, Part 2): 375.

Rivalan, P., R. Pradel, R. Choquet, M. Girondot, and A.-C. Prévot-Julliard. 2006. Estimating clutch frequency in the sea turtle *Dermochelys coriacea* using stop-over duration. Marine Ecology Progress Series 317: 285–295.

Robinson, N. J., S. E. Valentine, P. Santidrián Tomillo, V. S. Saba, J. R. Spotila, and F. V. Paladino. 2014. Multidecadal trends in the nesting phenology of Pacific and Atlantic leatherback turtles are associated with population demography. Endangered Species Research 24: 197–206.

Santidrián Tomillo, P., J. S. Suss, B. P. Wallace, K. D. Magrini, G. Blanco, F. V. Paladino, and J. R. Spotila. 2009. Influence of emergence success on the annual reproductive output of leatherback turtles. Marine Biology 156: 2021–2031.

Santidrián Tomillo, P., E. Velez, R. D. Reina, R. Piedra, F. V. Paladino, and J. R. Spotila. 2007. Reassessment of the leatherback turtle (*Dermochelys coriacea*) nesting population at Parque Nacional Marino Las Baulas, Costa Rica: effects of conservation efforts. Chelonian Conservation and Biology 6: 54–62.

Shillinger, G. L., E. Di Lorenzo, H. Luo, S. J. Bograd, E. L. Hazen, H. Bailey, and J. R. Spotila. 2012. On the dispersal of leatherback turtle hatchlings from Mesoamerican nesting beaches. Proceedings of the Royal Society B 279: 2391–2395. doi: 10.1098/rspb.2011.2348.

Sieg, A. E., E. Zandona, V. M. Izzo, F. V. Paladino, and J. R. Spotila. 2010. Population level "flipperedness" in the eastern Pacific leatherback turtle. Behavioural Brain Research 206: 135–138.

Stapleton, S. P., and K. L. Eckert. 2007. Nesting ecology and conservation biology of marine turtles in the Commonwealth of Dominica, West Indies: RoSTI 2007 Annual Project Report. Prepared by WIDECAST for the Ministry of Agriculture and the Environment (Forestry, Wildlife and Parks Division), Roseau, Dominica, West Indies. WIDECAST, Baldwin, MO, USA.

Stewart, K. R. 2007. Establishment and growth of a sea turtle rookery: the population biology of the leatherback in Florida. PhD diss., Duke University, Durham, North Carolina, USA.

Stewart, K. R., and P. H. Dutton. 2011. Paternal genotype reconstruction reveals multiple paternity and sex ratios in a breeding population of leatherback turtles (*Dermochelys coriacea*). Conservation Genetics 12: 1101–1113.

Tucker, A. D. 1988. A summary of leatherback turtle, *Dermochelys coriacea*, nesting at Culebra, Puerto Rico, from 1984–1987 with management recommendations. Unpublished report to the Fish and Wildlife Service, US Department of Interior. Fish and Wildlife Service, Washington, DC, USA.

Tucker, A. D., and K. V. Hall. 1984. Leatherback turtle (*Dermochelys coriacea*) nesting in Culebra, Puerto Rico, 1984. Unpublished report to the Fish and Wildlife Service, US Department of Interior. Fish and Wildlife Service, Washington, DC, USA.

Turtle Expert Working Group (TEWG). 2007. An assessment of the leatherback turtle population in the Atlantic Ocean. NOAA Technical Memorandum NMFS-SEFSC-555. NOAA, Miami, Florida, USA.

Veríssimo, D., D. A. Jones, R. Chaverri, and S. R. Meyer. 2012. Jaguar *Panthera onca* predation of marine turtles: conflict between flagship species in Tortuguero, Costa Rica. Oryx 46: 340–347.

Wallace, B. P., and V. S. Saba. 2009. Environmental and anthropogenic impacts on intra-specific variation in leatherback turtles: opportunities for targeted research and conservation. Endangered Species Research 7: 11–21.

Wallace, B. P., P. Sotherland, P. Santidrián-Tomillo, R. Reina, J. R. Spotila, and F. V. Paladino. 2007. Maternal investment in reproduction and its consequences in leatherback turtles. Oecologia 152: 37–47.

Whitmore, C. P., and P. H. Dutton. 1985. Infertility, embryonic mortality and nest-site selection in leatherback and green sea turtles in Suriname. Biological Conservation 34: 251–272.

Wyneken, J. 1997. Sea turtle locomotion: mechanisms, behavior, and energetics. In: P. L. Lutz and J. A. Musick (eds.), The biology of sea turtles. CRC Press, Boca Raton, FL, USA, pp. 165–198.

7

Egg Development and Hatchling Output of the Leatherback Turtle

PILAR SANTIDRIÁN TOMILLO
AND JENNIFER SWIGGS

Leatherback turtles (*Dermochelys coriacea*) exhibit the lowest hatching success of all sea turtle species. Globally, only about 50% of leatherback eggs complete development and hatch (Bell et al. 2004), compared to approximately 60–90 % in green turtles, *Chelonia mydas* (Broderick and Godley 1996; Antworth et al. 2006; Cheng et al. 2009); loggerhead turtles, *Caretta caretta* (Peters et al. 1994; Broderick and Godley 1996; Antworth et al. 2006); and solitary nesting olive ridley turtles, *Lepidochelys olivacea* (Clusella Trullas and Paladino 2007; Honarvar et al. 2008). Even though overall hatching success of leatherback clutches is low, there is variability among nesting sites around the world (table 7.1), as well as seasonal and interannual variability within the sites (Thomé et al. 2007; Santidrián Tomillo et al. 2009).

There are inconsistencies in the terms used to describe hatching of eggs and emergence of hatchlings, with different authors calculating success of clutches in various ways. For example, Eckert and Eckert (1990) estimated hatching success as the number of live hatchlings related to clutch size. However, more often, hatching success, hatch success, or hatching rate was calculated by dividing the number of hatchlings hatched (usually inferred from remaining eggshells) by the total number of eggs in a clutch (Whitmore and Dutton 1985; Leslie et al. 1996; Miller 1999). Shelled albumin gobs (SAGs), erroneously called yolkless eggs by some, may also have been included in calculations at times, despite not being real eggs because they lack yolk (Sotherland et al. 2003; chapter 12). Additionally, when calculating overall hatching success of leatherback clutches on a nesting beach, some studies included only clutches that survived to term, excluding those that had been eroded, inundated, or destroyed through predation (Eckert and Eckert 1990), whereas other studies may have included these clutches as having 0% hatching success. These differences complicate comparisons among nesting sites. Relocation of doomed clutches (clutches that will not hatch because of environmental factors such as flooding) to increase hatchling production is a common practice in sea turtle conservation projects, and may also affect the estimation of natural hatching success on a nesting beach. Relocated clutches generally result in lower hatching success than in situ ones (Eckert and Eckert

Table 7.1. Hatching success, emergence success, emergence rate, incubation period, incubation temperature, and natural loss of in situ nests of leatherback turtles around the world.

Nesting site	Ocean basin	Hatching success	Emergence success	Emergence rate	Incubation period (days)	Incubation temperature (°C)	Natural loss (%)
French Guiana	Atlantic	0.38[1]	–	–	–	27–32[2]	39[3]
USA, St. Croix	Atlantic	0.64[4], 0.47[5], 0.67[6]	–	–	64[4], 63[6]	–	45–60[7]
USA, Florida	Atlantic	0.49[8], 0.57[9]	0.49[9]	–	–	–	–
Suriname, Babunsanti	Atlantic	0.22–0.35[10]	–	–	61–65[10]	–	–
Suriname, Matapica	Atlantic	0.52[11], 0.58–0.64[10]	–	–	63–67[100], 64[12]	–	40[11]
Brazil	Atlantic	0.65[13]	–	–	68[13]	–	–
Venezuela	Atlantic	0.47[14]	–	–	59[14]	–	–
Costa Rica, Laguna Jalova	Caribbean	0.70[15]	–	–	61[15]	–	–
Costa Rica, Gandoca	Caribbean	–	0.41[16]	–	–	–	57[17]
Grenada, West Indies	Caribbean	–	–	–	–	30–33[18]*	–
Panama, Chiriqui	Caribbean	–	–	–	–	–	21[19]
Costa Rica, Tortuguero	Caribbean	0.14–0.47[20]	0.12–0.39[20]	–	–	–	20–25[20]
Indonesia, Papua	West Pacific	0.09–0.47[21]	–	–	61[21]	–	45[22]
Costa Rica, Playa Grande	East Pacific	0.47[23]	0.41[23]	0.84[23]	59.9[23]	30.6[23]	8[23]

References: (1) Caut et al. 2006, (2) Chevalier et al. 1999, (3) Mrosovsky 1983, (4) Eckert and Eckert 1990, (5) Garrett et al. 2010, (6) Boulon et al. 1996, (7) Eckert 1987, (8) Antworth et al. 2006, (9) Perrault et al. 2012, (10) Hilterman and Goverse 2007, (11) Whitmore and Dutton 1985, (12) Godfrey et al. 1996, (13), Thomé et al. 2007, (14) Hernández et al. 2007, (15) Hirth and Ogren 1987, (16) Chacón-Cheverri and Eckert 2007, (17) Chacón et al. 1996, (18) Houghton et al. 2007, (19) Ordoñez et al. 2007, (20) Troëng et al. 2007, (21) Tapilatu and Tiwari 2007, (22) Hitipeuw et al. 2007, (23) data from Playa Grande not previously published (seasons 2004–2005 to 2012–2013); (*) mean values.

1990; Garrett et al. 2010); the relocation may not always be justified and has the potential for altering hatchling output and sex ratios.

Emergence success is more often estimated as the number of fully emerged hatchlings related to clutch size (Leslie et al. 1996; Wallace et al. 2007). Therefore, this estimation is a combination of success during (1) egg development (eggs that complete development and hatch), and (2) hatchling emergence (percentage of hatchlings that emerge to the surface). Few studies on sea turtles have exclusively assessed the success of hatchlings during the emergence process (number of fully emerged hatchlings related to the number of hatchlings that hatched in the nest) using an emergence percentage (Matsuzawa et al. 2002) or emergence rate (Santidrián Tomillo et al. 2009).

Several biotic and abiotic factors influence egg development and emergence of hatchlings from the nest. Among biotic factors, maternal identity affects hatching success of leatherback clutches in Playa Grande, Costa Rica, where nests laid by some mothers exhibit higher hatching success than those laid by others (Rafferty et al. 2011). Supporting this, Perrault et al. (2012) reported that physiological parameters in the blood of leatherback mothers correlated with hatching and emergence success of eggs in Florida. Although clutch size affects hatching success in other sea turtles (Hewavishenthi and Parmenter 2002), the number of eggs does not influence hatching success of leatherback clutches in a hatchery (Wallace et al. 2007) or in situ conditions at Playa Grande (Santidrián Tomillo et al. 2009).

Some important abiotic factors that influence hatching success in the leatherback nest environment are temperature, PO_2, and PCO_2. Temperature may be the single most important variable affecting egg development and hatchling output in leatherback turtles. High temperature accelerates developmental rate, reduces hatching success and emergence rate, increases proportion of female hatchlings, and potentially affects fitness of hatchlings (fig. 7.1). Higher maximum temperatures in leatherback clutches at St. Croix reduces hatching success (Garrett et al. 2010) and high mean temperatures during incubation also reduces hatching success of clutches and emergence rate of hatchlings at Playa Grande, Costa Rica (Santidrián Tomillo et al. 2009).

Minimum PO_2 and maximum PCO_2 negatively affect hatching success of clutches at St. Croix (Garrett et al. 2010), but not at Playa Grande (Wallace et al 2004); the lack of effect at the latter is believed to be related to tidal pumping of gases (chapter 12). Physical location on the beach and position within the nest also influence development. In general, leatherback turtles tend to nest closer to the high tide line than other species, which nest farther up the beach (Whitmore and Dutton 1985), where there is higher risk of tidal inundation. Clutches that get overwashed have a lower hatching

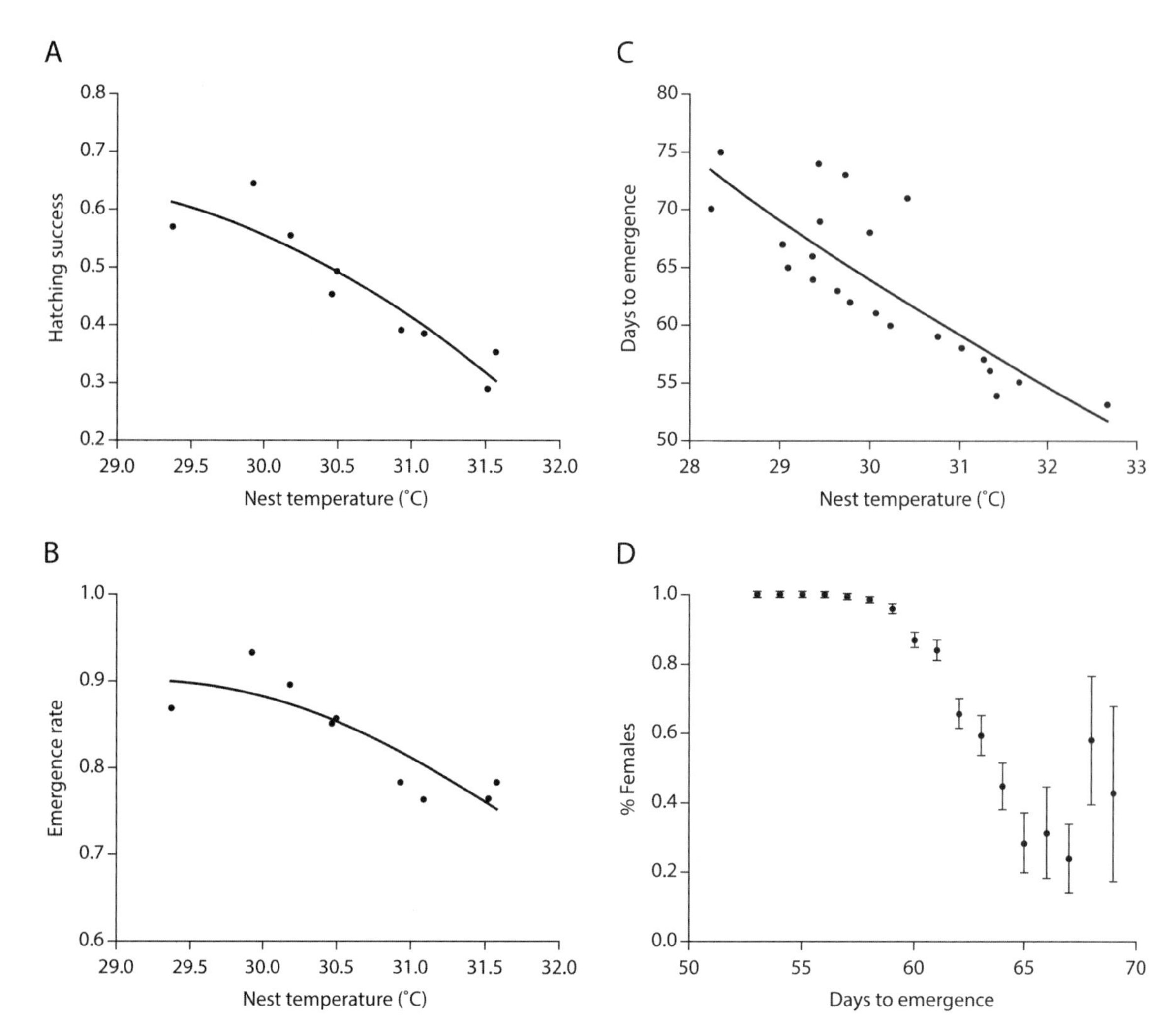

Fig. 7.1. Temperature affects egg development and hatchling output in the leatherback turtle. There is an inverse relationship between mean incubation temperature (°C) and (A) annual hatching success, (B) annual emergence rate, and (C) days to emergence (incubation period plus emergence time) in leatherback turtle clutches at Playa Grande, Costa Rica. Days to emergence (C) is a good predictor of (D) hatchling sex ratios (% females) for periods up to 65 days, because of the effect of temperature on sex ratios and developmental rate. For periods longer than 65 days, low fitness of hatchlings, barriers to emergence, or low numbers of emerging hatchlings may explain the lack of relationship; % females are presented as means (±SE) per incubation period. We included 9 seasons from 2004-2005 to 2012-2013 (n = 926 clutches for hatching success and emergence rate; n = 854 clutches for days to emergence). Relationships were significant in all cases ($R^2 = 0.85$, $p = 0.003$; $R^2 = 0.73$, $p = 0.021$; $R^2 = 0.70$, $p < 0.001$; $R^2 = 0.80$, $p < 0.001$ for figures a, b, c, and d respectively).

success than clutches located in safer areas (Hilterman and Goverse 2007; Caut et al. 2010). However, clutches located close to the high tide line that complete development may have a higher hatching success than clutches located further away (Santidrián Tomillo et al. 2009; Piedra 2011). Therefore, nesting close to the high tide line may constitute an advantage when nests do not get inundated. Within the nest, eggs in the center of the clutch are less successful than those in the periphery (Ralph et al. 2005).

Despite the large number of studies that analyzed egg development of leatherback turtles and the numerous factors known to reduce hatching success of their clutches, there is still a large proportion of variability in hatching success that remains unexplained (as indicated by statistical analyses). Therefore, the causes that drive low hatching success in leatherback turtles remain partially unknown. In this chapter, we review the current state of knowledge on egg development and hatchling emergence of leatherback turtles, and discuss implications for hatchling output and population recruitment.

Fertility, Developmental Arrest, and Oviposition

Development starts right after fertilization of the follicles in the female oviduct (Miller 1997). Low levels of fertility were once considered the cause of low hatching success in leatherback turtles because eggs with embryonic development not visible to the unaided eye were assumed to be "infertile" (Whitmore and Dutton 1985; Chan 1989). However, by using a dissecting microscope, Bell et al. (2004) found that approximately 93.3% of leatherback eggs at Playa Grande, Costa Rica, were fertilized. Although fertility seems to be generally high in leatherback turtles, there are differences among females. For example, turtles that nest late in the season at Playa Grande have overall lower fertility rates than turtles that nest earlier in the season (Bell et al. 2004).

Arrested embryonic development constitutes a reproductive strategy that provides some flexibility in time of nesting, allowing turtles to delay oviposition if environmental conditions are not optimal (Rafferty and Reina 2012). Preovipositional developmental arrest occurs in all turtle species (Andrews 2004). In sea turtles, development is temporally arrested in the female oviduct at middle gastrula (late gastrula in other turtles) and resumed approximately four or more hours after the egg is laid (Miller 1985; Andrews 2004), depending on the temperature. Eggs can tolerate relatively rough handling within the first five hours after oviposition (Chan et al. 1985; Eckert and Eckert 1990), but movement-induced mortality may occur if moved after this period without caution (Miller 1997).

Most unhatched leatherback eggs at Playa Grande died in early stages of development (Bell et al. 2004); according to Miller's stages of development (Miller 1985), these corresponded to developmental stages that occur inside the oviduct. Rafferty et al. (2011) suggested that a large proportion of eggs die in the gastrula stage before oviposition.

Developmental Stages and Hatching

A white spot develops on the sea turtle egg and is associated with the adherence of the vitelline membrane to the egg surface (Blanck and Sawyer 1981). In leatherback turtles this appears approximately four or five days after the egg is laid (Chan 1989). In field studies, the presence of a white spot indicates development. However, as mentioned above, the absence of the white spot does not indicate infertility. Although Miller (1985) identified 31 developmental stages in sea turtle eggs, field studies usually classify unhatched eggs into four stages (Leslie et al. 1996): stage 0 in which development is not visible, stage 1 in which blood vessels or signs of attachments to the shell are discernable, stage 2 in which black eyes of the embryo are visible but the embryo has no pigmentation, and stage 3 in which the embryo has pigmentation. For comparative purposes these stages can be further reduced to early (stages 0 and 1) versus late stages of development (stage 3). Stage 2 is found at very low frequency in nest excavations.

Predominance of dead embryos in early or late stages varies among nesting sites. Hirth and Ogren (1987) reported that a greater percentage of eggs have no visible development in Caribbean Costa Rica and therefore die in early stages of development (or were infertile). Likewise, Bell et al. (2004) reported a greater frequency of death in the early developmental stages in Pacific Costa Rica. However, this study did not include nests laid in the late part of the season when temperatures are highest and a great number of late-stage embryos are found dead in the nest (Santidrián Tomillo et al. 2009). Most mortality occurs in late developmental stages in leatherback clutches at St. Croix (Eckert and Eckert 1990; Garrett et al. 2010).

Among factors that can cause early and/or late mortality of developing embryos are temperature, PO_2, PCO_2, and contamination. High temperatures increased mortality both in early and late stages at Playa Grande (Santidrián Tomillo et al. 2009), and low levels of PO_2, high levels of PCO_2, and high temperatures increased late mortalities in St. Croix (Garrett et al. 2010). Patiño-Martínez et al. (2012) reported that eggs in the early stages of development are more sensitive to contact with eggshells contaminated with bacteria or fungi, which result in smaller hatchlings after hatching. However, the reasons why early or late mortalities are more frequent in some areas than in others are unknown. Differences within the nest environment affecting embryo mortalities may relate to overall climatic conditions and particular beach characteristics of the nesting sites.

At the end of development and before hatching, turtle embryos develop a caruncle, which is a toothlike structure located at the tip of the upper jaw; they use this to break the egg and hatch (Andrews 2004). After "pipping," hatchlings remain within the shell for several days before the process of emergence starts (Andrews 2004). Interestingly, Eckert and Eckert (1990) noticed increased numbers of dead, pipped hatchlings in relocated clutches when compared to in situ nests.

Turtle eggs hatch synchronously within a clutch (Spencer and Janzen 2011). The stability of the nest environment (temperature and humidity) at nest depth may explain the synchrony in leatherback and other sea turtles. Leatherback clutches are buried about 80 cm deep and daily fluctuations in temperature are minimal (Binckley et al. 1998; Drake and Spotila 2002). However, other reasons causing synchrony of hatching may be related to oxygen consumption (Thompson 1993) and the nature of developmental arrest, whereby eggs resume development within a few hours after being laid (Rafferty and Reina 2012). More recently, vibrations and communication among embryos in turtle eggs have also been proposed as cues for synchronous hatching (Spencer and Janzen 2011; reviewed by Rafferty and Reina 2012; Ferrara et al. 2013). Regardless of the mechanisms behind synchronous hatching, there are adaptive advantages to this grouping (Carr and Ogren 1959), such as the social facilitation among siblings during emergence and the reduced per capita risk of predation once on the beach (Santidrián Tomillo et al. 2010).

Emergence from Nest and the Run to the Ocean

Patiño-Martínez et al. (2010) observed that the first hatchlings to hatch waited on top of the group for other hatchlings before moving up in the sand column at Playa Playona, northwest Colombia. Once hatchlings start ascending from the egg chamber, they work as a team, hatchlings on the top of the group scratch the sand off the ceiling, consequently raising the floor (Carr and Ogren 1959). At Playa Playona, hatchlings worked together in burst periods that lasted on average 3.6 minutes and the average emergence time was 3.3 days (Patiño-Martínez et al. 2010).

Because the group effort facilitates the emergence process, the success of hatchlings (emergence rate) is greater in larger groups (Santidrián Tomillo et al. 2009). Additionally, emergence may be aided by air pockets in the sand created as SAGs dehydrate during incubation, coupled with reduced volume of hatched eggs (Patiño-Martínez et al. 2010). Leatherback turtles lay a larger quantity of SAGs than any other species (~1 kg per clutch), which are possibly production overruns (Wallace et al. 2007; chapter 12). Leatherback hatchlings likely benefit from the relatively high volume that SAGs occupy and the space and humidity they provide. Like the effect of temperature on egg development, high temperatures during emergence negatively affect hatchlings, reducing their overall success (Santidrián Tomillo et al. 2009).

The emergence process ends when hatchlings reach the surface of the sand and are exposed on the beach. This mainly occurs at night, with greatest frequency in the early part of the evening (Drake and Spotila 2002). Most leatherback hatchlings at Playa Grande (78–87%) emerge in the first group, although stragglers are frequently found in the morning after dawn (Santidrián Tomillo et al. 2010). At this location, hatchlings emerge when the sand temperature drops below 36°C, usually after dark (Drake and Spotila 2002). However, occasionally hatchlings complete emergence during the day. If temperatures are high (normal conditions on a sunny day) hatchlings will dehydrate, overheat and die since they start exhibiting uncoordinated movements at 33.6°C and reach their Critical Thermal Maxima (CTM) at 40.2°C (Drake and Spotila 2002).

Once hatchlings reach the surface they remain motionless and exposed for a period of time before running to the water. At Playa Grande, hatchlings spent on average 20 minutes visible on top of the nest and 17 minutes crawling to the ocean (Santidrián Tomillo et al. 2010). The time spent on the beach exposed to predators depends on the distance of the nest to the water and the hatchling's ability to move fast.

Hatchling Output

The seasonal hatchling output (total production of hatchlings in a season) per nesting female, or per nesting population, depends on the (1) reproductive investment of females (i.e., number of eggs in a clutch and number of clutches), (2) hatching success of eggs, (3) emergence rate of hatchlings, and (4) predation rates on eggs and hatchlings. Some females produce more hatchlings than others by nesting when conditions are optimal for egg development and hatchling emergence (Santidrián Tomillo et al. 2009), by laying larger clutches (Wallace et al. 2007) and by laying more clutches during the season (Santidrián Tomillo et al. 2009). Lifetime production of hatchlings will additionally depend on the total number of seasons a female leatherback turtle nests during her life, which will be affected by remigration intervals between nesting seasons. These intervals average two to four years (Girondot and Fretey 1996; Reina et al. 2002) but are highly variable in eastern Pacific populations (Saba et al. 2007). For more information on reproductive investment, see chapter 6.

Environmental variables related to beach dynamics further affect seasonal hatchling output. Leatherback nesting beaches can experience erosion, accretion, and tidal inundation. Beach erosion and inundation are a great threat to leatherback populations in the western

Pacific, such as in Papua Barat, Indonesia, Papua New Guinea, Solomon Islands, and Vanuatu (Tiwari et al. 2011). Relocation practices in Papua Barat, Indonesia save about 45% of nests from erosion (Hitipeuw et al. 2007). Likewise, high levels of erosion and inundation occur at locations in the Atlantic: 25% of nests were laid below the high tide line in 1985 in Laguna Jalova, Caribbean Costa Rica (Hirth and Ogren 1987), 40% in 1981 and 1982 in Krofajapasi Beach in Suriname (Whitmore and Dutton 1985), and about 20% in 1990 and 1991 in Tortuguero, Costa Rica (Leslie et al. 1996).

High levels of natural erosion occurring on several leatherback nesting beaches make relocation practices very efficient, because they can result in nearly a twofold increase in seasonal hatchling output. One of the best known examples of successful relocation practices is Sandy Point, St. Croix, where natural beach erosion caused 45–60% of nests to be lost before relocation programs started in 1982 (Eckert 1987; Eckert and Eckert 1990; Boulon et al. 1996). As a consequence, the number of nesting turtles increased substantially in the early 1990s (Dutton et al. 2005). Due to the low level of beach erosion in the eastern Pacific (estimated at Playa Grande, Costa Rica at approximately 8%; table 7.1) compared to beaches in other ocean basins, relocation practices have a less significant impact on hatchling output and future recruitment to the nesting population.

Egg relocations, however, can be detrimental if the clutches are moved from a safe location where they could have developed naturally. Relocated clutches often have a lower hatching success than in situ clutches (Eckert and Eckert 1990; Boulon et al. 1996; Hernández et al. 2007; Garrett et al. 2010) and the sex ratios of hatchlings may be altered (chapter 8). Additionally, egg infections from bacteria and fungi increase in reused hatcheries into which eggs are sometimes moved (Patiño-Martínez et al. 2012).

Other natural processes affecting hatchling output are clutch destruction by other female turtles in high-density areas, and egg and hatchling predation. Although clutch destruction is not likely to be a factor reducing hatchling output on most beaches due to current low numbers of nesting leatherbacks, it can have an effect on some highly populated beaches, such as Ya:lima:po-Awa:la, French Guiana, where about 21% of clutches are destroyed by other females (Girondot et al. 2002).

Wild animals, domestic animals, and people prey on leatherback eggs and/or hatchlings. Levels of predation vary from site to site and may be density dependent. Natural predators include mole crickets (Caut et al. 2006), coatis and black vultures (Hirth and Ogren 1987), caracaras and night herons (Santidrián Tomillo et al. 2010), jackals (Hughes 1996), lizards (Tapilatu and Tiwari 2007), ghostcrabs (Hirth and Ogren 1987; Santidrián Tomillo et al. 2010), and raccoons (authors' personal observation). Domestic animals such as dogs and pigs constitute an additional threat to eggs and hatchlings (Hughes 1996; Leslie et al. 1996; Hitipeuw et al. 2007; Ordoñez et al. 2007; Tapilatu and Tiwari 2007; Santidrián Tomillo et al. 2010). Finally, human predation (egg poaching) nearly collapsed the leatherback population of Playa Grande Costa Rica (Santidrián Tomillo et al. 2008) and did help extirpate the leatherback population of Terengganu, Malaysia (Chan and Liew 1996), where egg poaching reached about 90% and 100% respectively over several decades.

Climate Change and Hatchling Output

Prevailing climatic conditions on nesting beaches affect temperature and humidity in the nest environment, which in turn affect hatchling output and population dynamics of leatherback turtles. These effects are modulated through air temperature and precipitation (Saba et al. 2012; Santidrián Tomillo et al. 2012). We (Santidrián Tomillo et al. 2014) marked, monitored temperatures, and excavated 926 leatherback clutches at Playa Grande from 2004–2013. High temperatures reduced emergence success and female output (number of female hatchlings) (fig. 7.2). Output of female hatchlings increased with incubation temperature as it reached the upper end of the range of temperatures that produce both sexes (30 °C) and decreased afterward because high temperatures increased mortality of "female clutches." High temperatures reduced sex ratios from 85% female primary sex ratios to 79% secondary sex ratios due to hatchling mortality. Climate change may threaten populations by reducing overall hatchling output and increasing frequency of seasons with 100% female production, which will actually produce fewer female hatchlings.

Climatic conditions vary around the world. The northwest coast of Costa Rica, an area influenced by El Niño Southern Oscillation (ENSO), is particularly hot and dry during El Niño phases, but high levels of precipitation and cool temperatures during La Niña phases are favorable for hatching success and hatchling emergence (Santidrián Tomillo et al. 2012). However, other locations in the tropics may experience excessive levels of precipitation causing soil saturation and increased egg mortality; this has been recorded for loggerhead turtles (Ragotzkie 1959). Additionally, protracted rainfall has a cooling effect on nests, increasing male hatchling

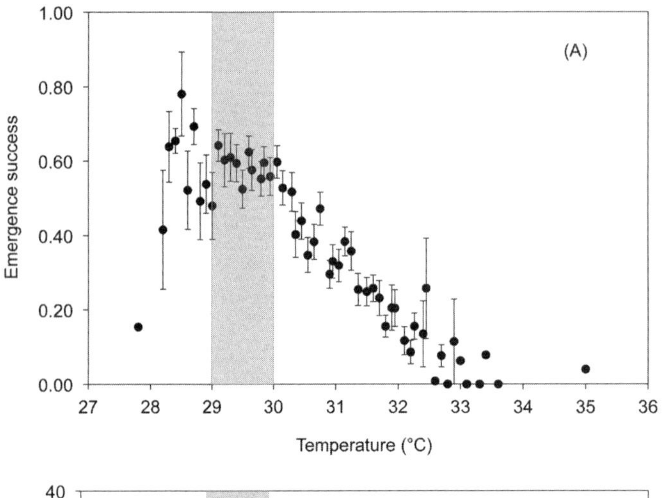

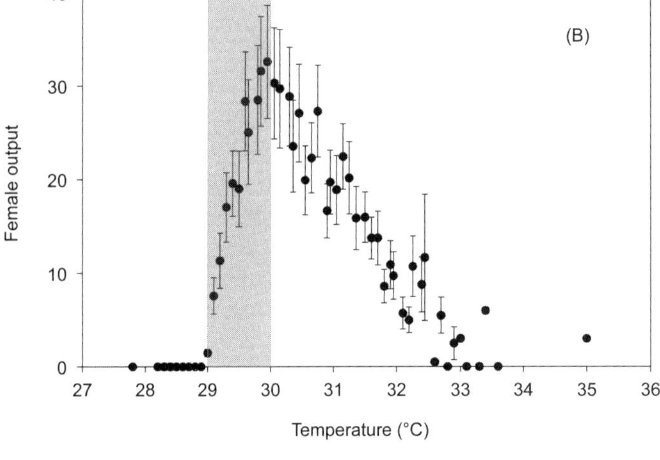

Fig. 7.2. Effect of nest temperature during the middle third of incubation on (A) emergence success, and (B) female hatchling output per nest of leatherback turtles at Playa Grande, Costa Rica, between 2004 and 2013. Values are mean ±SE. Gray area is transitional range of temperature for sex determination. From Santidrián Tomillo et al. 2014, with permission from Elsevier.

production (Houghton et al. 2007). We observed this on Playa Grande during the 2001–2002 nesting season when Hurricane Michelle sat over Central America and caused large amounts of rain for a week (Wallace et al. 2004). Clutch temperatures dropped about 6°C and they took approximately two weeks to recover to prehurricane levels. More male hatchlings were produced but there was no effect on overall hatching success.

Based on climate change projections from the Fourth Assessment report of the Intergovernmental Panel on Climate Change (IPCC), Santidrián Tomillo et al. (2012) project an approximate 50–60% decline in seasonal leatherback hatchling output (hatching success and emergence rate) in northwest Costa Rica through the twenty-first century, because climatic conditions are projected to get drier and warmer in this area. As a result of climate effects on eggs and hatchlings, Saba et al. (2012) estimate a 7% per decade decline in this nesting population for the same time period. Although, climate change may considerably reduce hatchling output in Pacific Costa Rica, other areas in the world may experience increased precipitation levels and, therefore, global hatchling output may not decrease substantially. However, if air temperatures are raised by several degrees due to human-caused global warming, it may become the main driving force affecting hatchling output. Additionally, increased rain on nesting beaches along the Inter-Tropical Convergence Zone may reduce beach temperatures and even drown clutches. This possibility requires additional research.

Climate change is becoming one of the main threats to leatherback populations due to effects on eggs and hatchlings. Outputs of leatherback hatchlings need to be assessed globally to determine high-risk areas where climate mitigation strategies may become necessary in the near future. Additionally, further research on the nest environment is crucial to better understand the causes behind low hatching success in leatherback turtles. As the climate gets warmer, increasing hatchling production may become the single most important conservation action to preserve leatherback turtle populations into the future.

ACKNOWLEDGMENTS

We would like to thank James Spotila, Frank Paladino, and the Leatherback Trust board and staff members for their leadership in research and conservation of leatherback turtles in Costa Rica. We also thank field managers and biologists who were instrumental in the data collection at all sites included in this chapter. Finally, we thank Spencer Roberts for reading early versions of the manuscript.

LITERATURE CITED

Andrews, R. M. 2004. Patterns of embryonic development. In: D. C. Deeming (ed.), Reptilian incubation: Environment, evolution and behaviour. Nottingham University Press, Nottingham, England, UK, pp. 75–102.

Antworth, R. L., D. A. Pike, and J. C. Stiner. 2006. Nesting ecology, current status, and conservation of sea turtles on an uninhabited beach in Florida, USA. Biological Conservation 130: 10–15.

Bell, B. A., J. R. Spotila, F. V. Paladino, and R. D. Reina. 2004. Low reproductive success of leatherback turtles, *Dermochelys coriacea*, is due to high embryonic mortality. Biological Conservation 115: 131–138.

Binckley, C. A., J. R. Spotila, K. S. Wilson, and F. V. Paladino. 1998. Sex determination and sex ratios of Pacific leatherback turtles, *Dermochelys coriacea*. Copeia 1998: 291–300.

Blanck, C. E., and R. H. Sawyer. 1981. Hatchery practices in relation to early embryology of the loggerhead sea turtle, *Caretta caretta* (Linné). Journal of Experimental Marine Biology and Ecology 49: 163–177.

Boulon, R. H., P. H. Dutton, and D. L. McDonald. 1996. Leatherback turtles (*Dermochelys coriacea*) on St. Croix, U.S. Virgin Islands, 1979–1995. Chelonian Conservation and Biology 2: 141–148.

Broderick, A. C., and G. J. Godley. 1996. Population and nesting ecology of the green turtle, *Chelonia mydas*, and the loggerhead turtle, *Caretta caretta*, in northern Cyprus. Zoology in the Middle East 13: 27–46.

Carr, A., and L. Ogren. 1959. The ecology and migrations of sea turtles, 3. *Dermochelys* in Costa Rica. American Museum Novitates 1958: 1–29.

Caut, S., E. Guirlet, and M. Girondot. 2010. Effect of tidal overwash on the embryonic development of leatherback turtles in French Guiana. Marine Environmental Research 69: 254–261.

Caut, S., E. Guirlet, P. Jouquet, and M. Girondot. 2006. Influence of nest location and yolkless eggs on the hatching success of leatherback turtle clutches in French Guiana. Canadian Journal of Zoology 84: 908–915.

Chacón, D., W. McLarney, C. Ampie, and B. Venegas. 1996. Reproduction and conservation of the leatherback turtle *Dermochelys coriacea* (Testudines: Dermochelyidae) in Gandoca, Costa Rica. Revista de Biología Tropical 44: 853–860.

Chacón-Chaverri, D., and K. L. Eckert. 2007. Leatherback sea turtle nesting at Gandoca beach in Caribbean Costa Rica: management recommendations from fifteen years of conservation. Chelonian Conservation and Biology 6: 101–111.

Chan, E. H. 1989. White spot development, incubation and hatching success of leatherback turtle (*Dermochelys coriacea*) eggs from Rantau Abang, Malaysia. Copeia 1989: 42–47.

Chan, E. H., and H. C. Liew. 1996. Decline of the leatherback population in Terengganu, Malaysia, 1956–1995. Chelonian Conservation and Biology 2: 196–203.

Chan, E. H., H. U. Salleh, and H. C. Liew. 1985. Effects of handling on hatchability of eggs of the leatherback turtle, *Dermochelys coriacea* (L.). Pertanika 8: 265–271.

Cheng, I. J., C. T. Huang, P. Y. Hung, B. Z. Ke, C. W. Kuo, and C. L. Fong. 2009. Ten years of monitoring the nesting ecology of green turtle, *Chelonia mydas*, on Lanyu (Orchid Island), Taiwan. Zoological Studies 48: 83–94.

Chevalier, J., M. H. Godfrey, and M. Girondot. 1999. Significant difference of temperature-dependent sex determination between French Guiana (Atlantic) and Playa Grande (Costa Rica, Pacific) leatherbacks (*Dermochelys coriacea*). Annales des Sciences Naturelles–Zoologie et Biologie Animale 20: 147–152.

Clusella Trullas, S., and F. V. Paladino. 2007. Micro-environment of olive ridley turtle nests deposited during an aggregated nesting event. Journal of Zoology 272: 367–376.

Drake, D. L., and J. R. Spotila. 2002. Thermal tolerances and the timing of sea turtle hatchling emergence. Journal of Thermal Biology 27: 71–81.

Dutton, D. L., P. H. Dutton, M. Chaloupka, and R. H. Boulon. 2005. Increase of a Caribbean leatherback turtle *Dermochelys coriacea* nesting population linked to long-term nest protection. Biological Conservation 126: 186–194.

Eckert, K. L. 1987. Environmental unpredictability and leatherback sea turtle (*Dermochelys coriacea*) nest loss. Herpetologica 43: 315–323.

Eckert, K. L., and S. A. Eckert. 1990. Embryo mortality and hatch success in *in situ* and translocated leatherback sea turtle *Dermochelys coriacea* eggs. Biological Conservation 53: 37–46.

Ferrara, C. R., R. C. Vogt, and R. S. Sousa-Lima. 2013. Turtle vocalizations as the first evidence of posthatching parental care in chelonians. Journal of Comparative Psychology 127: 24–32.

Garrett, K., B. P. Wallace, J. Garner, and F. V. Paladino. 2010. Variation in leatherback turtle nest environments: consequences for hatching success. Endangered Species Research 11: 147–155.

Girondot, M., and J. Fretey. 1996. Leatherback turtles, *Dermochelys coriacea*, nesting in French Guiana, 1978–1995. Chelonian Conservation and Biology 2: 204–208.

Girondot, M., A. D. Tucker, P. Rivalan, M. H. Godfrey, and J. Chevalier. 2002. Density-dependent nest destruction and population fluctuations of Guianan leatherback turtles. Animal Conservation 5: 75–84.

Godfrey, M. H., R. Barreto, and N. Mrosovsky. 1996. Estimating past and present sex ratios of sea turtles in Suriname. Canadian Journal of Zoology 74: 267–277.

Hernández, R., J. Buitrago, H. Guada, H. Hernández-Hamón, and M. Llano. 2007. Nesting distribution and hatching success of the leatherback, *Dermochelys coriacea*, in relation to human pressures at Playa Parguito, Margarita Island, Venezuela. Chelonian Conservation and Biology 6: 79–87.

Hewavishenthi, S., and C. J. Parmenter. 2002. Incubation environment and nest success of the flatback turtle (*Natator depressus*) from a natural nesting beach. Copeia 2002: 302–312.

Hilterman, M. L., and E. Goverse. 2007. Nesting and nest success of the leatherback turtle (*Dermochelys coriacea*) in Suriname, 1999–2005. Chelonian Conservation and Biology 6: 87–101.

Hirth, H., and L. H. Ogren. 1987. Some aspects of the ecology of the leatherback turtle *Dermochelys coriacea* at Laguna Jalova, Costa Rica. NOAA Technical Report NMFS 56. NOAA, Springfield, VA, USA.

Hitipeuw, C., P. H. Dutton, S. Benson, J. Thebu, and J. Bakarbessy. 2007. Population status and internesting movement of leatherback turtles, *Dermochelys coriacea*, nesting on the northwest coast of Papua, Indonesia. Chelonian Conservation and Biology 6: 28–37.

Honarvar, S., M. P. O'Connor, and J. R. Spotila. 2008. Density-dependent effects on hatching success of the olive ridley turtle, *Lepidochelys olivacea*. Oecologia 157: 221–230.

Houghton, J.D.R., A. E. Myers, C. Lloyd, R. S. King, C. Isaacs, and G. C. Hays. 2007. Protracted rainfall decreases temperature within leatherback turtle (*Dermochelys coriacea*) clutches in Grenada, West Indies: ecological implications for a species displaying temperature dependent sex determination. Journal of Experimental Marine Biology and Ecology 345: 71–77.

Hughes, G. R. 1996. Nesting of the leatherback turtle (*Dermochelys coriacea*) in Tongaland, KwaZulu-Natal, South Africa, 1963–1995. Chelonian Conservation and Biology 2: 153–159.

Leslie, A. J., D. N. Penick, J. R. Spotila, and F. V. Paladino. 1996. Leatherback turtle, *Dermochelys coriacea*, nesting and nest success in Tortuguero, Costa Rica, in 1990–1991. Chelonian Conservation and Biology 2: 159–168.

Matsuzawa, Y., K. Sato, W. Sakamoto, and K. A. Bjorndal. 2002. Seasonal fluctuations in sand temperature: effects on the incubation period and mortality of loggerhead sea turtle (*Caretta caretta*) pre-emergent hatchlings in Minabe, Japan. Marine Biology 140: 639–646.

Miller, J. D. 1985. Embryology of marine turtles. In: C. Gans, F. Billet, and P.F.A. Maderson (eds.), Biology of the Reptilia, Vol. 14. Wiley, New York, NY, USA, pp. 271–328.

Miller, J. D. 1997. Reproduction in sea turtles. In: P. L. Lutz and J. A. Musick (eds.), The biology of sea turtles, CRC Press, Boca Raton, FL, USA, pp. 51–80.

Miller, J. D. 1999. Determining clutch size and hatching success. In: K. L. Eckert, K. A. Bjorndal, F. A. Abreu-Grobois, and M. Donnelly (eds.), Research and management techniques for the conservation of sea turtles. IUCN/SSC Marine Turtle Specialist Group Publication No. 4. IUCN/SSC, Washington, DC, USA, pp. 124–129.

Mrosovsky, N. 1983. Ecology and nest-site selection of leatherback turtles *Dermochelys coriacea*. Biological Conservation 26: 47–56.

Ordoñez, C., S. Troëng, A. Meylan, P. Meylan, and A. Ruiz. 2007. Chiriqui beach, Panama, the most important leatherback nesting beach in Central America. Chelonian Conservation and Biology 6: 122–126.

Patiño-Martínez, J., A. Marco, L. Quiñones, E. Abella, R. M. Abad, and J. Diéguez-Uribeondo. 2012. How do hatcheries influence embryonic development of sea turtle eggs? Experimental analysis and isolation of microorganisms in leatherback turtle eggs. Journal of Experimental Zoology 317: 47–54.

Patiño-Martínez, J., A. Marco, L. Quiñones, and C. Calabuig. 2010. False eggs (SAGs) facilitate social post-hatching emergence behaviour in leatherback turtles *Dermochelys coriacea* (Testudines: Dermochelyidae). Revista de Biología Tropical 58: 943–954.

Perrault, J. R., D. L. Miller, E. Eads, C. Johnson, A. Merrill, L. J. Thompson, and J. Wyneken. 2012. Maternal health status correlates with nest success of leatherback sea turtles (*Dermochelys coriacea*) from Florida. PLoS ONE 7 (2): e3184. doi: 10.1371.

Peters, A., J. F. Verhoeven, and H. Strijbosch. 1994. Hatching and emergence in the Turkish Mediterranean loggerhead turtle, *Caretta caretta*: natural causes for egg and hatchling failure. Herpetologica 50: 369–373.

Piedra, R. 2011. Evaluación del éxito de incubación de los huevos de tortuga baula (*Dermochelys coriacea*) en dos áreas de anidación del Parque Nacional Marino Las Baulas de Guanacaste y su aporte a la conservación de la especie en el Pacífico Oriental Tropical. MS thesis, Universidad Nacional, Heredia, Costa Rica.

Rafferty, A. R., and R. D. Reina. 2012. Arrested embryonic development: a review of strategies to delay hatching in egg-laying reptiles. Proceedings of the Royal Society of London B 279: 2299–2308. doi: 10.1098/rspb.2012.0100.

Rafferty, A. R., P. Santidrián Tomillo, J. R. Spotila, F. V. Paladino, and R. D. Reina. 2011. Embryonic death is linked to maternal identity in the leatherback turtle (*Dermochelys coriacea*). PLoS ONE 6 (6): e21038. doi: 10.1371.

Ragotzkie R. 1959. Mortality of loggerhead turtle eggs from excessive rainfall. Ecology 40: 303–305.

Ralph, C. R., R. D. Reina, B. P. Wallace, P. R. Sotherland, J. R. Spotila, and F. V. Paladino. 2005. Effect of egg location and respiratory gas concentrations on developmental success in nests of the leatherback turtle, *Dermochelys coriacea*. Australian Journal of Zoology 53: 289–294.

Reina, R. D., P. A. Mayor, J. R. Spotila, R. Piedra, and F. V. Paladino. 2002. Nesting ecology of the leatherback turtle, *Dermochelys coriacea*, at Parque Nacional Marino Las Baulas, Costa Rica: 1988–1989 to 1999–2000. Copeia 2002: 653–664.

Saba, V. S., P. Santidrián Tomillo, R. D. Reina, J. R. Spotila, J. A. Musick, D. A. Evans, and F. V. Paladino. 2007. The effect of the El Niño Southern Oscillation on the reproductive frequency of eastern Pacific leatherback turtles. Journal of Applied Ecology 44: 395–404.

Saba, V. S., C. A. Stock, J. R. Spotila, F. V. Paladino, and P. Santidrián Tomillo. 2012. Projected response of an endangered marine turtle population to climate change. Nature Climate Change 2: 814–820.

Santidrián Tomillo, P., D. Oro, F. V. Paladino, R. Piedra, A. E. Sieg, and J. R. Spotila. 2014. High beach temperatures increased female-biased primary sex ratios but reduced output of female hatchlings in the leatherback turtle. Biological Conservation 176: 71–79.

Santidrián Tomillo, P., F. V. Paladino, J. S. Suss, and J. R. Spotila. 2010. Predation of leatherback turtle hatchlings during the crawl to the water. Chelonian Conservation and Biology 9: 18–25.

Santidrián Tomillo, P., V. S. Saba, G. S. Blanco, C. A. Stock, F. V. Paladino, and J. R. Spotila. 2012. Climate driven egg and hatchling mortality threatens survival of eastern Pacific leatherback turtles. PLoS ONE 7 (5): e37602. doi: 10.1371.

Santidrián Tomillo, P., V. S. Saba, R. Piedra, F. V. Paladino, and J. R. Spotila. 2008. Effects of illegal harvest of eggs on the population decline of leatherback turtles in Las Baulas Marine National Park, Costa Rica. Conservation Biology 22: 1216–1224.

Santidrián Tomillo, P., J. S. Suss, B. P. Wallace, K. D. Magrini, G. Blanco, F. V. Paladino, and J. R. Spotila. 2009. Influence of emergence success on the annual reproductive output of leatherback turtles. Marine Biology 156: 2021–2031.

Sotherland, P. R., R. D. Reina, S. Bouchard, B. P. Wallace, B. F. Franks, and J. R. Spotila. 2003. Egg mass, egg composition, clutch mass, and hatchling mass of leatherback turtles (*Dermochelys coriacea*) nesting at Parque Nacional Las Baulas, Costa Rica. Abstract. In: J. A. Seminoff (compiler), Proceedings of the Twenty-second Annual Symposium of Sea Turtle Biology and Conservation. NOAA Technical Memorandum NMFS-SEFSC-503. NOAA/NMFS, Miami, FL, USA, p. 31.

Spencer, R. J., and F. J. Janzen. 2011. Hatching behavior in turtles. Integrative and Comparative Biology 51: 100–110.

Tapilatu, R. F., and M. Tiwari. 2007. Leatherback turtle, *Dermochelys coriacea*, hatching success at Jamursba-Medi and Wermon beaches in Papua, Indonesia. Chelonian Conservation and Biology 6: 154–159.

Thomé, J.C.A., C. Paptistotte, L.M.P. Moreira, J. T. Scalfoni, A. P. Almeida, D. B. Rieth, and P.C.R. Barata. 2007. Nesting biology and conservation of the leatherback sea turtle (*Dermochelys coriacea*) in the State of Espírito Santo, Brazil, 1988–1989 to 2003–2004. Chelonian Conservation and Biology 6: 15–28.

Thompson, M. B. 1993. Oxygen consumption and energetics of development in eggs of the leatherback turtle, *Dermochelys coriacea*. Comparative Biochemistry Physiology A 104: 449–453.

Tiwari, M., D. L. Dutton, and J. A. Garner. 2011. Nest relocation. A necessary management tool for western Pacific leatherback nesting beaches. In: P. Dutton, D. Squires, and M. Ahmed (eds.), Conservation of Pacific sea turtles. University of Hawai'i Press, Honolulu, HI, USA, pp. 87–96.

Troëng, S., E. Harrison, D. Evans, A. De Haro, and E. Vargas. 2007. Leatherback turtle nesting trends and threats at Tortuguero, Costa Rica. Chelonian Conservation and Biology 6: 117–122.

Wallace, B. P., P. R. Sotherland, P. Santidrián Tomillo, R. D. Reina, J. R. Spotila, and F. V. Paladino. 2007. Maternal investment in reproduction and its consequences in leatherback turtles. Oecologia 152: 37–47.

Wallace, B. P., P. R. Sotherland, J. R. Spotila, R. D. Reina, B. F. Franks, and F. V. Paladino. 2004. Biotic and abiotic factors affect the nest environment of embryonic leatherback turtles, *Dermochelys coriacea*. Physiological and Biochemical Zoology 77: 423–432

Whitmore, C. P., and P. H. Dutton. 1985. Infertility, embryonic mortality and nest-site selection in leatherback and green sea turtles in Suriname. Biological Conservation 34: 251–272.

Sex Determination and Hatchling Sex Ratios of the Leatherback Turtle

CHRISTOPHER A. BINCKLEY
AND JAMES R. SPOTILA

Most vertebrates possess genetic sex determination (GSD) in which sex is determined at fertilization with the union of haploid gametes (Bull 1980). However, a wide diversity of reptiles, including all sea turtle species examined to date, exhibit temperature-dependent sex determination (TSD; see Wibbels 2003). Typically, temperatures experienced during the middle third of development, during development of gonads, determine the sex of sea turtle offspring (Binckley et al. 1998). Since first described by Charnier (1966), research on TSD has greatly expanded along four research lines: (1) its adaptive advantage, if any; (2) underlying physiological / genetic mechanisms; (3) ecological factors influencing nest temperatures and resulting patterns of hatchling sex ratios; and (4) the conservation / management implications of TSD, particularly in reference to human manipulation of nest temperatures at local and global scales.

This chapter reviews the current state of knowledge concerning TSD in leatherback turtles (*Dermochelys coriacea*) and TSD effects on hatchling sex ratios, and particularly emphasizes the third and fourth research lines noted above. Many excellent reviews of TSD in general and specifically TSD in sea turtles exist, and we avoid duplication where possible (Bull 1980; Standora and Spotila 1985; Mrosovsky 1994; Wibbels 2003, Wibbels 2008). Throughout, we address the surprisingly large number of basic and applied research questions that currently, and unfortunately, remain unanswered concerning leatherback TSD. Answering these questions is essential, given the continued human population growth that will cause anthropogenic impacts; these greatly modify leatherback nest temperatures and, hence, hatchling sex ratios. These same research gaps further hinder our ability to predict how changes in global climate will interact with leatherback TSD and affect sex ratios of future leatherback populations. Thus, we address the inevitable management and conservation issues related to leatherback TSD in a final section.

Patterns of TSD in Leatherback Turtles

In sea turtles low temperatures produce males, a narrow range of intermediate temperatures produces both sexes, and higher temperatures

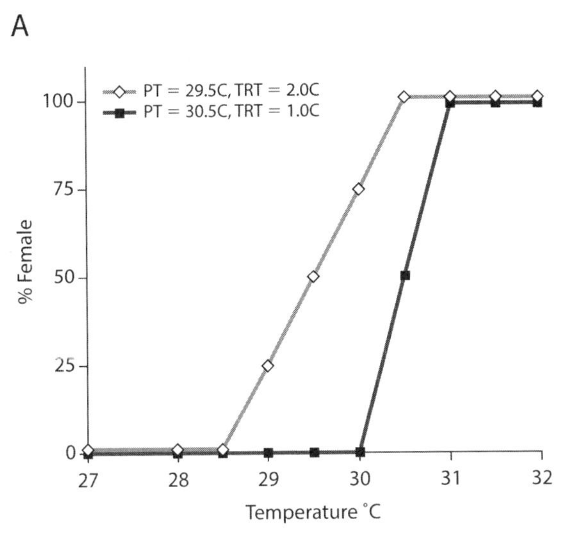

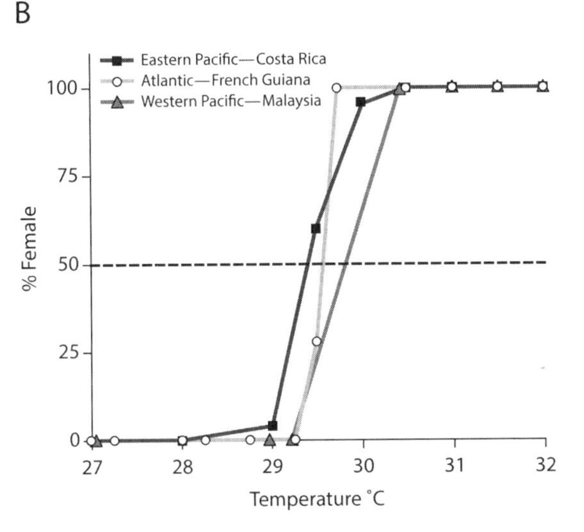

Fig. 8.1. A. Hypothetical TSD curves for two populations showing the percentage of female hatchlings produced at different temperatures. Populations differ in both the temperature producing a sex ratio of one to one (pivotal temperature: PT) and the range of temperatures causing a shift from all male to all female hatchlings (transitional range of temperatures: TRT). B. TSD curves from three leatherback populations: one nesting in Costa Rica (Binckley et al. 1998), another in French Guiana (Rimblot et al. 1985; Rimblot-Baly et al. 1987), and another in Malaysia (Chan and Liew 1995).

produce females (fig. 8.1). Graphical descriptions of changing sex ratios along temperature gradients are most often generated via controlled experiments in which eggs are incubated and sex of hatchlings is identified by histological examination of gonads. These experiments allow identification of two important parameters used for comparing TSD patterns within and among sea turtle species. The first is the range of temperatures producing both sexes and is called the transitional range of temperature (TRT) below which all hatchlings are male and above which all are female (fig. 8.1; see Mrosovsky and Pieau 1991). The second is the pivotal temperature (PT), the temperature within the TRT that produces a 1:1 sex ratio under constant temperature incubation (Mrosovsky and Pieau 1991). For example, figure 8.1A shows hypothetical TSD curves for two populations that differ in both TRT and PT. In the population represented by the grey line, eggs incubated below 28.5°C produce all males, those developing at 29.5°C produce an equal number of males and females (PT), while those above 30.5°C produce all females, thus giving this population a TRT of 2.0°C. The population represented by the black line has a PT of 30.5°C and a TRT of only 1.0°C as all males are produced below 30.0°C and only females above 31.0°C.

How PT varies within and among populations and species has received the most attention since this temperature produces a 1:1 sex ratio like that in GSD, and is under strong frequency-dependent selective pressure when deviations from balanced sex ratios occur (Binckley et al. 1998). Less well studied is variation in TRT within and among sea turtle populations and species, which is unfortunate since variation in this parameter could drastically affect hatchling sex ratios even if, for example, two populations have identical PTs but different TRT (Mrosovsky and Pieau 1991; Chevalier et al. 1999). Variation in PT occurs among female sea turtles from the same nesting population, conspecific females from different nesting populations, and the most variation occurs among females of different species (reviewed in Wibbels 2003). However, leatherbacks show only minor variation in the shape of their TSD curves among populations (Binckley et al. 1998) and females nesting in French Guiana have one of the narrowest TRT recorded for any reptile (Chevalier et al. 1999; Hulin et al. 2009). These results could be a consequence of the low number of leatherback populations examined for differences in TSD curves.

To date, leatherback TSD curves have been generated at only three nesting beaches using different incubation methodologies, number of eggs/females represented in the experiments, and the specific temperatures examined. Rimblot et al. (1985) and Rimblot-Baly et al. (1987) examined leatherback TSD from an Atlantic nesting colony located in French Guiana. They incubated 142 leatherback eggs at 10 different temperatures ranging from 27.0 to 32.0°C, with all males produced when temperatures were at and below 29.2°C, and all females from incubators set to 29.8°C and above. Eggs incubated at 29.5°C produced a mixed sex ratio of 28% female and 72% male. Binckley et al. (1998) examined leatherback TSD in the eastern Pacific nesting colony located at Parque Nacional Marino Las Baulas (PNMB), Costa Rica, to directly compare results from this population to those from French Guiana. They incubated 120 eggs at 8 different temperatures ranging from 28.0 to 33.0°C, with the last temperature reset to 31.5°C because no eggs developed at the highest temperature tested. Eggs incubated below 29.0°C produced primarily males (96%) with only a single female. Incubation temperatures above 30.0°C produced primarily females (96%) with only a single male. Eggs incubated at 29.5°C produced a mixed sex ratio of 40% male and 60% female. Chan and Liew (1995) determined the sex of 34 leatherback hatchlings from the western Pacific Rantau Abang turtle sanctuary in Malaysia. Eggs were incubated at four temperatures with means ranging from 27.0 to 30.4°C. Only male hatchlings were produced from temperatures at and below 29.2°C, and only females at the highest temperature tested. These results (fig. 8.1B) have been interpreted as showing either great similarities (Binckley et al. 1998) or differences among leatherback populations (Chevalier et al. 1999). We still favor the first interpretation, but both interpretations are informative and emphasize the need for future experiments that incubate leatherback eggs from genetically different leatherback populations in the same incubators.

The PTs often correspond to the thermal nesting environment of the turtle population being examined and vary both inter- and intraspecifically. These variations are hypothesized to avoid extreme bias in sex ratios, and differences are attributed to factors that influence nest temperature, including climate, nest site selection, and time of nesting season (Ewert et al. 1994). Generally, turtle species and populations that experience higher nest temperatures have higher PTs, as exemplified by desert tortoises (*Gopherus agassizii*) (Spotila et al. 1994) and olive ridleys (*Lepidochelys olivacea*) (Wibbels 2008). There are large differences in climate between different leatherback nesting beaches, However, estimated PTs for French Guiana, Costa Rica, and Malaysia differed by a mere 0.4°C (fig. 8.1B), ranging from 29.4 to 29.8°C, with the Costa Rica and French Guiana PT being almost identical. The higher PT for Malaysian leatherbacks reflects the total number and specific temperatures tested in this study. For example, if more incubators with mean temperatures at 29.5 to 30.0°C were used, a different TSD curve may have been produced. This is not a criticism of Chan and Liew (1995) but a lesson for future leatherback TSD studies as the number of specific temperatures tested will greatly influence the shape of the generated sex ratio curve (fig. 8.1). In addition, experiments are affected by the accuracy and precision with which temperatures are controlled and measured. Investigators typically do not use instrumentation that measures temperature to 0.00°C and do not calibrate their thermometers, thermistors, or thermocouples to that level of accuracy, so reports implying an accuracy of PT or TRT beyond +/−0.05°C, that is to the nearest 0.1°C, overstate their findings.

Interestingly, there are differences in TRT among leatherback populations (fig. 8.1B). The TRT for French Guiana is only 0.5°C compared to slightly greater than 1.0°C for Pacific Costa Rica and Malaysia. Given the relatively large number of temperatures tested in the Costa Rica and French Guiana experiments, their TRTs are significantly different (Chevalier et al. 1999). Within and among populations, differences in TRT have received almost no study in any species with TSD, including sea turtles (Chevalier et al. 1999). Perhaps this should be the primary focus of future leatherback TSD studies, given such slight differences in PT among populations whose eggs experience very different climates on their nesting beaches (Binckley et al. 1998).

Chevalier et al. (1999) suggested that differences in leatherback TRTs reflect differences in genetic diversity among nesting populations. This explanation, based on global phylogeography of leatherback turtles (Dutton et al. 1999; Dutton et al. 2007), provides a template for future studies that should focus on genetically distinct nesting populations, such as those located in Costa Rica, Trinidad, and St. Croix, to compare to the extensive TSD data from French Guiana. Global phylogeography also suggests the potential for differences among Pacific populations nesting in Malaysia, Solomon Islands, and Mexico to compare to the well-studied population at Playa Grande, Costa Rica. Such studies could be logically extended to further compare Atlantic and Pacific nesting populations. Such information is critically needed given the predicted future climate change (Saba et al. 2012; Santidrián Tomillo et al. 2012).

Even less well understood is how TSD curves might vary among female leatherbacks within the same nesting population. Investigators generally collect eggs from multiple females to determine if TSD is present (Dutton et al. 1985). If it is, investigators determine sex ratios generated at specific temperatures (Rimblot et al. 1985; Rimblot-Baly et al. 1987) and how TSD curves vary between different nesting populations (Binckley et al. 1998). No study has yet investigated if clutches from different leatherback females from the same nesting population vary in TSD curves. Mrosovsky et al. (2009)

provided an excellent example for hawksbill turtles (*Eretmochelys imbricata*), showing different clutches varying in both PT and TRT as documented in other species with TSD (reviewed in Wibbels 2003). Again, genetic analysis (Dutton et al. 1999; Dutton et al. 2007) provides a road map for examining within-population variation in leatherback TSD curves. For example, high haplotype and nucleotide diversity found in many Pacific leatherback populations (e.g., Malaysia and Solomon Islands) suggests that females from these nesting areas might vary greatly in their individual TSD curves. Such studies will be logistically easier to perform compared to among-population investigations given the large clutch size of leatherbacks, their multiple nesting events per year, and extended nesting seasons. These same parameters also allow for an ambitious future experiment where the same female is followed throughout a season to have her clutches collected, incubated, and sexed to see if any differences occur within the same individual. Such an exciting experiment has yet to be undertaken for any species with TSD.

Future studies of variation among and within leatherback populations and within individuals should focus on incubation temperatures between 29.0 and 30.0°C. An important caveat is that researchers collect eggs from different populations and from females in the same population, and hatch them in the same incubators to best control for differences in temperature that inevitably occur between experimental apparatus. This will be especially difficult for different populations that have different nesting seasons. Regardless, given the difficulty of accurately measuring temperature to even 0.1°C and the fact that these small differences can have large effects on leatherback hatchling sex ratios, controlling for temperature within the same incubators will be essential and should be attempted. One will be forced to take the same portable incubators from beach to beach to carry out the experiment. Such incubators can readily be built by hand using high quality temperature control units and materials such as insulating Styrofoam readily available in most countries (Binckley et al. 1998).

Patterns of Hatchling Sex Ratios in Leatherback Turtles

Numerous factors could potentially affect leatherback nest temperature and resulting hatchling sex ratios (Binckley et al. 1998). These vary from large-scale annual and seasonal differences in climate (patterns of precipitation and air temperature) during nesting seasons to mesoscale factors, such as nest placement on a beach (sun vs. shade, distance to high tide) or sand color (black vs. white). Smaller-scale factors affecting temperature at the individual nest include depth, egg position (bottom vs. top), and metabolic heat generated by developing embryos (Spotila et al. 1987; Godfrey et al. 1997; Binckley et al. 1998; Mickelson and Downie 2010).

Leatherback nests are deep compared to most egg laying species and this greatly reduces daily variation in temperature, allowing for more confidence when applying data from controlled incubation studies to predict hatchling sex ratios (Binckley et al. 1998). Leatherback nest site selection studies also suggest females avoid areas that are generally cooler, such as under vegetation (Mrosovsky et al. 1984; Godfrey et al. 1996; Binckley et al. 1998; Kamel and Mrosovsky 2004), and those clutches laid below the high tide line not only experience cooling effects (Houghton et al. 2007) but also drastically reduced clutch survivorship (Nordmoe et al. 2004; Fossette et al. 2008). Incubation studies indicate that most temperatures produce either all males or females, given the narrow TRT of leatherbacks (fig. 8.1B). Finally, there is only a narrow time range when nest temperatures determine hatchling sex; this corresponds to the middle third of development, known as the thermosensitive period (Mrosovsky and Pieau 1991).

Changes in nest temperature, regardless of cause, only modify hatchling sex ratios if temperatures are shifted into or out of the TRT during the thermosensitive period. For example, Binckley et al. (1998) reported differences in nest depth, metabolic heating, and temperature within leatherback nests at Playa Grande, in PNMB, Costa Rica. However, these had no effect on hatchling sex ratios since the majority of nest temperatures were well above 30.0°C and metabolic heating was most pronounced during the last third of development, after sex was already determined (but see Mickelson and Downie 2010). Although we are not attempting to trivialize the importance of mesoscale and nest-specific factors, we do suggest that overall data are lacking to indicate that the smaller-scale factors greatly modify the effects of larger-scale climatic processes, such as rainfall, on hatchling sex ratios. For example, Wallace et al. (2004) reported that a hurricane reduced leatherback nest temperatures by 4.0°C and about 20 days passed before temperatures returned to the earlier levels. It is unlikely that any behavioral or microclimatic variation could compensate for such a large temperature effect. This remains an area where future research is needed.

Mrosovsky et al. (1984) investigated sex ratios of hatchling leatherbacks in Suriname over two nesting seasons. Samples of hatchlings from natural nests ranged from 0.0–100.0% female. This encompassed a strong seasonal shift from predominately males during April when there was more precipitation, to a majority of female hatchlings in the drier month of June. Estimated hatchling sex ratios produced on the beach during two-week time periods ranged from 35.0 to 100.0% female, with strong seasonal shifts based on changes in precipitation (Mrosovsky et al. 1984). The beginning and end of the nesting season in Suriname are drier and produce more females compared to the wetter middle part when more males are produced. The overall sex ratio was 49.0% female with more female hatchlings produced during the drier April and May of 1981 compared to the same months in 1982 (Mrosovsky et al. 1984).

The Suriname study is informative for leatherbacks because it establishes the repeatable pattern of rainfall and nest temperatures being inversely related. Generally, higher rates of precipitation during the thermosensitive period will reduce nest temperatures and tend to produce male hatchlings, while reduced rainfall, or lack thereof, during the middle third of development has a strong feminizing effect (Mrosovsky et al. 1984; Godfrey et al. 1996; Binckley et al. 1998). For example, in a companion study at the same nesting beach examined by Mrosovsky et al. (1984), Dutton et al. (1985) reported a shift from 30 to 100% female hatchlings based on histological examination of hatchlings from natural nests; this shift corresponded to the end of the rainy season and beginning of the dry season (see also Hilterman and Goverse 2003). Godfrey et al. (1996) reported similar results during the 1993 nesting season. There were more females produced during the drier beginning and end of the season and more males produced during wetter middle months, with an overall sex ratio of 69.4% female. More importantly, Godfrey et al. (1996) used linear regression to relate precipitation to leatherback hatchling sex ratios and used these data to calculate past sex ratios for 14 nesting seasons. Estimated overall annual hatchling sex ratios varied from 35.0 to 70.0% female depending on seasonal rainfall, with a mean of 53.6% for all seasons combined.

Generally, studies of leatherback nest or sand temperatures from Atlantic beaches show cooler temperatures compared to some Pacific sites, most often due to differences in rainfall. For example, using sand temperature profiles, Leslie et al. (1996) estimated hatchling sex ratios at Tortuguero, Costa Rica, of 38.0% to 70.8% female over two nesting seasons that included dramatic cooling effects of rainfall. Livingstone (2006) reported a mean temperature of 28.3°C during the middle third of incubation from 45 nests on the north coast of Trinidad. Also, in an unpublished report, Livingstone (2007) documented a mean temperature of 29.1°C from 11 nests in southwestern Gabon. These two studies suggest production of male hatchlings in both areas. However, there are notable exceptions and this general pattern of cooler nest temperatures on Atlantic nests could just reflect the low number of nesting colonies from which leatherback nest temperature data exist. Houghton et al. (2007) provided leatherback nest temperature data from eight nests in Grenada, suggesting that primarily females were produced, yet periods of rainfall reduced nest temperatures by up to 2.0°C. Temperature reductions of this magnitude could induce male hatchling production if they occur during the thermosensitive period, depending on the temporal duration of the reduction. Mickelson and Downie (2010) estimated 100.0% female production from 10 nests in Tobago, West Indies, with metabolic heating during the thermosensitive period pushing nest temperatures up to 31.0–32.0°C. Patiño-Martínez et al. (2012) reported highly female-biased hatchling sex ratios (83–97.0%) for Atlantic nesting beaches in Colombia. These studies, combined with those from Suriname, suggest that production of males at several Atlantic beaches is a common phenomenon, but that other beaches produce mainly females during certain years. This is in stark contrast to the predominance of female-biased hatchling sex ratios reported in most other sea turtle species (Wibbels 2003), making some Atlantic leatherback populations unique in this regard and vastly different from some conspecific populations nesting in the Pacific.

Chan and Liew (1995) documented sand temperatures above 30.0°C in Malaysia regardless of beach station, zone, nest depth, and importantly, over a six month period. These data strongly suggest production of female hatchlings at least for the year of study. This is similar to the findings of Binckley et al. (1998) who used both nest temperatures and histological examinations to estimate hatchling sex ratios at Playa Grande in PNMB, on the Pacific coast of Costa Rica. Estimates ranged from 73.4% to 100.0% female depending on season. Recently, Sieg et al. (2011) expanded the Playa Grande data set by providing nine more years of data (1998–2006). The annual estimated sex ratio was above 90.0% for each year with one exception. The 1999 nesting season experienced the most precipitation recorded in the region in the last 50 years (A. Sieg, unpublished data) producing the only year with a male-biased sex ratio (48.0% female). The combined estimated hatchling sex ratio from Binckley et al. (1998) and Sieg et al. (2011) covers 12 nesting seasons and is 91.1% female. A recent study by Steckenreuter (2010) provided the first temperature data from nesting beaches in Papua

New Guinea, and shows some of the lowest nest temperatures and highest precipitation levels experienced by leatherbacks nests. Not surprisingly their estimated sex ratio from three nests is 19.1% female and only 7.7% female over one nesting season.

Taken together, hatchling sex ratio studies in leatherbacks show the full range from mostly males (Steckenreuter 2010) to almost all females (Sieg et al. 2011) being produced; these results were largely dependent on patterns of precipitation and how precipitation interacts with the number of nesting females, and therefore, the number of clutches laid during each month of the season. Figure 8.2 demonstrates this with nesting distribution and precipitation data from different leatherback beaches. In Suriname the general pattern was more females at the beginning and end of the nesting season with more males during middle months following patterns of rainfall. At Playa Grande, Costa Rica, most nests were laid from November to January and hence their sex was determined from December to February when virtually no rain occurred (fig. 8.2A). At this nesting beach, male-producing temperatures occurred primarily from September to October as these months received the majority of precipitation, but few nesting females. Even the nests laid in October will have their sex determined by November rainfall, which at Playa Grande was similar in rainfall to female producing months in Suriname (fig. 8.2B). We also include data from Papua New Guinea (Steckenreuter 2010) for comparison, because this study suggested the fewest number of female hatchlings generated to date; relatively high rainfall amounts most likely contributed to these findings (fig. 8.2B). The consequences of regional differences in precipitation on generating annual variation in hatchling sex ratios are shown in figure 8.2C, where a predominately female-producing beach (Playa Grande, Costa Rica) received substantially less rainfall during the nesting season compared to Suriname where hatchling sex ratios were closer to 1:1.

Long-term data, combined with controlled incubation experiments and histological examination of gonads, exist only from Suriname (Godfrey et al. 1996) and Pacific Costa Rica (Binckley et al. 1998; Sieg et al. 2011). Clearly, more temperature monitoring is needed at a variety of different beaches over much longer time scales. Since work conducted in the 1980s and 1990s, we have noticed a trend of researchers quantifying temperatures of fewer nests over shorter time periods. This pattern must be reversed given predicted changes to the global climate. Furthermore, more studies that relate climate to past sex ratios (Godfrey et al. 1996; Hays et al. 2003) are essential and currently cannot be constructed with sex ratio estimates of leatherback hatchlings that only span one or two seasons and that also assume TSD curves similar to other populations.

Leatherback Turtle Sex Determination and Climate Change

Leatherback nesting beaches are a thermal mosaic at the global scale that currently produce a variety of sex ratios from mostly male to all female. Much has been written concerning how global warming will change climate and affect species with TSD (Mitchell and Janzen 2010), including sea turtles (Hawkes et al. 2007; Hawkes et al. 2009; Hulin et al. 2009; Poloczanska et al. 2009; Fuentes et al. 2011; Patiño-Martínez et al. 2012; Saba et al. 2012; Santidrián Tomillo et al. 2012), and we refer the reader to these excellent studies. Increases in global temperature will affect leatherback hatchling fitness (Mickelson and Downie 2010), sex ratios (Binckley et al. 1998; Patiño-Martínez et al. 2012; Santidrián Tomillo et al. 2012), and survival of hatchlings (Saba et al. 2012; chapter 16).

Rising global temperatures are predicted to increase the proportion of female leatherback hatchlings and reduce hatching success and fitness primarily at sites already experiencing female-producing temperatures (Binckley et al. 1998; Mickelson and Downie 2010; Patiño-Martínez et al. 2012; Santidrián Tomillo et al. 2012). In response to a warming climate, selection could favor leatherback females that nest in cooler microhabitats, nest during cooler parts of the year, or differ in their TSD curve (PT and TRT). However, females nesting in warmer areas (Playa Grande) do not differ in nest site selection compared to those that nest on cooler beaches (Suriname). In addition, leatherbacks' long generation time (15–20 years) may preclude rapid selection for a change in PT, TRT, or nesting seasonality and behavior. Mrosovsky et al. (1984) did report annual shifts in the timing of the nesting season that affected hatchling sex ratios. This has not been reported for other nesting beaches (Binckley et al. 1998), but quantifying this seemingly basic aspect of leatherback natural history is of great importance for predicting future changes to hatchling sex ratios.

Long-term data sets are urgently needed to provide a baseline of natural sex ratios produced at nesting beaches as the climate changes. For example, the existence of other male-producing beaches (Steckenreuter 2010) needs to be identified given their increased importance for population viability (Hawkes et al. 2007). Finally, much research concerning climate change and TSD has emphasized temperature, but changing pre-

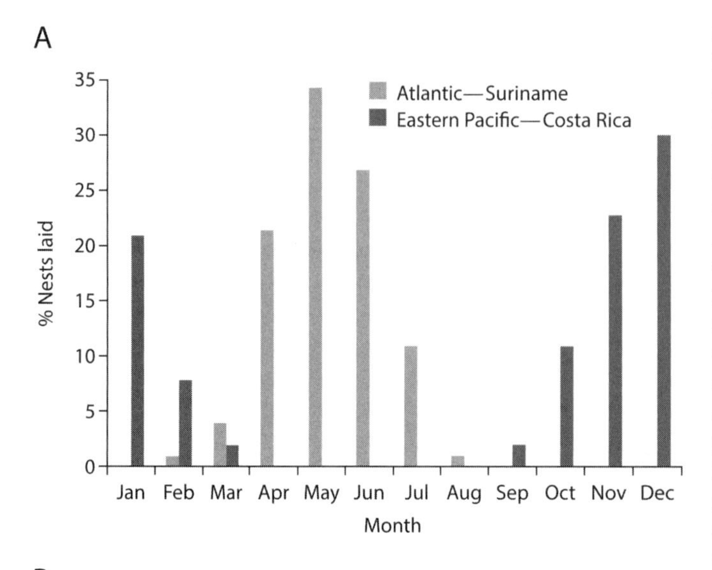

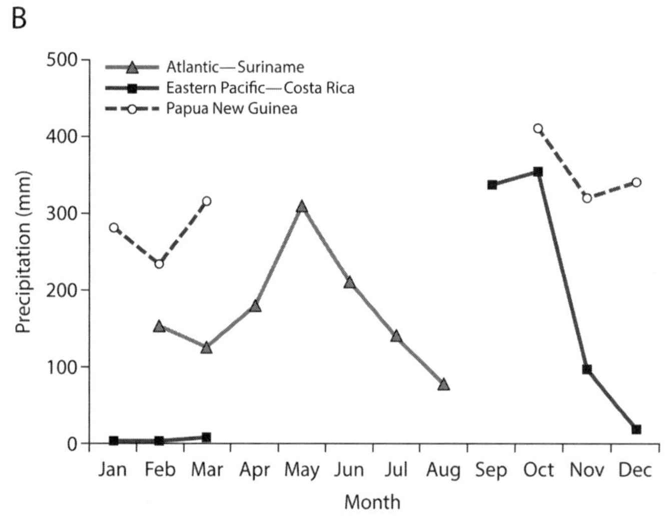

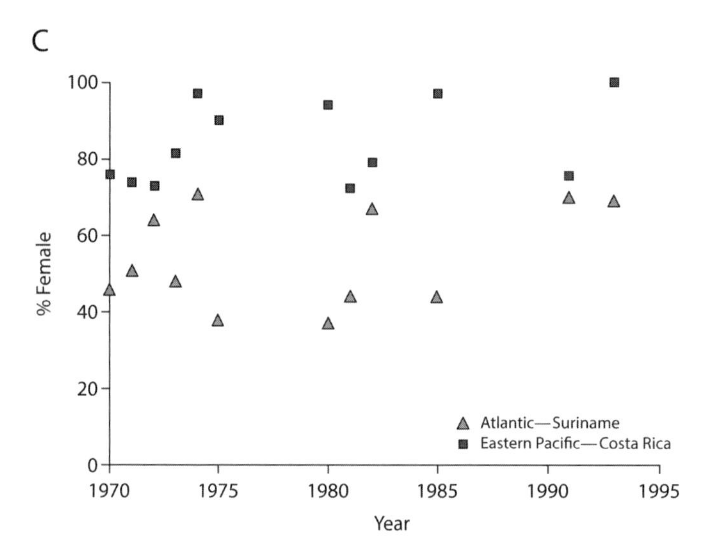

Fig. 8.2. A. Mean percentage of leatherback nests laid in Suriname (gray bars, data from Mrosovsky et al. 1984) and Costa Rica (black bars, data from Reina et al. 2002) during each of the main months of the nesting season. B. Mean monthly precipitation at Suriname (gray triangles, data from Mrosovsky et al. 1984), Costa Rica (black squares, J. R. Spotila, unpublished data), and Papua New Guinea (white circles, data from Steckenreuter et al. 2010). C. Estimated sex ratios of hatchlings from Suriname (53.6% female, Godfrey et al. 1996) and Costa Rica (91.1% female, Binckley et al. 1998; Sieg et al. 2011) for 22 years reflect differences in rainfall among the nesting beaches. This is exemplified by production of only 7.7% female hatchlings in Papua New Guinea in 2007, which received about 40% more precipitation than Suriname nesting beaches (Steckenreuter et al. 2010; not included on fig. 8.2C).

cipitation patterns (Houghton et al. 2007) will be just as important (Saba et al. 2012), further emphasizing our need to quantify long-term leatherback sex ratios across the globe. This is exemplified by research (Santidrián Tomillo et al. 2012; Santidrián Tomillo et al. 2014) documenting increased mortality of leatherback eggs at Playa Grande, Costa Rica, during the warmest and driest months of the nesting season that produce female hatchlings. Thus, this beach, and perhaps others, could have its female-biased sex ratios reduced, but at the high cost of increased egg mortality. Monitoring this phenomenon will be critical given predictions of increased temperatures and reduced precipitation at Playa Grande under most climate change scenarios (Santidrián Tomillo et al. 2012, Santidrián Tomillo et al. 2014). Monitoring different phenomenon such as the increased precipitation projected in Papua New Guinea, where at least one beach already produces many male hatchlings, will be equally critical (Steckenreuter 2010).

Leatherback hatcheries may become a common management tool as nesting beaches heat up due to climate change. Hatcheries are one of the most controversial conservation topics in sea turtle biology (Morreale et al. 1982; Sieg et al. 2011) and they inevitably change sex ratios (Sieg et al. 2011). Leatherback hatcheries modify nest temperatures and sex ratios (Dutton et al. 1985; Sieg et al. 2011). Besides rainfall, few factors consistently change hatchling sex ratios as do leatherback hatcheries. They generally reduce temperatures and produce male hatchlings (Sieg et al. 2011). The mechanisms for this effect are not well known and more research is needed to improve our understanding of this phenomenon. That is because production of more male hatchlings may be necessary in managed populations given the effect of global warming on regional temperatures and rainfall, and consequently on beach temperatures (Saba et al. 2012, Santidrián Tomillo et al. 2012).

We are not promoting the use of hatcheries but hypothesize their future importance given the above facts. If they do become more commonly used, it raises the uncomfortable and unavoidable question of what sex ratios should be produced. Such questions are particularly germane for beaches that already and naturally produce almost all females, as exemplified at Playa Grande, Costa Rica (Binckley et al. 1998; Sieg et al. 2011). Given this site's importance as one of the last remaining nesting colonies in the eastern Pacific, questions concerning whether more males should deliberately be produced in hatcheries with a warmer and drier future climate need to be addressed so that mitigation measures can be implemented in a timely fashion.

ACKNOWLEDGMENTS

We would like to thank all past, present, and future investigators of leatherback sex determination for their dedication to understanding this important phenomenon. Our research was supported by grants from the Earthwatch Institute, the Leatherback Trust, the Betz Chair of Environmental Science at Drexel University, and the Shrey Chair of Biology at Purdue University. This chapter was supported by the Betz Chair and The Leatherback Trust.

LITERATURE CITED

Binckley, C. A., J. R. Spotila, K. S. Wilson, and F. V. Paladino. 1998. Sex determination and sex ratios of Pacific leatherback turtles, *Dermochelys coriacea*. Copeia 1996: 291–300.

Bull, J. J. 1980. Sex determination in reptiles. Quarterly Review of Biology 55: 3–21.

Chan, E. H., and H. C. Liew. 1995. Incubation temperatures and sex-ratios in the Malaysian leatherback turtle *Dermochelys coriacea*. Biological Conservation 74: 169–174.

Charnier, M. 1966 Action de la temperature sur la sex-ratio chez l'embryon d'*Agama agama* (Agamidae, Lacertilien). Comptes Rendus des Séances de la Société de Biologie de l'Ouest Africain, Paris 160: 620–622.

Chevalier, J., M. H. Godfrey, and M. Girondot. 1999. Significant difference of temperature-dependent sex determination between French Guiana (Atlantic) and Playa Grande (Costa-Rica, Pacific) leatherbacks (*Dermochelys coriacea*). Annales des Sciences Naturelles 20: 147–152.

Dutton, P. H., B. W. Bowen, D. W. Owens, A. Barragán, and S. K. Davis. 1999. Global phylogeography of the leatherback turtle (*Dermochelys coriacea*). Journal of the Zoological Society of London 248: 397–409.

Dutton, P. H., C. Hitipeuw, M. Zein, S. R. Benson, G. Petro, J. Pita, V. Rei, et al. 2007. Status and genetic structure of nesting populations of leatherback turtles (*Dermochelys coriacea*) in the western Pacific. Chelonian Conservation and Biology 6: 47–53.

Dutton, P. H., C. P. Whitmore, and N. Mrosovsky. 1985. Masculinization of leatherback turtle *Dermochelys coriacea* hatchlings from eggs incubated in styrofoam boxes. Biological Conservation 31: 249–264.

Ewert, M. A., D. R. Jackson, and C. E. Nelson. 1994. Patterns of temperature dependent sex determination in turtles. Journal of Experimental Zoology 270: 3–15.

Fossette, S., L. Kelle, M. Girondot, E. Goverse, M. L. Hilterman, B. Verhage, B. de Thoisy, and J.-Y. Georges. 2008. The world's largest leatherback rookeries: a review of conservation-oriented research in French Guiana / Suriname and Gabon. Journal of Experimental Marine Biology and Ecology 356: 69–82.

Fuentes, M. B., C. J. Limpus, and M. Hamann. 2011. Vulnerability of sea turtle nesting grounds to climate change. Global Change Biology 17: 140–153.

Godfrey, M. H., R. Barreto, and N. Mrosovsky. 1996. Estimating past and present sex ratios of sea turtles in Suriname. Canadian Journal of Zoology 74: 267–277.

Godfrey, M. H., R. Barreto, and N. Mrosovsky. 1997. Metabolically generated heat of developing eggs and its potential effect on sex ratio of sea turtle hatchlings. Journal of Herpetology 31: 616–619.

Hawkes, L. A., A. C. Broderick, M. H. Godfrey, and B. J. Godley. 2007. Investigating the potential impacts of climate change on a marine turtle population. Global Change Biology 13: 923–932.

Hawkes, L. A., A. C. Broderick, M. H. Godfrey, and B. J. Godley. 2009. Climate change and marine turtles. Endangered Species Research 7: 137–154.

Hays, G. C., A. C. Broderick, F. Glen, and B. J. Godley. 2003. Climate change and sea turtles: a 150-year reconstruction of incubation temperatures at a major marine turkey rookery. Global Change Biology 9: 642–646.

Hilterman, M. L., and E. Goverse. 2003. Aspects of nesting and nest success of the leatherback turtle (*Dermochelys coriacea*) in Suriname, 2002. Guianas Forests and Environmental Conservation Project (GFECP) Technical Report. World Wildlife Fund Guianas / Biotopic Foundation, Amsterdam, the Netherlands.

Houghton, J. D., R. Meyers, A. E. Myers, C. Lloyd, R. S. King, C. Isaacs, and G. C. Hays. 2007. Protracted rainfall decreases temperature within leatherback turtle (*Dermochelys coriacea*) clutches in Grenada, West Indies: ecological implications for a species displaying temperature dependent sex determination. Journal of Experimental Marine Biology and Ecology 345: 71–77.

Hulin, V., V. Delmas, M. Girondot, M. H. Godfrey, and J. M. Guillon. 2009. Temperature-dependent sex determination and global change: Are some species at greater risk? Oecologia 160: 493–506.

Kamel, S. J., and N. Mrosovsky. 2004. Nest site selection in leatherbacks, *Dermochelys coriacea*: individual patterns and their consequences. Animal Behaviour 68: 357–366.

Leslie, A. J., D. N. Penick, J. R. Spotila, and F. V. Paladino. 1996. Leatherback turtle, *Dermochelys coriacea*, nesting and nest success at Tortuguero, Costa Rica, in 1990–1991. Chelonian Conservation and Biology 2: 159–168.

Livingstone, S. R. 2006. Sea turtle ecology and conservation on the North Coast of Trinidad, West Indies. PhD diss., University of Glasgow, UK.

Livingstone, S. R. 2007. Leatherback nest ecology in the Gamba Complex: implications for a successful hatchery and sustainable conservation, Gabon, Central Africa. Rufford Foundation Small Grants for Nature Conservation Report. www.rufford.org/files/Gabon_leatherbacks_livingstone_2007_0.pdf.

Mickelson, L. E., and J. R. Downie. 2010. Influence of incubation temperature on morphology and locomotion performance of leatherback (*Dermochelys coriacea*) hatchlings. Canadian Journal of Zoology 88: 359–368.

Mitchell, N. J., and F. J. Janzen. 2010. Temperature-dependent sex determination and contemporary climate change. Sexual Development 4: 129–140.

Morreale, S. J., G. J. Ruiz, J. R. Spotila, and E. A. Standora. 1982. Temperature dependent sex determination: current practices threaten conservation of sea turtles. Science 216: 1245–1247.

Mrosovsky, N. 1994. Sex ratios of sea turtles. Journal of Experimental Zoology 270: 16–27.

Mrosovsky, N., P. H. Dutton, and C. P. Whitmore. 1984. Sex ratios of two species of sea turtle nesting in Suriname. Canadian Journal of Zoology 62: 2227–2239

Mrosovsky, N., S. J. Kamel, C. E. Diez, and R. P. van Dam. 2009. Methods of estimating natural sex ratios of sea turtles from incubation temperatures and laboratory data. Endangered Species Research 8: 147–155.

Mrosovsky, N., and C. Pieau. 1991. Transitional range of temperature, pivotal temperatures and thermosensitive stages for sex determination in reptiles. Amphibia-Reptila 12: 169–179.

Nordmoe, E. D., A. E. Sieg, P. R. Sotherland, J. R. Spotila, F. V. Paladino, and R. D. Reina. 2004. Nest site fidelity of leatherback turtles at Playa Grande, Costa Rica. Animal Behaviour 68: 387–394.

Patiño-Martínez, J., A. Marco, L. Quiñones, and L. Hawkes. 2012. A potential tool to mitigate the impacts of climate change to the Caribbean leatherback sea turtle. Global Change Biology 18: 401–411.

Poloczanska, E. S., C. J. Limpus, and C. G. Hays. 2009. Vulnerability of marine turtles to climate change. Advances in Marine Biology 56: 151–211.

Reina, R. D., P. A. Mayor, J. R. Spotila, R. Piedra, and F. V. Paladino. 2002. Nesting ecology of the leatherback turtle, *Dermochelys coriacea*, at Parque Nacional Marino Las Baulas, Costa Rica: 1988–1989 to 1999–2000. Copeia 2002: 653–664.

Rimblot, F., J. Fretey, N. Mrosovsky, J. Lescure, and C. Pieau. 1985. Sexual differentiation as a function of the incubation temperature of eggs in the sea turtle *Dermochelys coriacea* (Vandelli, 1761). Amphibia-Reptilia 6: 83–92.

Rimblot-Baly, F., J. Lescure, J. Fretey, and C. Pieau 1987. Sensibiliteé à la température de la différenciation sexuelle chez la Tortue Luth, *Dermochelys coriacea* (Vandelli, 1761); application des données del'incubation artificielle à l'étude de la sex-ratio dans la nature. Annales des Sciences Naturelles Paris 8: 277–290.

Saba, V. S., C. A. Stock, J. R. Spotila, F. V. Paladino, and P. Santidrián Tomillo. 2012. Projected response of an endangered marine turtle to climate change. Nature Climate Change 2: 814–820. doi: 10.1038/nclimate1582.

Santidrián Tomillo, P., D. Oro, F. V. Paladino, R. Piedra, A. Sieg, and J. R. Spotila. 2014. High beach temperatures increase female-biased primary sex ratios but reduce output of female hatchlings in the leatherback turtle. Biological Conservation 176: 71–79.

Santidrián Tomillo, P, V. S. Saba, G. S. Blanco, C. S. Stock, F. V. Paladino, and J. R. Spotila. 2012. Climate driven egg and hatchling mortality threatens survival of eastern Pacific leatherback turtles. PloS One 7(5): e37602. doi: 10.1371/journal.pone.0037602.

Sieg, A. E., C. A. Binckley, B. P. Wallace, P. S. Tomillo, R. D. Reina, F. V. Paladino, and J. R. Spotila. 2011. Sex ratios of leatherback turtles: hatchery translocation decreases metabolic heating and female bias. Endangered Species Research 15: 195–204.

Spotila, J. R., E. A. Standora, S. J. Morreale, and G. J. Ruiz. 1987. Temperature dependent sex determination in the green turtle (*Chelonia mydas*): effects on the sex ratio on a natural nesting beach. Herpetologica 43: 74–81.

Spotila J. R., L. C. Zimmerman, and C. A. Binckley. 1994. Effects of incubation conditions on sex determination, hatching success, and growth of hatchling desert tortoises, *Gopherus agassizii*. Herpetological Monographs 8: 104–116.

Standora, E. A., and J. R. Spotila. 1985. Temperature dependent sex determination in sea turtles. Copeia 1985: 711–722.

Steckenreuter, N. P. 2010. Male-biased primary sex ratio of leatherback turtles (*Dermochelys coriacea*) at the Huon Coast, Papua New Guinea. Chelonian Conservation and Biology 9: 123–128.

Wallace, B. P., P. R. Sotherland, J. R. Spotila, R. D. Reina, B. F. Franks, and F. V. Paladino. 2004. Biotic and abiotic factors affect the nest environment of embryonic leatherback turtles, *Dermochelys coriacea*. Physiological and Biochemical Zoology 77: 423–432.

Wibbels, T. 2003. Critical approaches to sex determination in sea turtles. In: P. L. Lutz, J. A. Musick, and J. Wyneken (eds.), The biology of sea turtles, Vol. II. CRC Press, Boca Raton, FL, USA, pp. 103–134.

Wibbels, T. 2008. Sex determination and sex ratio in ridley sea turtles. In: P. Plotkin (ed.), Biology and conservation of ridley sea turtles. John Hopkins University Press, Baltimore, MD, USA, pp. 167–189.

Part III POPULATION STATUS AND TRENDS

Leatherback Turtle Populations in the Atlantic Ocean

MARC GIRONDOT

Leatherback turtles (*Dermochelys coriacea*) are the most widely distributed sea turtles and span tropical to subarctic oceanic waters worldwide. They range from the Arctic Circle (Goff and Lien 1988) to south of Africa (Luschi et al. 2003). Distribution in the Atlantic reflects the ancestry of this species, which separated from other marine turtles 50 ± 6.6 million years ago (Near et al. 2005), and that has a homeothermic capacity allowing them to have higher body temperatures than water (Paladino et al. 1990; Bostrom and Jones 2007; Bostrom et al. 2010).

The Atlantic Ocean began to form 130 million years ago, when continents that formed the ancestral super continent Pangaea were drifting apart. Collision between South and North American tectonic plates began around 15 million years ago and the Atlantic and Pacific Oceans were definitively separated 3.5 million years ago (Coates et al. 1992). At the same time, the Caribbean Sea and adjacent Gulf of Mexico formed. The Mediterranean Sea formed from the Neotethyan Ocean (popularly called the Tethys Sea) that began to form in the Late Jurassic (Dercourt et al. 1993; Smith et al. 1994). During the Late Cretaceous, when modern sea turtles were flouishing, the Neotethyan Ocean communicated with the North Atlantic via connections with the North Sea and through the Polish Trough and Paris Basin; this resulted in the formation of a large number of small and large European islands (Weishampel and Jianu 2011). After this period the Neotethyan Ocean was in regular contact with the Atlantic basin. By 20 million years ago, in the Miocene, the modern Mediterranean Sea would have been recognizable. Then, about 5.6 million years ago, the Mediterranean Sea became isolated and mostly dried up through evaporation. Atlantic waters rapidly refilled the Mediterranean 5.33 million years ago in an event known as the Zanclean flood (Garcia-Castellanos et al. 2009). The Adriatic, Ionian, Aegean, and Tyrrhenian Seas are part of the Mediterranean Sea, and the Gulf of Mexico is discussed together with the Caribbean Sea.

Distribution in Water

Leatherbacks in the Atlantic belong to three different regional management units, or RMUs (Wallace et al. 2010). These include the Atlantic Northwest, Atlantic Southeast, and Indian Southwest.

Mediterranean Leatherbacks (Atlantic Northwest RMU)

The first leatherback turtle known to science originated in the Mediterranean Sea south of France (Rondelet 1554). The first valid taxonomic description was made by Vandelli (1761) in a letter to Linnaeus from a specimen in the Mediterranean, but the exact locality is unknown (Fretey and Bour 1980). Leatherbacks are now rare (Casale et al. 2003), but are known from almost every area in the Mediterranean (Casale and Margaritoulis 2010), and exhibit some intraseasonal distribution patterns (Karaa et al. 2013). Their relative absence in the southeastern Mediterranean is probably due to sampling bias in this region. Nesting is absent or exceptional in the Mediterranean (Lescure et al. 1989); therefore, most leatherbacks are of Atlantic origin (Casale et al. 2003). Little is known about leatherbacks in the Mediterranean (origin, movement, foraging).

Caribbean Leatherbacks (Atlantic Northwest RMU)

Leatherbacks occur throughout the Caribbean Sea. Most observations are of adults that nest in this region (Eckert 2001), but a few juveniles have been reported (Eckert 2002). Leatherbacks forage throughout the Caribbean and Gulf of Mexico (Fossette, Girard, et al. 2010). At the border between the Caribbean and the South Atlantic, a huge number of leatherbacks are caught in fishing nets (Lee Lum 2006). Some are females that nest in Trinidad and Tobago and some are females from Suriname and French Guiana rookeries that migrate back and forth from southern nesting to northern foraging grounds (Girondot et al. 2007).

North Atlantic Leatherbacks (Atlantic Northwest RMU)

Persistent coastal foraging occurs in Canadian waters (James et al. 2006) and west of France (Fretey and Girondot 1996; Dell'Amico and Morinière 2010). Other coastal foraging occurs along Florida (Hoffman and Fritts 1982; Schroeder and Thompson 1987) and in upwelling zones close to Mauritania, in West Africa (Eckert 2006; Fossette, Hobson, et al. 2010). The use of North Atlantic pelagic waters by leatherbacks is strongly dependent on oceanographic conditions (Fossette, Hobson, et al. 2010) and has been described in detail by Bailey et al. (2012).

South Atlantic Leatherbacks (Atlantic Southeast and Indian Southwest RMUs)

Distribution of leatherbacks in the South Atlantic is less well known than for the North Atlantic. Coastal foraging areas exist south of Rio de Janeiro and along the Brazilian (Barata et al. 2004; Kotas et al. 2004) and Uruguayan (Fallabrino et al. 2000; López-Mendilaharsu et al. 2009) coasts. As nesting abundance of leatherbacks is very low in Brazil (Thomé et al. 2007), most of these turtles probably originate from rookeries on the Central African coast (Witt et al. 2011). The Angolan, Namibian, and South African coasts are foraging areas for leatherbacks from Gabon and the Indian Ocean (Luschi et al. 2006; Elwen and Leeney 2011; Witt et al. 2011).

Nesting Sites in the Atlantic Ocean and Caribbean Sea

WIDECAST (Wider Caribbean Sea Turtle Conservation Network) provides an atlas of leatherback nesting in the Caribbean Sea (Dow et al. 2007), and SWOT (State of the World's Sea Turtles) provides a spatial database of nesting around the world (DiMatteo et al. 2013). Here I synthesize these databases to get a concise view of leatherback nesting.

Brazil (Atlantic Southeast RMU)

Leatherbacks nest in Espírito Santo State in southern Brazil (Marcovaldi and Marcovaldi 1999). The annual number of nests varies between 6 (in 1993–1994) and 92 (in 2002–2003). Between 1995–1996 and 2003–2004, the annual number of nests increased by 20.4% per year. Leatherback nesting in Brazil occurs mostly around the austral summer, from October to February, but a slight nesting peak is also observed in May–June (fig. 9.1; Thomé et al. 2007).

French Guiana and Suriname (Atlantic Northwest RMU)

These countries are one of the hotspots for leatherback nesting. Discovered more than 40 years ago (Pritchard 1969; Schulz 1971), these nesting grounds have been monitored since then and have generated one of the longest time series of sea turtle nesting in the world (Fossette et al. 2007; Girondot et al. 2007).

Leatherback nesting regularly occurs from the eastern end of French Guiana around Cayenne westward to Braamspunt at the confluence of the Commewijne and Suriname rivers in Suriname. It encompasses ap-

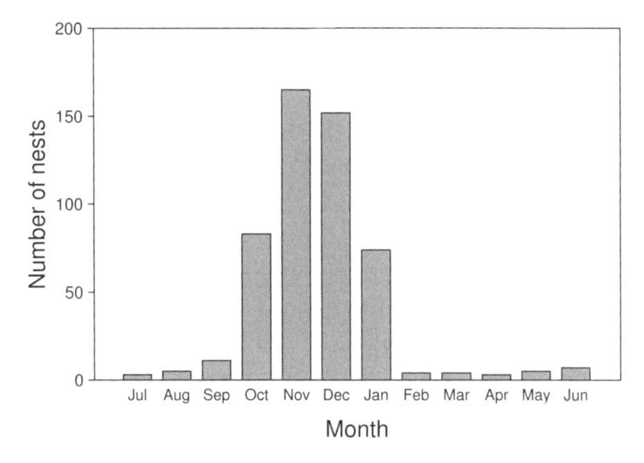

Fig. 9.1. Seasonality of nesting of leatherback turtles in State of Espirito Santo, Brazil, 1988-1989 through 2003-2004. Peak nesting is in November and December. From Thomé et al. 2007, with permission of Allen Press Publishing Services.

proximately 600 km of coastline with highly dynamic beaches that appear or disappear according to the displacement of large mud banks of Amazonian origin (Marchand et al. 2003). Principal nesting areas are separated into five zones: (1) Matapica area in central coastal Suriname, (2) Galibi Natural Reserve area in eastern Suriname near the border with French Guiana (Baboensanti and remote nesting beaches), (3) French Guiana beaches in and near the estuary of the Maroni and Mana rivers (Ya:lima:po-Awa:la and Pointe Kawana), (4) western oceanic beaches of French Guiana (from Pointe-Isère nesting beach), and (5) Kourou and Cayenne area in eastern French Guiana (Girondot et al. 2007).

Lack of strong fidelity of leatherback females to nesting beaches in this region, due to rapid changes in beach distribution (Schulz 1975; Fretey 1980), hampered a clear vision of the status of the population during the first 20 years of monitoring. Turtles were shifting from one beach to the other while monitoring was always taking place on the same beaches. This explains why leatherbacks from French Guiana were thought to have declined in the 1990s (Chevalier, Desbois, and Girondot 1999; Ferraroli et al. 2000). However, turtles were actually just shifting to Surinamese beaches. Recent compilation of data from both French Guiana and Suriname indicated that the population is stable (Girondot et al. 2007). However, interannual fluctuation in the number of nests is huge, from 10,000 to more than 60,000 per year (fig. 9.2). This makes detecting and deciphering a population trend difficult.

Leatherbacks in this region generally do not nest every year, but show extreme variability between nesting seasons. Up to 10% of females nest in consecutive years, whereas the mode is three years between nesting seasons. The number of clutches laid per female during a nesting season varies (Rivalan et al. 2006), and the number of clutches laid by females nesting after two or three years may be a trade-off between these two components of reproductive output (Rivalan et al. 2005). Some females deposit very few clutches on the monitored beach (one or two), while others lay many clutches, of up to seven or eight (Briane et al. 2007). It is not known if females lay other clutches on un-monitored beaches. Such heterogeneity is not included in demographic models. The annual survival rate for higher-mode nesting females is 0.910 (95% confidence interval: 0.864–0.942; Rivalan 2004). The nesting season in French Guiana and Suriname is bimodal with a large peak in May–June (Girondot et al. 2006) and a slight peak in December (Chevalier, Talvy, et al. 1999).

Guyana (Atlantic Northwest RMU)

The most important leatherback nesting area in Guyana is the North West District, especially Almond Beach, also known as Shell Beach, close to Trinidad and Tobago and the Paria Peninsula in Venezuela. There is little recent information on leatherbacks in this region. During 2006–2008, the number of nests markedly increased (P. Pritchard, personal communication), but the precise magnitude of this increase is unknown (Turtle Expert Working Group 2007). It could be on the same magnitude as the increase in Trinidad, and Paria Peninsula in Venezuela (see below).

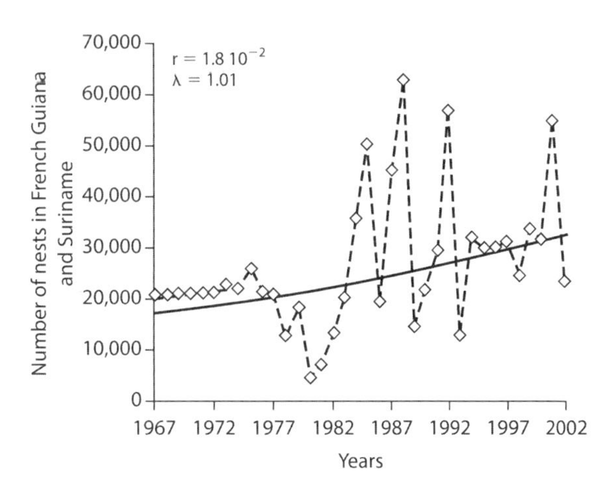

Fig. 9.2. Leatherback turtle nesting trends in French Guiana and Suriname. Rate of population growth is r, and λ is finite rate of increase. From Girondot et al. 2007, with permission of Allen Press Publishing Services.

Trinidad and Tobago (Atlantic Northwest RMU)

Trinidad has one of the three largest leatherback nesting colonies in the world. The nesting population has grown at an average rate of 1.8% per year since 1993, and in 2007 and 2008, 11,097 and 5,963 leatherbacks nested. Between 1999 and 2008 the number of leatherbacks nesting on Matura Beach ranged from approximately 1,500 to almost 6,000 in a single season (Eckert and Sammy 2008). In 2008, the total number of nests on the Matura, Grand Rivière, and Fishing Pond beaches was 49,731 (95% confidence interval: 47,832–51,630). Tobago Island hosts only sporadic nesting on the Atlantic and the Caribbean beaches (Forestry Division [Government of the Republic of Trinidad and Tobago], et al. 2010). The nesting season spans from March to August with a peak in May–June (Eckert and Sammy 2008).

Venezuela (Atlantic Northwest RMU)

The Paria Peninsula in eastern Venezuela is the most important nesting site in that country. In 2006 there were around 500 nests with more than 50% on Playa de Cipara and Playa de Querepare (Médicci et al. 2010). From 2000 to 2006, the number of nesting females on these two beaches almost doubled. Paria Peninsula is very close to Trinidad, but the relationship between the females in these two areas is unknown. The nesting season is concentrated in May–June.

Colombia (Gulf de Urabá), Panama, and Costa Rica (Atlantic Northwest RMU)

Nesting leatherbacks occur quasicontinuously along these coasts. Therefore, the entire region from the Gulf of Urabá, Colombia, to the Northern Caribbean coast of Costa Rica is included in this section. Extensive monitoring of leatherback turtle nesting in the Gulf of Urabá, Colombia, and the Caribbean coast of southern Panama identified three important coastal stretches totalling 18.9 km that hold 98.5–98.7% of nesting activity. Activity at all sites included 6,254 nests in 2006, and 7,509 in 2007; 90.9% and 86.2% of those attempts, respectively, resulted in clutch deposition (Patiño-Martínez et al. 2008).

Chiriquí Beach, in the Comarca Ngöbe-Buglé on the Caribbean coast of Panama, is the most important nesting beach for leatherbacks in Caribbean Central America. In the 2004 nesting season there were 3,077 nests and a minimum of 234 individual leatherbacks (Ordoñez et al. 2007). Aerial and track survey results indicated 5,759–12,893 nests per year between the San Juan River and Chiriquí Beach (Troëng et al. 2004). There were 199–1,623 nests per year at Tortuguero, Costa Rica, from 1995 to 2006. A Bayesian regression model suggested that leatherback nesting at that site decreased by 67.8% between 1995 and 2006 (Troëng et al. 2007). The most recent data indicated as few as 200 nests per year during 2008–2010 (Atkinson et al. 2011).

When considering this entire region's nesting population as a single management unit, it is not decreasing, due to the high abundance (an order of magnitude higher than Tortuguero) of nesting in Panama. The peak of the nesting season occurs in May–June (Patiño-Martínez et al. 2008).

Caribbean Islands (Atlantic Northwest RMU)

Some leatherbacks nest on almost every island of the Caribbean Sea (Dow et al. 2007), but generally at a very low level. For example, nesting occurs on both St. Christopher and Nevis—also known as St. Kitts and Nevis (Eckert and Honebrink 1992) and on Barbados (Horrocks 1987), but nesting at these sites has not been quantified.

US VIRGIN ISLANDS

There was a significant increase in the number of females nesting annually, from approximately 18–30 in the 1980s to 186 in 2001, with a corresponding increase in annual hatchling production from approximately 2,000 to over 49,000 at Sandy Point on St. Croix. An analysis with an open robust-design, capture-mark-recapture model of tagging data from 1991 to 2001 estimated annual survival probability as 0.89 (95% confidence interval: 0.87–0.92) and a population increase of 13% per year. Considered together with DNA fingerprinting that identified local mother-daughter relations, the increase in the size of the nesting population was the result of beach protection and egg relocation (Dutton et al. 2005). Peak nesting period occurred from middle to late May (Boulon et al. 1996).

PUERTO RICO

There is a biennial nesting pattern for leatherbacks at Culebra, but a decrease in the number of nests over time. However, there is an increase in nesting activities in nearby locations such as Fajardo, on mainland Puerto Rico (Diez et al. 2010). It is possible that the decreasing number of nests at Culebra Island is due to emigration of nesting females from there to nearby nesting areas, including St. Croix. Recent molecular studies suggested a regional stock interchange among different nesting beaches such as Culebra, US Virgin Islands, and possibly others in the Antilles, including the east coast of mainland Puerto Rico (Dutton et al. 2005). In addition, nesting leatherbacks tagged on Culebra Island nest elsewhere in the region, and vice versa (K. L. Eckert et al. 1989; Dutton et al. 2005). Peak of nesting occurred between 15 April and 15 June (Diez et al. 2010).

GUADELOUPE

Guadeloupe is located in the lesser Antilles arc. More than 150 beaches are located on the main islands (Grande Terre and Basse Terre) and adjacent smaller islands, including Petite Terre, Marie Galante, Les Saintes, and La Désirade (Delcroix et al. 2011). In 2007, there were 527 (95% confidence interval 240–870) nests on 45 monitored beaches, and in 2008, there were 180 (95% confidence interval 64–432) nests on 59 monitored beaches. Nesting is dispersed all around the islands without high aggregations, which makes global estimation very difficult (Delcroix et al. 2013). Similar situations exist on some other Caribbean islands (Dominique, Martinique) but no quantitative data are published. The peak of nesting occurs on Guadeloupe in the end of April (Delcroix et al. 2013).

Florida and Southeast USA (Atlantic Northwest RMU)

Leatherbacks nesting along the Atlantic coast of Florida was low during the 1990s (Calleson et al. 1998) but greatly increased since 2000 (Stewart et al. 2011). Number of nesting females is low (< 1000 females) compared to the spatial scale of monitoring. The recent discovery of the northernmost leatherback nest in North Carolina (Rabon et al. 2003) could be related to an expansion of leatherback nesting to the north. Nesting season peaks in middle to late May (Stewart et al. 2011).

Africa (Atlantic Southeast RMU)

Leatherbacks nest along the Atlantic coast of Africa from Mauritania at 21°N (Márquez 1990; Fretey et al. 2007) to Angola at 17°S (Weir et al. 2007). Nesting takes place in all countries in between (Fretey 2001; Fretey et al. 2007); however, the exact geographical limits of nesting are unknown. Nesting occurs at low frequency along 7,000 km of coast and quantification is missing for most locations. The highest known aggregation of nesting is in the Gabon-Congo biogeographical complex (Witt et al. 2009).

GABON

Leatherback nesting in Gabon was discovered in the late 1980s by Nicole Girardin, a French teacher in Libreville, the capital of Gabon (Fretey and Girardin 1988). The coast of Gabon, which covers nearly 885 km, has few rocky areas and consists mainly of long, sandy beaches. Nesting activity takes place along three main sites, from north to south (Witt et al. 2009):

- The beaches of Pongara National Park, which are 25 km long.
- Gamba Complex of Protected Areas, which is a beach 5.75 km long, situated near the airport of Gamba.
- Mayumba National Park, which is 95 km long.

The total number of clutches laid was between 36,185 and 126,480 during the 2006–2007 nesting season, making Gabon perhaps the largest leatherback nesting colony in the world (table 9.1; chapters 10 and 11). Nesting season is centered on December (Witt et al. 2009).

CONGO-BRAZZAVILLE

The Republic of the Congo is located south of Gabon and has a 170 km coastline, the majority of which comprises sandy beaches in front of mangroves. Nesting season is centered on December (Godgenger et al. 2009).

Leatherback nesting activity has been monitored in Congo since the 2003–2004 nesting season, making this the longest time series for leatherbacks on the African coast. The number of nests on beaches in the southern half of Congo is between 100 and 500 annually (Godgenger et al. 2009).

Phenology of Nesting Season

For Atlantic leatherbacks, the peak of nesting season occurs in May–June for rookeries in the Northern Hemisphere and in December for rookeries in the Southern Hemisphere. However, more than the latitude of the rookery, probably the latitude of the foraging area is the determinant of nesting phenology. The bimodal nesting seasons observed in French Guiana (Chevalier, Talvy, et al. 1999) and in Brazil (Thomé et al. 2007), with one peak of nesting in December and one peak of nesting in May, suggest a more complex foraging strategy than currently thought for these populations. I propose that some of these females forage in the Northern Hemisphere while others forage in the Southern Hemisphere.

Genetic and Ecological Distinctiveness of Atlantic Populations

Dutton et al. (2013) did a genetic analysis of stock structure for leatherback turtles (*Dermochelys coriacea*), combining 17 microsatellite loci and 763 bp (base pairs) of the mitochondrial DNA (mtDNA) control region from

Table 9.1. Summary of nesting at major leatherback rookeries across the Atlantic Ocean and Caribbean Sea.

Location	Nests per year	Source
Costa Rica and northern Panama	5,759–12,893	Troëng et al. (2004)
Colombia and southeast Panama	5,689–6,470	Patiño-Martínez et al. (2008)
French Guiana and Suriname	13,291–63,294	Girondot et al. (2007)
Gabon	31,185–126,480	Witt et al. (2009)
Trinidad and Tobago	6,000 females	Turtle Expert Working Group (2007)

1,417 individuals belonging to 9 nesting sites (Brazil, Costa Rica, French Guiana and Suriname, Gabon, Ghana, Trinidad, Florida, St. Croix in the US Virgin Islands, and South Africa) in the Atlantic and southwestern Indian Ocean. Population differentiation was significant between all pairs and genetic fixation (F_{ST}) values ranged from 0.034 to 0.676 and 0.004 to 0.205 for mtDNA and microsatellite data respectively, suggesting that male-mediated gene flow is not as widespread as previously assumed. All nine nesting colonies may be considered demographically independent populations for the purpose of conservation. However, this interpretation is not consistent with the observation that females nesting in French Guiana or Suriname also nest in Trinidad (Girondot et al. 2007). It illustrates the timescale difference in population genetic studies for long-lived individuals (> 10,000 years) and population monitoring (< 20 years). The Trinidad, and French Guiana and Suriname populations were differentiated only using microsatellites. A more precise analysis of genetic structuration of populations is needed for northwestern populations; these include Martinique and Guadeloupe in Caribbean Sea, Ya:lima:po in the west of French Guiana and Cayenne in the east (Canestrelli et al. 2013). The genetics of the recent population in the east of French Guiana seems to be a copy of the more ancient one in the west, but with lower genetic diversity. We propose that the new population from Cayenne arose from a few nesting females from the west that colonized the newly formed beach during the middle 1990s. Although population size of nesting leatherbacks in French Guiana and Suriname since the 1980s has been variable, but always large, observations of leatherbacks in this area before 1950 were sporadic or absent. The number of nesting females was probably extremely low (Chevalier et al. 1998). This is consistent with the observation of a recent very large increase in the effective size of this population detected using coalescence analysis (Canestrelli et al. 2013).

Only a few leatherbacks nest in Brazil, but high densities of pelagic individuals occur along the southern and southeastern coast. Diversity of the mtDNA control region partially supports a close association between nesting and pelagic turtles from Brazil and points to a complex origin for pelagic individuals along the Brazilian coast (Vargas et al. 2008).

Use of stable isotope methodology to characterize different leatherback populations is more relevant to the time scale of conservation biology. Stable isotope ratios for nitrogen and carbon have been studied in nesting leatherbacks from both Saint Croix, US Virgin Islands (Wallace et al. 2006) and French Guiana (Caut et al. 2008). Red blood cells showed no difference for either $\partial^{15}N$ and $\partial^{13}C$ composition ($t = 0.98$, df $= 39$, $p = 0.32$; and $t = 0.04$, df $= 23$, $p = 0.96$, respectively; combined probability: $c^2 = 2.29$, df $= 2$; $p = 0.31$) but yolk did show differences in $\partial^{15}N$ and $\partial^{13}C$ composition ($t = 3.46$, df $= 52$, $p = 0.001$; and $t = 9.73$, df $= 67$, $p < 10^{-13}$, respectively; combined probability: $c^2 = 77.06$, df $= 2$; $p < 10^{-16}$). The biological meaning of these differences is unclear. It could be related to differences in foraging area where lipids used in egg yolks are acquired.

Biometry of Mediterranean and Atlantic Leatherbacks

National Marine Fisheries Service (2001) and Eckert (2002) considered specimens less than 145 cm CCL (curved carapace length) to be juveniles. However, reproductive females as small as 105–125 cm CCL occur at most leatherback rookeries and their nests produce viable hatchlings (Stewart et al. 2007). Thus, size is not a good indicator of the reproductive status of individuals, and age of maturity for leatherbacks is unknown (Jones et al. 2011). On 23 September 1988, the largest leatherback turtle recorded was found dead in fishing gear in Gwynedd, Wales. This adult male weighed 916 kg and measured (CCL) 256.5 cm (Eckert and Luginbuhl 1988).

The biometry of individuals caught at sea is very difficult to interpret. This is because they represent a mix of unknown proportion of juveniles, subadults, and adults (Lescure et al. 1989; Casale et al. 2003; Innis et al. 2010).

A recent review of carapace length for 20 leatherback populations (Stewart et al. 2007) gave a mean CCL of 154.8 cm (SD 4.4 cm). Mean CCL is not significantly different among oceanic basins (GLM—generalized linear model—with oceanic basin effect, $p > 0.5$). However, the morphology of leatherbacks renders most length or width measures uncertain because the caudal part of the carapace is often damaged and morphology of the carapace makes curved measurements very subjective (Godfrey et al. 2001). The small differences in size distributions among rookeries could be due to differences in methodology used to measure females rather than true differences of size.

Body mass is a less ambiguous biometrical measure for leatherbacks. Mean body mass (±SD) for 49 nesting females in French Guiana was 408.1 ± 63.4 kg (Georges and Fossette 2006). In Tortuguero, Costa Rica, mean body mass (±SD) for 22 females was 346.8 ± 55.4 kg (Leslie et al. 1996). Mean body mass (±SD) for 127 nesting leatherbacks at St. Croix, US Virgin Island was 327.9 ± 44.3 kg (Boulon et al. 1996). Mean body mass is significantly different between French Guiana and both Caribbean rookeries ($t = 4.09$, df = 46, $p < 0.0001$ and $t = 8.12$, df = 67, $p < 10^{-9}$; t-test with different variance). Mean body masses are not significantly different between the two Caribbean rookeries ($t = 1.50$, df = 26, $p = 0.14$). The reason for differences between the French Guiana and Caribbean rookeries is unknown, but females lose mass during the nesting season even if some females forage (Fossette et al. 2008). If the body mass is not measured during the first nesting event, differences between rookeries could be simply due to differences in capture probability. If capture probability is lower in one study area, females will be weighed on average later in their annual nesting cycle and will exhibit a lower mass. This hypothesis was confirmed by data on mass loss during inter-nesting at Sandy Croix, US Virgin Islands (S. A. Eckert et al. 1989).

Mean body mass (±SD) for 23 leatherbacks captured off Canada or stranded on Canada shorelines was 392.5 ± 104.3 kg (James et al. 2007), 33% heavier than turtles of similar length on nesting grounds. The large standard deviation as compared to data obtained from nesting beaches could result from inclusion of males, females, and juveniles in the sample obtained at sea, which increases variation.

Threats to Atlantic Leatherbacks

The list of putative factors negatively influencing leatherback demography includes direct turtle and egg harvest, egg predation, loss or degradation of nesting beach habitat, fisheries bycatch, pollution, and large-scale changes in oceanographic conditions and nutrient availability (Wallace et al. 2011). Recently, fisheries bycatch, in particular that from longline fisheries, has received increased attention and has been proposed as a primary source of turtle mortality. Catch data from over 40 nations and bycatch data from 13 international observer programs were integrated into a single model to extrapolate the global pelagic longline bycatch per oceanic basin (Lewison et al. 2004); in 2000, an estimated 30,000 to 60,000 leatherbacks were caught in Atlantic longline fisheries, and between 250 to 10,000 in Mediterranean longlines.

Preliminary data also suggest that bycatch from gill nets and trawl fisheries is equally high or higher than longline bycatch, with far higher mortality rates (Lee Lum 2006). Until gill net and trawl fisheries are subject to the same level of scrutiny given to pelagic longlines, our understanding of the overall impact of fisheries bycatch on vulnerable sea turtle populations will be incomplete (Lewison and Crowder 2007). Exact fishing procedure and pressure is lacking for most rookeries (Dunn et al. 2010).

High levels of leatherback bycatch occur close to major rookeries. In Africa, in September and October 2005, there were 2 leatherbacks among 98 dead turtles along 145 km of southern Gabon and northern Congo. Between 16 and 24 October 2006, 6 dead leatherbacks were stranded within 40 km immediately south of Mayumba in southern Gabon. Leatherbacks had premortem amputations of a front flipper or multiple fractures of the cranium, both consistent with human-induced trauma, as could occur during removal from a net (Parnell et al. 2007). In Congo-Brazzaville, the conservation program Rénatura managed, since 2005, a release program with local fishermen. In 2006, 170 leatherbacks were released alive in exchange for material to repair fishing nets. This particular number of turtles saved was important, because the number of Congo-Brazzaville leatherback nests in the 2005–2006 nesting season was around 500 (Godgenger et al. 2009). In Trinidad, leatherback deaths in the northeast region of the country probably exceeded 1,000 egg-bearing female turtles every year (Eckert and Eckert 2005; Lee Lum 2006). In French Guiana and Suriname, the main threat is due to bycatch in gill nets. In front of the Maroni river mouth, illegal fishing activity is very difficult to control; each night several Surinamese boats fish within the French Exclusive Economic Zone (EEZ) with nets 5 km in length and 4 m in height, with 10 cm meshes. There were more than 50 leatherback carcasses stranded in 1998 along the beaches of French Guiana and Suriname (Chevalier et al. 1998). Due to the strong northwesterly current in this region, the true number of dead leatherbacks in these nets is probably higher. Furthermore, legal fishermen in French Guiana also often interact with leatherbacks: 1,192 ± 429 leatherbacks were caught in legal fisheries in 2004 and this rose to 1268 ± 410 in 2005 (Delamare 2005). The mortality rate for these captured and released leatherbacks is unknown.

Another major threat for leatherbacks is plastic pollution of the ocean; plastic is reported in 34% of dissections of stranded leatherbacks, and blockage of the gut by plastic is mentioned in some accounts (Mrosovsky et al. 2009). For example, a leatherback dead in a fishing net in Azores had six pieces of soft plastic together with a hard plastic belt and a small hard plastic cap in the anterior part of the intestine (Barreiros and Barcelos 2001). A tightly compacted piece of plastic (15 × 25 cm) blocked the entrance to the small intestine of the largest leatherback found so far, but the reason of death was due to entanglement in fishing gear (Eckert and Luginbuhl 1988). In some cases females can expel plastic debris out of their body via the cloaca (Plot and Georges 2010).

Atlantic Leatherback Abundance Trends

The population abundance trend in the Atlantic varies among rookeries. Nesting declined slightly from 1995 to 2003 on three Caribbean Costa Rican beaches, including Tortuguero, Pacuare, and Gandoca (Troëng et al. 2004), with further declines in recent years (Atkinson et al. 2011). However, leatherbacks from these Costa Rican beaches belong to the same RMU as the one in Panama, which could compensate for this decline. The trend detected in Congo-Brazzaville was positive between 2003–2004 to 2006–2007 (Godgenger et al. 2009), then negative after very low nesting seasons in 2009–2010 and 2010–2011, but was again positive in 2011–2012 (A. Girard, N. Bréheret, and M. Girondot, unpublished data). The trend seems to be negative in Gabon, but long time series are unavailable (Godgenger et al. 2008).

For other rookeries, the trend is strongly positive. For example, researchers in Florida (Stewart et al. 2011), US Virgin Islands (Dutton et al. 2005), and Trinidad (Eckert and Sammy 2008) reported a demographic explosion of leatherback nesting. The Brazilian rookery, while minor, also showed a positive trend between 1995–1996 and 2003–2004 (Thomé et al. 2007). The trend for eastern French Guiana is also highly positive (Canestrelli et al. 2013). The exact situation in the west of French Guiana and in Suriname is unknown because recent data from Suriname are lacking.

Conclusions

Leatherbacks occur throughout Atlantic, Mediterranean, and Caribbean waters in a highly dispersed, pelagic pattern, which renders any spatially explicit conservation measure extremely difficult to define. On the other hand, coastal foraging aggregations off eastern Canada, western France, West Africa, and South America are documented by satellite data, and three major nesting rookeries are well monitored in French Guiana / Suriname, Gabon / Congo-Brazzaville, and Trinidad. These feeding and nesting aggregations provide opportunities for conservation, such as regulation of fishing activities during specific periods of the year. However, leatherback nesting aggregations are not stable in space or time, and new major aggregations can develop quite quickly, as recently seen in Trinidad and eastern French Guiana. For each of these aggregations (nesting and foraging), we lack a clear, ranked diagnosis of threats. There is a natural tendency to pinpoint visible threats, such as presence of logs on the beaches—for example, on the Gabonese coast (Laurance et al. 2008)—or artificial lightning (Deem et al. 2007) impeding nesting activities and hatchling dispersal, without assessment of their relative demographic effects. On the other hand, water pollution (Perrault et al. 2011) or coastal development (e.g., port building in Africa) could be of major importance. More effort is needed to quantify the relative importance of threats and determine appropriate conservation measures.

ACKNOWLEDGMENTS

The author thanks the editors of this book, James R. Spotila and Pilar Santidrián Tomillo, for proposing this review. Anna Han helped with the English. This review was made possible by the work of dozens of organizations all around the Atlantic and hundreds of field workers. They are enthusiastically thanked here and encouraged for their future indispensable contributions.

LITERATURE CITED

Atkinson, C., D. Nolasco del Aguila, and E. Harrison. 2011. Report (unpublished) on the 2010 leatherback program at Tortuguero, Costa Rica. Sea Turtle Conservancy, San Pedro, Costa Rica.

Bailey, H., S. Fossette, S. J. Bograd, G. L. Shillinger, A. M. Swithenbank, J.-Y. Georges, P. Gaspar, et al. 2012. Movement patterns for a critically endangered species, the leatherback turtle (*Dermochelys coriacea*), linked to foraging success and population status. PLoS One 7: e36401 doi: 10.1371.

Barata, P.C.R., E.H.S.M. Lima, M. Borges-Martins, J. T. Scalfoni, C. Bellini, and S. Siciliano. 2004. Records of the leatherback sea turtle (*Dermochelys coriacea*) on the Brazilian coast, 1969–2001. Journal of the Marine Biological Association of the UK 84: 1233–1240.

Barreiros, J. P., and J. Barcelos. 2001. Plastic ingestion by a leatherback turtle *Dermochelys coriacea* from the Azores (NE Atlantic). Marine Pollution Bulletin 42: 1196–1197.

Bostrom, B. L., and D. R. Jones. 2007. Exercise warms adult leatherback turtles. Comparative Biochemistry and Physiology, Part A 147: 323–331.

Bostrom, B. L., T. T. Jones, M. Hastings, and D. R. Jones. 2010. Behaviour and physiology: the thermal strategy of leatherback turtles. PLoS One 5: e13925. doi: 10.1371.

Boulon, R. H., P. H. Dutton, and D. L. McDonald. 1996. Leatherback turtles (*Dermochelys coriacea*) on St. Croix, U.S. Virgin Islands: fifteen years of conservation. Chelonian Conservation and Biology 2: 141–147.

Briane, J. P., P. Rivalan, and M. Girondot. 2007. The inverse problem applied to the observed clutch frequency of leatherbacks from Yalimapo Beach, French Guiana. Chelonian Conservation and Biology 6: 63–69.

Calleson, T. J., G. O. Bailey, and H. L. Edmiston. 1998. Rare nesting occurrence of the leatherback sea turtle, *Dermochelys coriacea*, in northwest Florida, USA. Herpetological Review 29: 14–15.

Canestrelli, D., É. Molfetti, S. Torres Vilaça, J.-Y. Georges, V. Plot, E. Delcroix, et al. 2013. Recent demographic history and present fine-scale structure in the Northwest Atlantic Leatherback (*Dermochelys coriacea*) turtle population. PLoS One 8: e58061. doi: 10.1371.

Casale, P., and D. Margaritoulis. 2010. Sea turtles in the Mediterranean: Distribution, threats and conservation priorities. IUCN-MTSG, Gland, Switzerland.

Casale, P., P. Nicolosi, D. Freggi, M. Turchetto, and R. Argano. 2003. Leatherback turtles (*Dermochelys coriacea*) in Italy and in the Mediterranean basin. Herpetological Journal 13: 135–140.

Caut, S., S. Fossette, E. Guirlet, E. Angulo, K. Das, and M. Girondot. 2008. Isotope analysis reveals foraging area dichotomy for Atlantic leatherback turtles. PLoS One 3: e1845. doi: 10.1371.

Chevalier, J., B. Cazelles, and M. Girondot. 1998. Apports scientifiques à la stratégie de conservation des tortues luths en Guyane française. JATBA, Revue d'éthnobiologie 40: 485–507.

Chevalier, J., X. Desbois, and M. Girondot. 1999. The reason of decline of leatherback turtles (*Dermochelys coriacea*) in French Guiana: an hypothesis. In: C. Miaud and R. Guyétant (eds.), Proceedings of the Ninth Ordinary General Meeting of the Societas Europaea Herpetologica, Le Bourget du Lac, France. Societas Europea Herpetologica, Le Bourget du Lac, France, pp. 79–87.

Chevalier, J., G. Talvy, S. Lieutenant, S. Lochon, and M. Girondot. 1999. Study of a bimodal nesting season for leatherback turtles (*Dermochelys coriacea*) in French Guiana. In: H. Kalb and T. Wibbels (eds.), Proceedings of the Nineteenth Annual Symposium on Sea Turtle Biology and Conservation, South Padre Island, TX, USA. NOAA Technical Memoorandum NMFS-SEFSC-443. NOAA, Miami, FL, USA, pp. 264–267.

Coates, A. G., J.B.C. Jackson, L. S. Collins, T. M. Cronin, H. J. Dowsett, L. M. Bybell, P. Jung, and J. A. Obando. 1992. Closure of the Isthmus of Panama: the near-shore marine record of Costa Rica and western Panama. Geological Society of America Bulletin 104: 814–828.

Deem, S. L., F. Boussamba, A. Z. Nguema, G. P. Sounguet, S. Bourgeois, J. Cianciolo, and A. Formia. 2007. Artificial lights as a significant cause of morbidity of leatherback sea turtles in Pongara National Park, Gabon. Marine Turtle Newsletter 116: 15–17.

Delamare, A. 2005. Estimation des captures accidentelles de tortues marines par les fileyeurs de la pêche côtière en Guyane. Unpublished report. Diplôme d'Agronomie Approfondie, spécialisation Halieutique, Agrocampus Rennes.

Delcroix, E., S. Bédel, G. Santelli, and M. Girondot. 2013 Monitoring design for quantification of marine turtle nesting with limited human effort: a test case in the Guadeloupe Archipelago. Oryx 48: 95–105.

Delcroix, E., F. Guiougou, S. Bédel, G. Santelli, A. Goyeau, L. Malglaive, T. Guthmüller, et al. 2011. Le programme Tortues marines Guadeloupe: bilan de 10 années de travail partenarial. Bulletin de la Société Herpétologique de France 139–140: 21–35.

Dell'Amico, F., and P. Morinière. 2010. Observations de tortues marines en 2008 et 2009 (Côtes atlantiques françaises). Annales de la Société des Sciences Naturelles de la Charente-Maritime 10: 69–76.

Dercourt, J., L.-E. Ricou, and B. Vrielynck (eds.). 1993. Atlas Tethys palaeoenvironmental maps. Gauthier-Villars, Paris, France.

Diez, C. E., R. Soler, G. Olivera, A. White, T. Tallevast, N. Young, and R. P. Van Dam. 2010. Caribbean leatherbacks: results of nesting seasons from 1984–2008 at Culebra Island, Puerto Rico. Marine Turtle Newsletter 127: 22–23.

DiMatteo, A, E. Fujioka, B. Wallace, B. Hutchinson, J. Cleary, and P. Halpin. 2013. SWOT Database Online. http://seamap.env.duke.edu/swot. Accessed 10 May 2013.

Dow, W., K. Eckert, M. Palmer, and P. Kramer. 2007. An atlas of sea turtle nesting habitat for the wider Caribbean region. WIDECAST Technical Report No. 6. WIDECAST, Beaufort, North Carolina, USA.

Dunn, D. C., K. Stewart, R. Bjorkland, M. Haughton, S. Singh-Renton, R. Lewison, L. Thorne, and P. N. Halpin. 2010. A regional analysis of coastal fishing effort and intensity in the wider Caribbean. Fisheries Research 102: 60–68.

Dutton, D. L., P. H. Dutton, M. Chaloupka, and R. H. Boulon. 2005. Increase of a Caribbean leatherback turtle *Dermochelys coriacea* nesting population linked to long-term nest protection. Biological Conservation 126: 186–194.

Dutton, P. H., S. E. Roden, K. R. Stewart, E. LaCasella, M. Tiwari, A. Formia, J. C. Thomé, et al. 2013. Population stock structure of leatherback turtles (*Dermochelys coriacea*) in the Atlantic revealed using mtDNA and microsatellite markers. Conservation Genetics 14: 625–636.

Eckert, K. L. 2001. Status and distribution of the leatherback turtle, *Dermochelys coriacea*, in the wider Caribbean Region. In: K. L. Eckert and F.A.A. Grobois (eds.), Proceedings of the regional meeting, Marine turtle conservation in the wider Caribbean region: A dialogue for effective regional management. WIDECAST, IUCN/SSC/MTSG, WWF, and the UNEP CEP. WIDECAST, Baldwin, MO, USA, pp. 24–31.

Eckert, K. L., S. A. Eckert, T. W. Adams, and A. D. Tucker. 1989. Inter-nesting migrations by leatherback sea turtles (*Dermochelys coriacea*) in the West Indies. Herpetologica 45: 190–194.

Eckert, K. L., and T. D. Honebrink. 1992. WIDECAST sea turtle recovery action plan for St. Kitts and Nevis, CEP Technical Report No. 17. UNEP Caribbean Environment Programme, Kingston, Jamaica.

Eckert, K. L., and C. Luginbuhl. 1988. Death of a giant. Marine Turtle Newsletter 43: 2–3.

Eckert, S. A. 2002. Distribution of juvenile leatherback sea turtle *Dermochelys coriacea* sightings. Marine Ecology-Progress Series 230: 289–293.

Eckert, S. A. 2006. High-use oceanic areas for Atlantic leatherback sea turtles (*Dermochelys coriacea*) as identified using satellite telemetered location and dive information. Marine Biology 149: 1257–1267.

Eckert, S. A., and K. L. Eckert. 2005. Strategic plan for eliminating the incidental capture and mortality of leatherback turtles in the coastal gillnet fisheries of Trinidad and Tobago. Proceedings of a National Consultation, Ministry of Agriculture, Land and Marine Resources, Republic of Trinidad and Tobago, in collaboration with WIDECAST. WIDECAST, Baldwin, MO, USA.

Eckert, S. A., K. L. Eckert, P. Ponganis, and G. L. Kooyman. 1989. Diving and foraging behavior of leatherback sea turtles (*Dermochelys coriacea*). Canadian Journal of Zoology 67: 2834–2840.

Eckert, S. A., and D. Sammy. 2008. Trinidad's leatherback sea turtles. Research report (unpublished) 2008, Trinidad. Nature Seekers, Matura, Trindad and Tobago.

Elwen, S. H., and R. H. Leeney. 2011. Interactions between leatherback turtles and killer whales in Namibian waters, including possible predation. South African Journal of Wildlife Research 41: 205–209.

Fallabrino, A., A. Bager, A. S. Estrades, and F. Achaval. 2000. Current status of marine turtles in Uruguay. Marine Turtle Newsletter 87: 4–5.

Ferraroli, S., S. Eckert, J. Chavelier, M. Girondot, L. Kelle, and Y. LeMaho. 2000. Marine behavior of leatherback turtles nesting in French Guiana for conservation strategy. NOAA Technical Memorandum NMFS-SEFSC-477. NOAA, Miami, FL, USA, pp. 283–285.

Forestry Division (Government of the Republic of Trinidad and Tobago), Save our Seaturtles-Tobago, and Nature Seekers. 2010. WIDECAST Sea Turtle Recovery Action Plan for Trinidad & Tobago. Karen L. Eckert (ed.), CEP Technical Report No. 49. United Nations Caribbean Environment Programme, Kingston, Jamaica.

Fossette, S., S. Ferraroli, H. Tanaka, Y. Ropert-Coudert, N. Arai, K. Sato, Y. Naito, et al. 2007. Dispersal and dive patterns in gravid leatherback turtles during the nesting season in French Guiana. Marine Ecology Progress Series 338: 233–247.

Fossette, S., P. Gaspar, Y. Handrich, Y. Le Maho, and J.-Y. Georges. 2008. Dive and beak movement patterns in leatherback turtles (*Dermochelys coriacea*) during inter-nesting intervals in French Guiana. Journal of Animal Ecology 77: 236–246.

Fossette, S., C. Girard, M. López-Mendilaharsu, P. Miller, A. Domingo, et al. 2010. Atlantic leatherback migratory paths and temporary residence areas. PLoS One 5: e13908. doi: 10.1371.

Fossette, S., V. J. Hobson, C. Girard, B. Calmettes, P. Gaspar, J.-Y. Georges, and G. C. Hays. 2010. Spatio-temporal foraging patterns of a giant zooplanktivore, the leatherback turtle. Journal of Marine Systems 81: 225–234.

Fretey, J. 1980. Délimitation des plages de nidification des Tortues marines en Guyane française. Comptes Rendus de la Société de Biogéographie 496: 173–191.

Fretey, J. 2001. Biogeography and conservation of marine turtles of the Atlantic coast of Africa. CMS Technical Series Publication 6. UNEP/CMS Secretariat, Bonn, Germany.

Fretey, J., A. Billes, and M. Tiwari. 2007. Leatherback, *Dermochelys coriacea,* nesting along the Atlantic coast of Africa. Chelonian Conservation and Biology 6: 126–129.

Fretey, J., and R. Bour. 1980. Redécouverte du type de *Dermochelys coriacea* (Vandelli) (Testudinata, Dermochelyidae). Bollettino di Zoologia 47: 193–205.

Fretey, J., and N. Girardin. 1988. La nidification de la tortue luth, *Dermochelys coriacea* (Vandelli, 1761) (*Chelonii, Dermochelyidae*) sur les côtes du Gabon. Journal of African Zoology 102: 125–132.

Fretey, J., and M. Girondot. 1996. Première observation en France métropolitaine d'une tortue Luth, *Dermochelys coriacea* baguée en Guyane. Annales de la Société des Sciences Naturelles de la Charente-Maritime 8: 515–518.

Garcia-Castellanos, D., F. Estrada, I. Jiménez-Munt, C. Gorini, M. Fernàndez, J. Vergés, and R. De Vicente. 2009. Catastrophic flood of the Mediterranean after the Messinian salinity crisis. Nature 462: 778–781.

Georges, J.-Y., and S. Fossette. 2006. Estimating body mass in the leatherback turtle *Dermochelys coriacea*. Marine Ecology Progress Series 318: 255–262.

Girondot, M., M. H. Godfrey, L. Ponge, and P. Rivalan. 2007. Modeling approaches to quantify leatherback nesting trends in French Guiana and Suriname. Chelonian Conservation and Biology 6: 37–46.

Girondot, M., P. Rivalan, R. Wongsopawiro, J. P. Briane, V. Hulin, S. Caut, E. Guirlet, and M. H. Godfrey. 2006. Phenology of marine turtle nesting revealed by a statistical model of the nesting season. BMC Ecology 6: 11.

Godfrey, M. H., O. Drif, and M. Girondot. 2001. Two alternative approaches to measuring carapace length in leatherback sea turtles. Abstract. In: M. S. Coyne and R. D. Clark (eds.), Proceedings of the Twenty-first Annual Symposium on Sea Turtle Biology and Conservation, NOAA Technical memorandum NMFS-SEFSC-528. NOAA, Miami, FL, USA, p. 177.

Godgenger, M. C., N. Bréheret, G. Bal, K. N'Damité, A. Girard, and M. Girondot. 2009. Nesting estimation and analysis of threats for critically endangered leatherback *Dermochelys coriacea* and endangered olive ridley *Lepidochelys olivacea* marine turtles nesting in Congo. Oryx 43: 556–563.

Godgenger, M. C., A. Gibudi, and M. Girondot. 2008. Activités de ponte des tortues marines sur l'Ouest africain. Rapport d'étude pour Protomac, Gabon. Université Paris Sud, AgroParisTech. CNRS et Protomac, Orsay, France.

Goff, G. P., and J. Lien. 1988. Atlantic leatherback turtles, *Dermochelys coriacea,* in cold water off Newfoundland and Labrador. Canadian Field-Naturalist 102: 1–5.

Hoffman, W., and T. H. Fritts. 1982. Sea turtle distribution along the boundary of the Gulf Stream current off eastern Florida. Herpetologica 38: 405–409.

Horrocks, J. A. 1987. Leatherbacks in Barbados. Marine Turtle Newsletter 41: 7.

Innis, C., C. Merigo, K. Dodge, M. Tlusty, M. Dodge, B. Sharp, A Myers, et al. 2010. Health evaluation of leatherback turtles (*Dermochelys coriacea*) in the Northwestern Atlantic during direct capture and fisheries gear disentanglement. Chelonian Conservation and Biology 9: 205–222.

James, M., S. Sherrill-Mix, K. Martin, and R. Myers. 2006. Canadian waters provide critical foraging habitat for leatherback sea turtles. Biological Conservation 133: 347–357.

James, M. C., S. A. Sherrill-Mix, and R. A. Myers. 2007. Population characteristics and seasonal migrations of leatherback sea turtles at high latitudes. Marine Ecology Progress Series 337: 245–254.

Jones, T. T., M. D. Hastings, B. L. Bostrom, D. Pauly, and D. R. Jones. 2011. Growth of captive leatherback turtles, *Dermochelys coriacea*, with inferences on growth in the wild: implications for population decline and recovery. Journal of Experimental Marine Biology and Ecology 399: 84–92.

Karaa, S., I. Jribi, A. Bouain, M. Girondot, and M. N. Bradaï. 2013. On the occurrence of leatherback turtles *Dermochelys coriacea* (Vandelli, 1761), in Tunisian waters (Central Mediterranean Sea) (*Testudines: dermochelydae*). Herpetozoa 26: 65–75.

Kotas, J. E., S. dos Santos, V. G. de Azevedo, B.M.G. Gallo, and P.C.R. Barata. 2004. Incidental capture of loggerhead (*Caretta caretta*) and leatherback (*Dermochelys coriacea*) sea turtles by the pelagic longline fishery off southern Brazil. Fishery Bulletin 102: 393–399.

Laurance, W. F., J. M. Fay, R. J. Parnell, G. P. Sounguet, A. Formia, and M. E. Lee. 2008. Does rainforest logging threaten marine turtles? Oryx 42: 246–251.

Lee Lum, L. 2006. Assessment of incidental sea turtle catch in the artisanal gillnet fishery in Trinidad and Tobago, West Indies. Applied Herpetology 3: 357–368.

Lescure, J., M. Delaugerre, and L. Laurent. 1989. La nidification de la tortue Luth, *Dermochelys coriacea* (Vandelli, 1761) en Méditerranée. Bulletin Société Herpetologie de France 50: 9–18.

Leslie, A. J., D. N. Penick, J. R. Spotila, and F. V. Paladino. 1996. Leatherback turtle, *Dermochelys coriacea*, nesting and nest success at Tortuguero, Costa Rica, in 1990–1991. Chelonian Conservation and Biology 2: 159–168.

Lewison, R. L., and L. B. Crowder. 2007. Putting longline bycatch of sea turtles into perspective. Conservation Biology 21: 79–86.

Lewison, R. L., S. A. Freeman, and L. B. Crowder. 2004. Quantifying the effects of fisheries on threatened species: the impact of pelagic longlines on loggerhead and leatherback sea turtles. Ecology Letters 7: 221–231.

López-Mendilaharsu, M., C.F.D. Rocha, P. Miller, A. Domingo, and L. Prosdocimi. 2009. Insights on leatherback turtle movements and high use areas in the southwest Atlantic Ocean. Journal of Experimental Marine Biology and Ecology 378: 31–39.

Luschi, P., J.R.E. Lutjeharms, P. Lambardi, R. Mencacci, G. R. Hughes, and G. C. Hays. 2006. A review of migratory behaviour of sea turtles off southeastern Africa. South African Journal of Science 102: 51–58.

Luschi, P., A. Sale, R. Mencacci, G. R. Hughes, J.R.E. Lutjeharms, and F. Papi. 2003. Current transport of leatherback sea turtles (*Dermochelys coriacea*) in the ocean. Proceedings of the Royal Society B 270: S129-S132.

Marchand, C., E. Lallier-Verges, and F. Baltzer. 2003. The composition of sedimentary organic matter in relation to the dynamic features of a mangrove-fringed coast in French Guiana. Estuarine, Coastal and Shelf Science 56: 119–130.

Marcovaldi, M. Â., and G. G. Marcovaldi. 1999. Marine turtles of Brazil: the history and structure of Projeto TAMAR-IBAMA. Biological Conservation 91: 35–41.

Márquez, M. R. 1990. Sea turtles of the world. An annotated and ilustrated catalogue of sea turtles species known to date. UN Food and Agriculture Organization, Rome, Italy.

Médicci, M. R., J. Buitrago, and H. J. Guada. 2010. Biología reproductiva de la tortuga cardón (*Dermochelys coriacea*) en playas de la Península de Paria, Venezuela, durante las temporadas de anidación 2000–2006. Interciencia 35: 263–270.

Mrosovsky, N., G. D. Ryan, and M. C. James. 2009. Leatherback turtles: the menace of plastic. Marine Pollution Bulletin 58: 287–289.

National Marine Fisheries Service. 2001. Stock assessments of loggerhead and leatherback sea turtles and an assessment of the impact of the pelagic longline fishery on the loggerhead and leatherback sea turtles of the Western North Atlantic, NOAA Technical Memorandum NMFS-SEFSC-455. NOAA, Miami, FL, USA.

Near, T. J., P. A. Meylan, and H. B. Shaffer. 2005. Assessing concordance of fossil calibration points in molecular clock studies: an example using turtles. American Naturalist 165: 137–146.

Ordoñez, C., S. Troëng, A. Meylan, P. Meylan, and A. Ruiz. 2007. Chiriqui Beach, Panama, the most important leatherback nesting beach in Central America. Chelonian Conservation and Biology 6: 122–126.

Paladino, F. V., M. O. O'Connor, and J. R. Spotila. 1990. Metabolism of leatherback turtles, gigantothermy, and thermoregulation of dinosaurs. Nature 344: 858–860.

Parnell, R., B. Verhage, S. L. Deem, H. V. Leeuwe, T. Nishihara, C. Moukoula, and A. Gibudi. 2007. Marine turtle mortality in Southern Gabon and Northern Congo. Marine Turtle Newsletter 116:12–14.

Patiño-Martínez, J., A. Marco, L. Quiñones, and B. Godley. 2008. Globally significant nesting of the leatherback turtle (*Dermochelys coriacea*) on the Caribbean coast of Colombia and Panama. Biological Conservation 141: 1982–1988.

Perrault, J., J. Wyneken, L. J. Thompson, C. Johnson, and D. L. Miller. 2011. Why are hatching and emergence success low? Mercury and selenium concentrations in nesting leatherback sea turtles (*Dermochelys coriacea*) and their young in Florida. Marine Pollution Bulletin 62: 1671–1682.

Plot, V., and J.-Y. Georges. 2010. Plastic debris in a nesting leatherback turtle in French Guiana. Chelonian Conservation and Biology 9: 267–270.

Pritchard, P.C.H. 1969. Sea turtles of the Guianas. Bulletin of the Florida Sate Museum, Biological Science 13: 85–140.

Rabon, D. R., Jr., S. A. Johnson, R. Boettcher, M. Dodd, M. Lyons, S. Murphy, S. Ramsey, S. Roff, and K. Stewart. 2003. Confirmed leatherback turtle (*Dermochelys coriacea*) nests from North Carolina, with a summary of leatherback nesting activities north of Florida. Marine Turtle Newsletter 101: 4–8.

Rivalan, P. 2004. La dynamique des populations de tortues luths de Guyane française: recherche des facteurs impliqués et application à la mise en place de stratégies de conservation. PhD diss., Université Paris Sud, France.

Rivalan, P., R. Pradel, R. Choquet, M. Girondot, and A.-C. Prévot-Julliard. 2006. Estimating clutch frequency in the sea turtle *Dermochelys coriacea* using stopover duration. Marine Ecology Progress Series 317: 285–295.

Rivalan, P., A.-C. Prévot-Julliard, R. Choquet, R. Pradel, B. Jacquemin, J-P. Briane, and M. Girondot. 2005. Trade-off between current reproduction investment and delay until next reproduction in the leatherback sea turtle. Oecologia 145: 564–574.

Rondelet, G. 1554. De Testudinibus. Libri de piscibus marinis, in quibus verae piscium effigies expressae sunt, Lugduni, apud Matthiam Bonhomme, pp. 443–453.

Schroeder, B. A., and N. B. Thompson. 1987. Distribution of the loggerhead turtle, *Caretta caretta*, and the leatherback turtle, *Dermochelys coriacea*, in the Cape Canaveral, Florida area: results of aerial surveys. In: W. N. Witzell (ed.), Ecology of east Florida sea turtles, Cape Canaveral, Florida. NOAA Technical Report NMFS 53. NOAA, Miami, FL, USA, pp. 45–53.

Schulz, J. P. 1971. Nesting beaches of sea turtles in west French Guiana. Proceedings of the Koninklijke Nederlandse Akademie van Wetenschappen C 74: 398–404.

Schulz, J. P. 1975. Sea turtles nesting in Suriname. Zoologische Verhandelingen 143: 1–143.

Smith, A. G., D. G. Smith, and B. M. Funnell. 1994. Atlas of Mesozoic and Cenozoic coastlines. Cambridge University Press, Cambridge, UK.

Stewart, K., C. Johnson, and M. H. Godfrey. 2007. The minimum size of leatherbacks at reproductive maturity, with a review of sizes for nesting females from the Indian, Atlantic and Pacific Ocean basins. Herpetological Journal 17: 123–128.

Stewart, K., M. Sims, A. Meylan, B. Witherington, B. Brost, and L. B. Crowder. 2011. Leatherback nests increasing significantly in Florida, USA; trends assessed over 30 years using multilevel modeling. Ecological Applications 21: 263–273.

Thomé, J.C.A., C. Baptistotte, L.M.P. Moreira, J. T. Scalfoni, A. P. Almeida, D. B. Rieth, and P.C.R. Barata. 2007. Nesting biology and conservation of the leatherback sea turtle (*Dermochelys coriacea*) in the State of Espírito Santo, Brazil, 1988–1989 to 2003–2004. Chelonian Conservation and Biology 6: 15–27.

Troëng, S., D. Chacón, and B. Dick. 2004. Possible decline in leatherback turtle *Dermochelys coriacea* nesting along the coast of Caribbean Central America. Oryx 38: 395–403.

Troëng, S., E. Harrison, D. Evans, A. De Haro, and E. Vargas. 2007. Leatherback turtle nesting trends and threats at Tortugero, Costa Rica. Chelonian Conservation and Biology 6: 117–122.

Turtle Expert Working Group. 2007. An assessment of the leatherback turtle population in the Atlantic Ocean. NOAA Technical Memorandum NMFS-SEFSC-555. NOAA, Miami, FL, USA.

Vandelli, D. 1761. Epistola de Holothurio, et Testudine coriacea ad Celeberrinum Carolum Linnaeum. Conzatti, Padua, Italy.

Vargas, S. M., F.C.F. Araújo, D. S. Monteiro, S. C. Estima, A. P. Almeida, L. S. Soares, and F. R. Santos. 2008. Genetic diversity and origin of leatherback turtles (*Dermochelys coriacea*) from the Brazilian Coast. Journal of Heredity 99: 215–220.

Wallace, B. P., A. D. DiMatteo, A. B. Bolten, M. Y. Chaloupka, B. J. Hutchinson, F. A. Abreu-Grobois, J. A. Mortimer, et al. 2011. Global conservation priorities for marine turtles. PLoS One 6: e24510. doi:10.1371/journal.pone.0024510.

Wallace, B. P., A. D. DiMatteo, B. J. Hurley, E. M. Finkbeiner, A. B. Bolten, M. Y. Chaloupka, B. J. Hutchinson, et al. 2010. Regional management units for marine turtles: a novel framework for prioritizing conservation and research across multiple scales. PLoS One 5: e15465. doi: 10.1371.

Wallace, B. P., J. A. Seminoff, S. S. Kilham, J. R. Spotila, and P. H. Dutton. 2006. Leatherback turtles as oceanographic indicators: stable isotope analyses reveal a trophic dichotomy between ocean basins. Marine Biology 149: 953–960.

Weir, C. R., T. Ron, M. Morais, and A.D.C. Duarte. 2007. Nesting and at-sea distribution of marine turtles in Angola,

West Africa, 2000–2006: occurrence, threats and conservation implications. Oryx 41: 224–231.

Weishampel, D. B., and C.-M. Jianu. 2011. Transylvanian dinosaurs. Johns Hopkins University Press, Baltimore, MD, USA.

Witt, M. J., E. Augowet Bonguno, A. C. Broderick, M. S. Coyne, A. Formia, A. Gibudi, G. A. Mounguengui Mounguengui, et al. 2011. Tracking leatherback turtles from the world's largest rookery: assessing threats across the South Atlantic. Proceedings of the Royal Society B 278: 2338–2347.

Witt, M. J., B. Baert, A. C. Broderick, A. Formia, J. Fretey, A. Gibudi, C. Moussounda, et al. 2009. Aerial surveying of the world's largest leatherback turtle rookery: a more effective methodology for large-scale monitoring. Biological Conservation 142: 1719–1727.

10

Leatherback Turtle Populations in the Pacific Ocean

SCOTT R. BENSON,
RICARDO F. TAPILATU,
NICOLAS PILCHER,
PILAR SANTIDRIÁN TOMILLO,
AND LAURA SARTI MARTÍNEZ

The leatherback turtle, *Dermochelys coriacea*, is globally listed as vulnerable under the International Union for the Conservation of Nature (IUCN) criteria (Wallace et al. 2013), but trends and status differ markedly among basins (Sarti Martínez 2000; Eckert et al. 2012). While populations in the Atlantic appear largely stable or increasing (Turtle Expert Working Group 2007; Stewart et al. 2011), populations in the Pacific Ocean basin have declined precipitously during the last several decades, including declines of more than 90% in Mexico and Costa Rica (Sarti Martínez et al. 2007; Santidrián Tomillo et al. 2007; Santidrián Tomillo et al. 2008) and 78% in Papua Barat, Indonesia (Tapilatu et al. 2013). Therefore, the Pacific populations are listed as critically endangered (Wallace et al. 2013). This is particularly notable because Pacific nesting populations once represented the largest breeding populations of leatherbacks in the world (Spotila et al. 2000). Land-based threats have included overharvesting of eggs, coastal development, beach erosion, lethal sand temperatures, predation of eggs and hatchlings by introduced predators, and harvesting of adult females for meat (e.g., Tapilatu and Tiwari 2007; Tiwari et al. 2011; Sarti Martínez et al. 2007; see Eckert et al. 2012 for overview). At-sea causes of mortality include incidental bycatch in diverse industrial and artisanal fisheries, killing of free-swimming animals for food or bait, and possibly ingestion of marine pollution such as plastic bags (e.g., NMFS and USFWS 1998; Sarti Martínez 2000; Eckert et al. 2012).

Within the Pacific basin, leatherbacks are known to inhabit a wide range of coastal and pelagic waters in tropical and temperate ecosystems. They are found from the equator to subpolar regions in both hemispheres, although nesting activity is confined to tropical and subtropical latitudes. Major nesting populations of leatherbacks are located on both sides of the Pacific basin. Genetic studies and movement data (Dutton et al. 2000; Dutton et al. 2007; Shillinger et al. 2008; Benson et al. 2011; chapter 2) have confirmed the existence of three genetically and demographically distinct subpopulations: eastern Pacific, Malaysian, and western Pacific (fig. 10.1), although the Malaysian population is now considered functionally extinct (Chan and Liew 1996). The level of research and monitoring activities has differed among populations, and each will be discussed separately in the pages that follow.

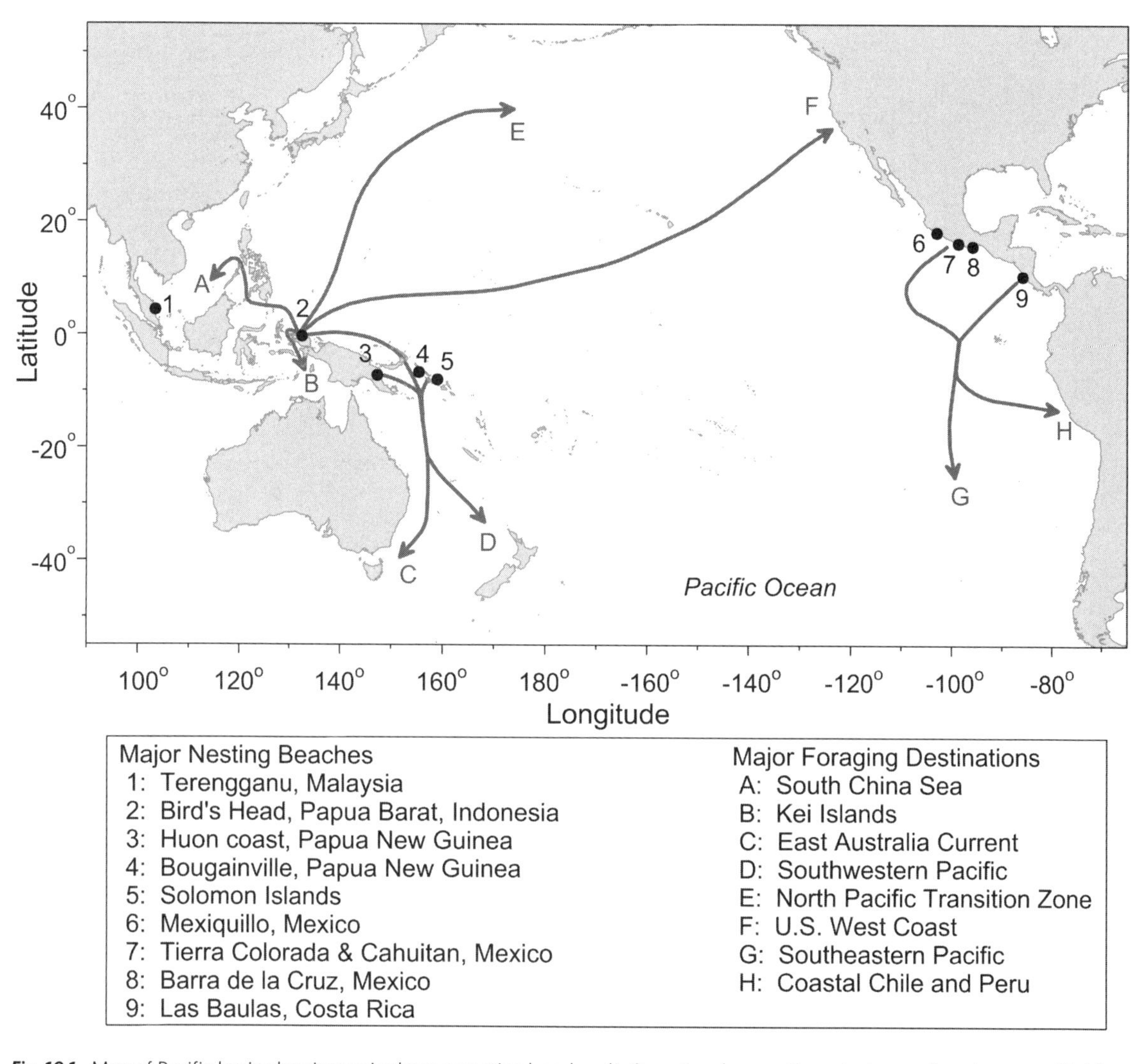

Fig. 10.1. Map of Pacific basin showing major known nesting beaches (1–9; past and present), major known foraging areas (A–H), and stylized movement patterns from nesting beaches to foraging grounds. Sources: Eckert and Sarti (1997), Mast (2006), Shillinger et al. (2008), and Benson et al. (2011).

Eastern Pacific Leatherback Population

Leatherback nesting in the eastern Pacific ranges from the southern tip of Baja California, Mexico, to Panama (Mast 2006; Seminoff and Wallace 2012). The nesting season extends from October/November through March, with a peak in December/January (Eckert et al. 2012; Sarti et al. 2003). Limited information is available on the total number of nesting females or total population size prior to the 1980s, and early estimates of total Pacific nesting activity and number of females varied widely. As recently as 1971, no areas of concentrated nesting activity were known (Pritchard 1971), but reports began to emerge of thousands of nesting leatherbacks at beaches along the Mexican Pacific coast (Márquez et al. 1981; Fritts et al. 1982), and the first eastern Pacific estimate was over 87,000 females (Pritchard 1982). Information on population trends is most comprehensive for leatherback nesting beaches in Costa Rica and Mexico (Sarti Martínez et al. 2007; Santidrián Tomillo et al. 2007).

Mexico

Pritchard conducted the first comprehensive aerial survey of leatherback nests along the Pacific coast of Mexico in 1981 (Pritchard 1982), resulting in an estimate of 75,000 nesting females within this region. Nests were so dense that Pritchard (1982) considered this to be a minimum estimate. Beginning in 1982, systematic

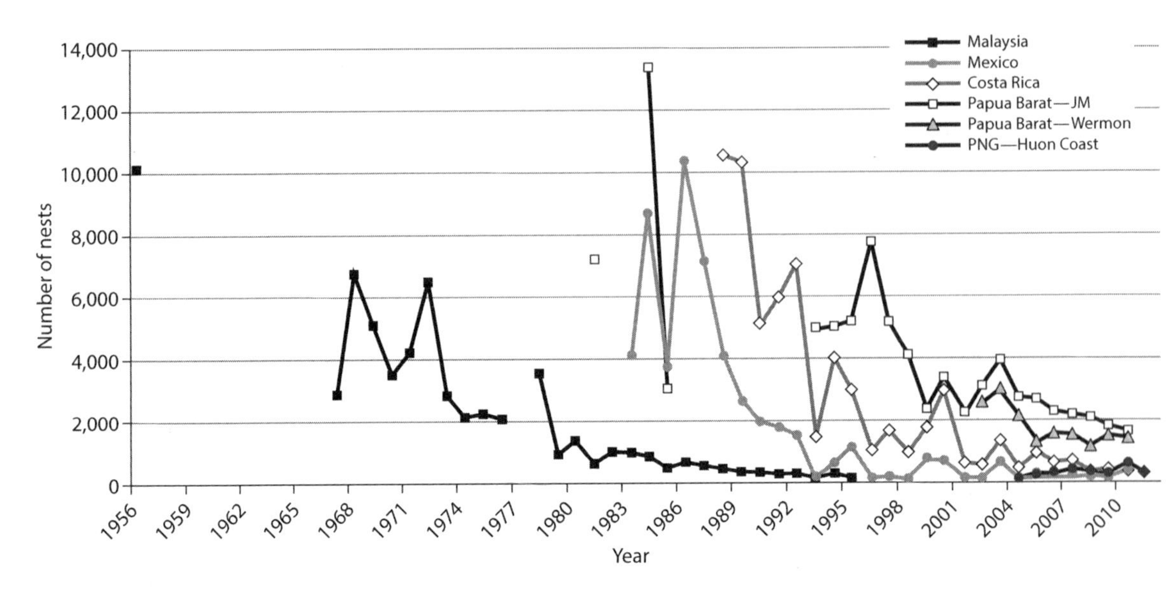

Fig. 10.2. Nesting trends of leatherback turtles (*Dermochelys coriacea*) at major nesting beaches in the Pacific, 1956-2011. Sources: Chan and Liew (1996) for Malaysia; Sarti Martínez et al. (2007) and L. Sarti Martínez (unpublished data) for Mexico; Santidrián-Tomillo et al. (2007) and F. V. Paladino and J. R. Spotila (unpublished data) for Costa Rica; Tapilatu et al. (2013) for Papua Barat; and N. Pilcher et al. (unpublished data) for PNG, Huon Coast.

monitoring and conservation activities began at nesting beaches in Mexico (Sarti Martínez et al. 2007). Standardized monitoring methods were implemented in 1997 at four index beaches spanning a total of 64 km (fig. 10.1), which were estimated to encompass about 42% of the total leatherback nesting activity on the Mexican Pacific coast. Several secondary beaches, spanning about 150 km in combined length and including an additional 31% of the total leatherback nesting activity, were monitored less frequently since 1982. Peak nest counts along a single 4 km stretch of beach at Mexiquillo included 5,000 nests in 1985–1986, resulting in a total estimate for the entire 18 km index site of over 10,000 nests. A marked decline in nest counts occurred at several beaches during the 1993–1994 season, followed by continued declines and a low of only 120 nests recorded across all four index beaches during 2002–2003 (fig. 10.2). Despite intensive conservation efforts to protect eggs and increase hatchling success during the past decade, this decline had not been reversed as of 2011 (fig. 10.2).

Central America

A similar pattern of sharp population decline occurred at the largest known nesting beach complex spanning 6 km of coastline in Parque Nacional Marino Las Baulas (PNMB), Nicoya Peninsula, Costa Rica (including Playa Grande, Playa Langosta, and Playa Ventanas). Monitoring began in 1988, and these beaches support about 85–90% of leatherback nesting on the Pacific coast of Costa Rica (Santidrián Tomillo et al. 2007). A population of over 1,500 leatherbacks that nested during the 1988–1989 season declined to only 100 individuals by 2006–2007 (Spotila et al. 1996; Spotila et al. 2000; Santidrián Tomillo et al. 2007; Santidrián Tomillo et al. 2008; fig. 10.2). For comparison with other nesting beaches for which only nest counts are available, this decline would correspond to a change from about 10,500 nests to 700 nests, based on a clutch frequency of 7 per season (Reina et al. 2002). As with the Mexican nesting populations, the steepest declines occurred during the 1990s.

Additional beaches in Costa Rica with lower levels of nesting activity (see Mast 2006 for details) include Ostional National Wildlife Refuge (59 nests in 2004, 4 nests in 2012); Caletas (24 nests in 2004); and San Miguel, Guanacaste (2 nests in 1999, 1 nest in 2000, no nests since 2001). The lengths of these beaches (3–7 km) are comparable to other major nesting beaches along the Central American Pacific coast, but historical nesting counts are not available for the evaluation of trends or past importance to the eastern Pacific leatherback population. Substantial nesting activity was also documented in 1989–1993 at Playa Naranjo in Santa Rosa National Park, Costa Rica (just north of the Nicoya Peninsula), with 466–1,212 leatherback crawls reported per season (Araúz-Almengor and Morera-Avila 1994). Recent nest count data for Playa Naranjo have not been published, but a 1999 aerial survey designed to identify major leatherback nesting areas along the entire Central American Pacific coast (Sarti et al. 2000) found only 11 nests in Costa Rica outside of the Nicoya Peninsula. Some additional nesting activity was documented during the 1999 aerial survey in Nicaragua (61 nests), Guatemala (6), El Salvador (4), and Panama (4). Sarti et al. (2000) concluded that the lack of any new major nesting beaches confirmed the decline of the eastern Pacific leatherback population, rather than a potential shift of nesting activity to new beaches. Since 2002, monitoring efforts at three major Nicaraguan nesting beaches identified 48 distinct females and documented up to 420 nests annually (Urteaga et al. 2012) Prior to 2002, nearly 100% of the eggs were poached, but conservation efforts are now protecting about 94% of the nests (Urteaga et al. 2012).

Causes of Decline and Conservation Efforts

Causes of the decline are documented most thoroughly for leatherbacks nesting at the main index beaches at PNMB, Costa Rica, and in Mexico. The dominant factor appears to be egg harvesting for consumption, although mortality of adult animals at sea and on nesting beaches contributed, and has become increasingly important as population sizes have decreased. Illegal harvest of eggs is the primary cause of population collapse at PNMB (Santidrián Tomillo et al. 2008). Although egg harvesting by local inhabitants who lived adjacent to the beaches occurred as early as the 1950s, systematic large-scale poaching occurred during the 1970s when newly constructed roads provided access to people from distant villages and cities. Illegal plunder of eggs continued for 16 years, removing an estimated 90% of eggs, until Parque Nacional Marino Las Baulas was established in 1991.

In Mexico, Sarti Martínez et al. (2007) reviewed the likely causes of the sharp decline in nesting activity at Mexiquillo during the 1993–1994 season. They identified (1) intensive egg harvest, slaughter of adults on the beaches, and bycatch at sea; (2) natural fluctuations in the reproductive biology of leatherbacks; or (3) movement by nesting females to un-monitored beaches as possible causes. To address these factors, "Proyecto Láud" (Leatherback Project) was developed to coordinate monitoring and management activities at multiple primary and secondary nesting beaches in Mexico (Sarti Martínez et al. 2007; Sarti and Barragán 2011). Investigators collected data to assess the size of the nesting population, including replicated aerial surveys of the entire Mexican coast to ensure that previously unknown nesting aggregations were not overlooked. They calculated reproductive parameters, such as average estimated clutch frequency (ECF) and average clutch interval (CI) and compared them to those at two other leatherbacks nesting beaches: PNMB, Costa Rica (where nesting has also declined), and St. Croix, US Virgin Islands in the Caribbean, where nesting activity was increasing. The effort spanned 214 km of Mexican coastline and included relocation of clutches when in situ incubation was unsafe due to likely poaching or predation.

Although some movement of turtles between various index sites occurred, aerial surveys did not locate any large, previously unknown aggregations of nesting leatherbacks, thereby refuting the hypothesis of possible movement to unknown areas as a cause of decline at index beaches. Reproductive parameters were similar to values reported from the declining nesting population in Costa Rica (Sarti Martínez et al. 2007; Spotila et al. 2000). The proportion of remigrant turtles (nesting turtles returning from previous seasons) for both of these declining populations (22–25%) was much lower than for the increasing population at St. Croix during the same period (52%). Estimated mortality of female leatherbacks nesting at PNMB, Costa Rica is 22–25% (Reina et al. 2002; Santidrián Tomillo et al. 2007). The combined results of these studies indicated substantial at-sea mortality of adult eastern Pacific leatherback turtles. Fishery bycatch of at least 1,500 animals per year was documented in a variety of gill net and longline fisheries in the North Pacific and off Central and South America (Spotila et al. 2000). However, the source population of these leatherbacks was not determined until genetic and telemetry studies were conducted beginning in the late 1990s, revealing that eastern Pacific leatherbacks are subject to bycatch in fisheries of the eastern tropical Pacific and off South America (fig. 10.1; Eckert and Sarti 1997; Dutton et al. 2000; Shillinger et al. 2008).

Although little is known about historic bycatch rates and the total number of leatherbacks killed in these areas, intentional and incidental takes of leatherbacks have occurred in waters off Central and South America since at least the 1970s (Brown and Brown 1982; Alfaro-Shigueto et al. 2007; Saba et al. 2008). Total leatherback mortality in coastal fisheries may have been substantial, and Eckert and Sarti (1997) estimated that a minimum of 2,000 leatherbacks were killed annually in gill net fisheries off Chile and Peru, based on data collected during the 1980s and 1990s. In a single Peruvian port (Pucusana), 200 adult and subadult leatherbacks were killed during the 1978 season alone (Brown and Brown 1982), and this fishery was not banned until 1995 (Morales and Vargas 1996). Bycatch also occurred in longline fisheries for swordfish off Chile and Peru; however, recent conservation efforts have reduced the number of turtles killed in these fisheries (Donoso and Dutton 2010) or show promise for future reductions (Alfaro-Shigueto et al. 2012).

Ongoing conservation efforts for eastern Pacific leatherbacks include active programs to protect nesting beaches and enhance reproductive output through increased hatchling survival (e.g., Arauz et al. 2003; Santidrián Tomillo et al. 2007; Santidrián Tomillo et al. 2008; Sarti Martínez et al. 2007). No indication of a reversal of the declining population trend is yet apparent (fig. 10.2); however, recent simulations (Saba et al. 2012) have highlighted the need to continue these programs to mitigate the projected adverse effects of climate change on eastern Pacific leatherback reproduction.

Malaysian Leatherback Population

The most dramatic decline in a Pacific leatherback population occurred at Terengganu, Malaysia, where there were over 10,000 nests annually during the 1950s and the population was reduced to less than 1% of its historic size by 1995 (Chan and Liew 1996). The primary cause was the nearly complete removal of eggs for many decades, although bycatch in fisheries that rapidly expanded during the 1970s accelerated the decline (Chan et al. 1988). As late as 1984–1985, when the population had already collapsed dramatically (Hamann, Ibrahim, and Limpus 2006), hundreds of leatherback deaths per year occurred in Malaysian fisheries. Low-level nesting activity along the adjacent coast of southeastern Thailand also ceased by the 1980s (Mortimer 1988). Similarly, a nesting population of leatherbacks in Vietnam that was estimated to include 500 females laying 10–20 nests per night prior to the 1960s, declined to a small remnant population with 10–20 nests per year by 2002 (Hamann, Kuong, et al. 2006). Only rare sporadic nesting has occurred at Terengganu over the last decade, and the Malaysian population appears to be functionally extinct (Liew 2011).

Western Pacific Leatherback Population

Monitoring activities in most areas of the western Pacific are relatively recent, and comparatively less is known about historically important nesting areas, status, and trends. The records are further confounded by changes in place names and jurisdictional boundaries during the past decades (e.g., the Indonesian province formerly known as Irian Jaya is currently comprised of two provinces named Papua and Papua Barat, and common-use village names have changed over time). Below we use current naming conventions, which may differ from those reported in the cited sources.

Papua Barat, Indonesia

Salm (1981) published the first indication of a significant nesting population in the western Pacific region outside Malaysia; this was based on an August aerial survey to assess four reputedly large leatherback nesting sites being considered for "reserve" designation on the north coast of Bird's Head Peninsula, Papua Barat, Indonesia. Leatherback nesting activity documented during the aerial survey was lower than expected, and one beach had no indication of leatherback nesting activity. Salm (1981) considered it likely that locals from adjacent villages had collected all the eggs, and concluded that turtle populations at those four beaches were already drastically depleted by egg and turtle collecting. However, in other areas on the north coast of Bird's Head Peninsula, Salm (1981) discovered previously undocumented nesting beaches that contained thousands of leatherback nests. His account was eerily similar to the description provided by Pritchard (1982) from Mexico, with nest densities so great that a precise count was impossible. At the time, Salm (1981) did not provide location details out of concern that public disclosure prior to protection would be detrimental. Follow-up studies during the 1980s and 1990s indicated that these large nesting populations were located along the less developed coastal beaches of northern Bird's Head Peninsula, at Jamursba-Medi Beach (Bhaskar 1985).

Systematic monitoring of leatherbacks, primarily in the form of annual nest counts, began during the early 1990s on the north coast of the Bird's Head Peninsula (Hitipeuw et al. 2007). Within this region, nesting occurs mainly at Jamursba-Medi, a complex of three beaches that span 18 km, and Wermon, a smaller 6 km beach approximately 30 km east of Jamursba-Medi. The primary nesting season at Jamursba-Medi occurs during May–September, while nesting occurs year-round at Wermon with peaks in July and December. Hitipeuw et al. (2007) provided the first assessment of trends at Jamursba-Medi between 1984 and 2004, concluding that the estimated number of nesting females declined from a peak of 2,303–3,036 in 1984 (based on nest counts by Bhaskar [1987]) to 667–879 during 2004. They also reported that nesting at Wermon during two seasons in 2002–2004 was only slightly lower than nesting at Jamursba-Medi, with year-round nesting and a second peak during January. However, beach erosion and predation by pigs and dogs caused the loss of 28% of nests at Wermon (Hitipeuw et al. 2007). Thus, although the leatherback nesting population at Bird's Head had not experienced the collapse observed at Malaysian and eastern Pacific rookeries, declines and population impacts were evident.

Follow-up studies at Bird's Head have increased monitoring activities at Jamursba-Medi and Wermon to identify trends in the total number of nesting females in this population. Tapilatu et al. (2013) applied correction factors to partial nest counts going back as far as 1984, based on more comprehensive data collected between 2004 and 2011. At Bird's Head, the total estimated number of nests has undergone a steady and sustained decline at both beaches, averaging about 5.9% per year since 1984. At Jamursba-Medi, total nest counts declined by 78.3%, from 14,522 nests in 1984 to 1,596 in 2011. A shorter time series at Wermon revealed a decline of 62.8% between 2002 and 2011 (from 2,994 to 1,096 nests). The most recent numbers of females nesting annually, for both beaches combined, were estimated to be 382 during the boreal summer 2011, and 93 during the austral summer of 2011–2012, based on estimated clutch frequency and clutch interval (Tapilatu et al. 2013). Thus, the last remaining significant nesting population of Pacific leatherbacks is also at risk of imminent collapse unless effective conservation efforts are implemented immediately.

Papua New Guinea

During a comprehensive review of marine turtles in Papua New Guinea (PNG) covering all areas except Morobe, Northern, and Gulf Provinces, Spring (1982a) reported regular, but low-density leatherback nesting activity along the north coast of PNG and on several islands including Manus, Long, New Britain, New Ireland, and Normanby. Occasional nesting was reported at Bougainville and Woodlark Islands. Village surveys indicated that population declines were already underway in many areas, primarily because of changes in village life brought about through the introduction of new technologies (e.g., outboard motors) and a cash economy (Spring 1982a, 1982b), which increased access to beaches and incentives for egg harvest. Local villagers in many areas regularly consumed eggs and nesting female leatherbacks. Substantial nesting activity was documented in Morobe Province, including a few hundred females along a 15 km beach at Busama (Maus Buang), and greater nesting activity was reported at Lababia village about 30 km to the south, within the Kamiali Wildlife Management Area (Quinn and Kojis 1985; Bedding and Lockhart 1989). Extensive and nearly complete egg harvest occurred at Busama, with up to 70% of leatherback eggs taken to markets in the nearby city of Lae, with another 20% consumed locally (Quinn and Kojis 1985). Peak nesting occurs during December–January.

Aerial surveys to assess leatherback nesting in PNG during January–February of 2004–2006 confirmed that the largest concentration of nesting activity within PNG occurred along the Morobe Province where there were up to 320 nests (Benson et al. 2007). A monitoring program for nesting leatherbacks began during 1999 and expanded to include additional beaches along the Huon coast during 2005–2007. Since 2006–2007, nesting activity has been relatively stable with 200–500 nests per year and the greatest level of nesting at two beaches near the city of Lae (Busama and Labu Tale) and at the Kamiali Wildlife Management Area (Pilcher 2012; fig 10.2).

Solomon Islands

Nesting of leatherbacks in Solomon Islands takes place along numerous isolated beaches on several islands (McKeown 1977; Vaughan 1981), peaking during November–January. Activity across all beaches varies from a few nests per season to over 20 nests in a single night, totaling a minimum of several hundred nests throughout the entire Solomon Island archipelago per season (Vaughan 1981; Dutton et al. 2007). The greatest annual nesting activity (summarized by Dutton et al. 2007) occurs on Santa Isabel (640–717 nests), Choiseul (50 nests), and Rendova and Tetepare (123 nests), but some nesting also occurs on most of the other major islands. Killing of nesting leatherbacks was common throughout Solomon Islands, and eggs were also consumed regularly. Declines were suspected as early as the 1970s, particularly at beaches near villages (Vaughan 1981). Breakdown of traditional customs, increasing population size, better access to nesting beaches, and lack of enforcement of laws protecting turtles were listed as contributing factors where declines were observed (Vaughan 1981; Leary and Laumani 1989).

Other South Pacific Islands

Low levels of nesting occur on several islands of Vanuatu, with 31 nests documented during November–February 2002–2003 on Epi Island (Petro et al. 2007). The islands of Fiji also have occasional leatherback nesting, but most early accounts involved capture and

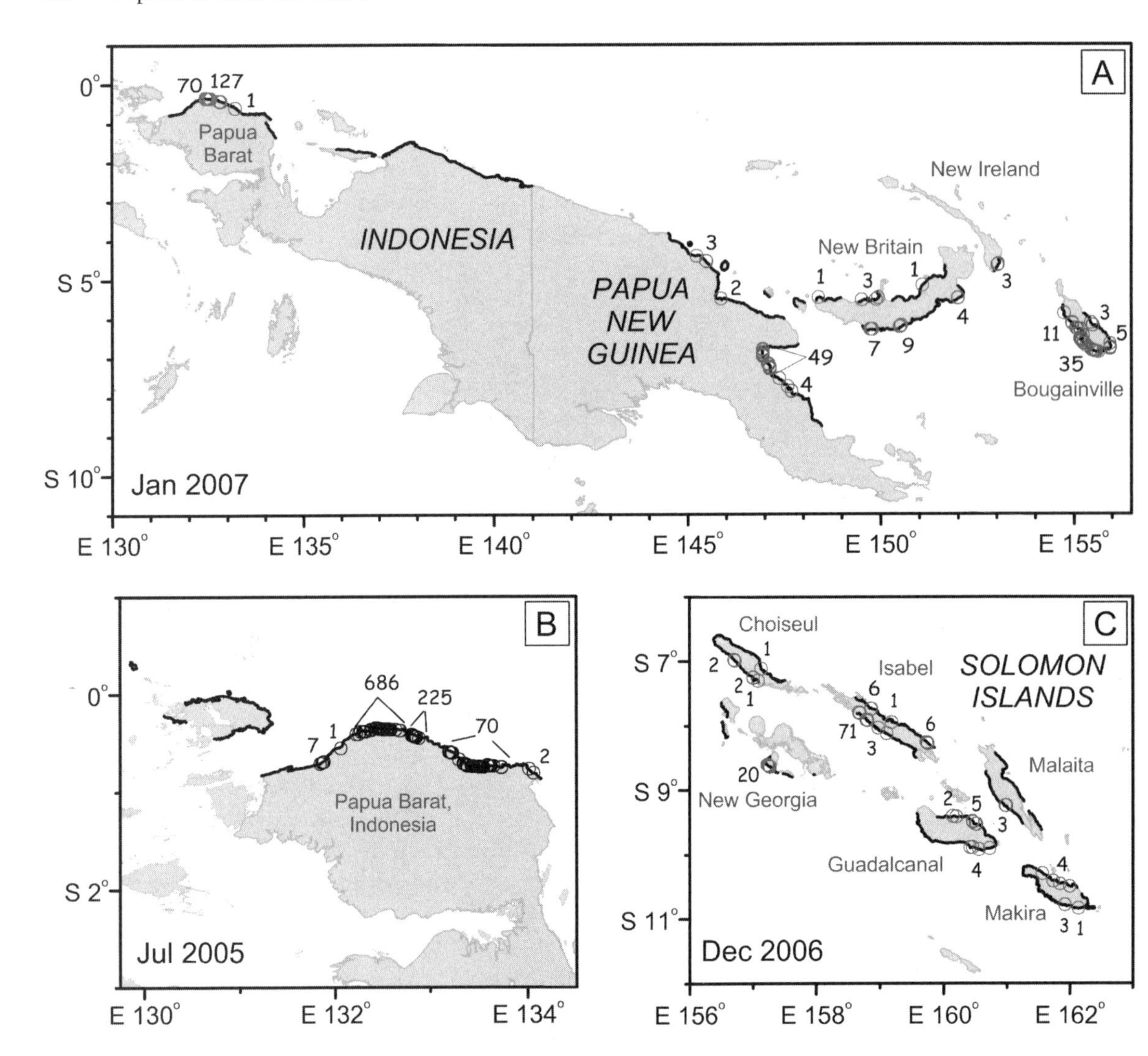

Fig. 10.3. Leatherback (*Dermochelys coriacea*) nests detected during tri-national aerial surveys conducted in (A) Papua and Papua Barat (Indonesia) and Papua New Guinea during January 2007, (B) Papua Barat during July 2005, and (C) Solomon Islands during December 2006. Black lines indicate aerial survey coverage and gray circles show areas of nesting activity with the number of nests shown. From Benson et al. (2012), and S. Benson (unpublished data).

killing of the animals (Guinea 1993). Nesting of leatherbacks is rare in Australia, but isolated historical records exist and report December–February nesting along the eastern coast of Queensland, in New South Wales, and in the Northern Territories (Limpus 2009).

Multinational Assessments

A summary of nest counts at 28 western Pacific nesting sites (Dutton et al. 2007) indicated there were 5,067–9,176 nests annually, with the majority of nesting activity occurring within the Jamursba-Medi and Wermon beaches of Papua Barat, and most remaining activity taking place on Huon Coast beaches of PNG and on Santa Isabel Island, Solomon Islands. Although the authors of that study suggested caution when deriving the number of turtles from nest counts, they estimated a minimum range of 844–3,294 females nesting annually. A coordinated, tri-national aerial survey to assess regional leatherback nesting activity in PNG, Solomon Islands, and along the north coast of Bird's Head, Papua Barat, was completed during December 2006–January 2007 (Benson et al. 2012). Although turtle nests are only visible from an aircraft for a period of time that depends upon tides, weather, and overall beach activity, such broad-scale aerial surveys are effective at identifying major nesting beaches within a region and documenting the relative importance of various beaches (Sarti

Martínez et al. 2007; Benson et al. 2012). Combined with a separate aerial survey in Papua Barat during July 2005 (fig. 10.3), the tri-national surveys confirmed that Bird's Head Peninsula in Papua Barat is the primary remaining nesting beach during both winter and summer seasons. Secondary, lower-level, boreal winter nesting occurs at a few beaches in PNG (Huon Coast and Bougainville; Kinch et al. 2012) and Solomon Islands, with scattered reports of occasional nesting elsewhere (e.g., Yapen and Waigeo Islands during boreal summer). No previously unknown nesting areas were identified. Leatherback nesting appears to be seasonally limited to those regions where beaches are in suitable condition. The seasonal monsoons affect wind, rain, and ocean current patterns, causing sand erosion and accretion to modify beach morphology markedly throughout the year (Benson et al. 2007; Hitipeuw et al. 2007).

Causes of Decline and Conservation Efforts

Kaplan (2005) conducted a Pacific-wide risk assessment of leatherback turtles that included consideration of multiple sources of mortality, including egg collection, killing of adults on nesting beaches and foraging grounds, and bycatch of turtles at sea. There is a long history, spanning many human generations, of harvesting sea turtles and their eggs for local subsistence use in the western Pacific region (Spring 1982b; Bhaskar 1987). Leatherback turtles have been an important part of the culture of indigenous populations through harvest of eggs and of adults (Suarez and Starbird 1995; Hitipeuw et al. 2007). With the introduction of a cash economy and motorized boats during the 1980s, the harvest of eggs and adults expanded to provide a source of income beyond the subsistence needs of local villages (e.g., Spring 1982b; Betz and Welch 1992); this involved nearby urban markets. This increased harvest pressure on leatherback populations caused sharp declines during the 1980s and beyond (Hitipeuw et al. 2007). Continued at-sea harvests of foraging leatherbacks and nesting females (e.g., in the Kei Islands, Indonesia; Benson et al. 2011) contributed to this trend.

Concern over the observed declines resulted in local programs to eliminate egg harvesting at key nesting beaches (Hitipeuw et al. 2007) and efforts to raise local awareness of the need to prevent the killing of adults. However, local customs vary and the success of such educational programs has been mixed. The taking of eggs and adults still occurs in many areas of western Pacific, especially where there are no active conservation programs (e.g., Kinch et al. 2012; Benson et al. 2011). Furthermore, recovery has remained hampered by other factors, including depredation by feral pigs and hunting dogs, loss of nests through beach erosion, and lethal sand temperatures leading to high rates of hatching failure (Tapilatu and Tiwari 2007).

At-sea bycatch of western Pacific leatherbacks occurs in a variety of gill net, trawl, and longline fisheries (as described below), but little is known about the total magnitude or full geographic extent of this source of mortality, in part because the movements of western Pacific leatherbacks are poorly understood. Recently, integrated telemetry studies identified movements of western Pacific leatherbacks into the North Pacific, southeastern Pacific, and Indo-Pacific tropical seas (fig. 10.1) and revealed fidelity to specific foraging regions (Benson et al. 2011; Seminoff et al. 2012). Several of the turtles tagged in Papua Barat, Indonesia, were known or suspected to have been killed in fisheries operating off Japan, Philippines, and Malaysia (Benson et al. 2011).

Historically, significant leatherback bycatch occurred in the North Pacific high seas drift net fishery, which expanded rapidly during the late 1970s and was banned in 1992 by United Nations resolution. Wetherall et al. (1993) estimated that over 750 leatherback turtles were killed in Japanese, Korean, and Taiwanese drift net fisheries during the 1990–1991 season. Although complexities of the fishery make extrapolation to a total mortality estimate very difficult, these estimates suggest that 5,000–10,000 leatherbacks may have been killed between the late 1970s and 1992. Based on current knowledge of movement patterns (Benson et al. 2011), all or most of these animals caught unintentionally would have originated from western Pacific nesting beaches during the boreal summer nesting period. Thus, high seas, drift net fishery bycatch was likely a significant contributor to the population declines observed at western Pacific nesting beaches during the 1980s and 1990s.

Additional detailed bycatch data are available for US drift net and longline fisheries in the central and eastern Pacific, indicating that tens of leatherbacks were killed or injured annually during the 1990s (Julian and Beeson 1998; McCracken 2000) and that there were markedly lower rates following implementation of sea turtle protection measures in the early 2000s. Genetic analysis indicated almost all of these killed and injured turtles originated from the western Pacific population (Dutton et al. 2000).

Bycatch data for the South Pacific regions within the known range of western Pacific leatherbacks (Benson et al. 2011) are more limited. Molony (2005) provides multinational turtle bycatch data for the 1990–2004 purse seine fishery and the deep, shallow, and albacore longline fisheries operating between 15° N and 31° S, indicating that an average of about 100 leatherbacks were killed per year. In Australia, bycatch records exist for pelagic longline fisheries (Stobutzki et al. 2006; Robins et al. 2002), prawn trawls off Queensland and Northern Territory, gill net fisheries off Queensland and Tasmania (Limpus 2009), and pot gear off Tasmania. Although no overall leatherback mortality estimates are available for Australian fisheries, gill net bycatch is reported as widespread (Limpus 2009). In particular, anecdotal reports of leatherback takes in Tasmanian tuna gill net fisheries may be of concern (Limpus 2009).

Pacific-wide Synthesis

The critically endangered status of Pacific leatherback populations is a reflection of conservation problems and challenges that globally affect large marine organisms whose ranges span entire ocean basins. Problems and successes in leatherback conservation have been determined by myriad factors operating at local, regional, and international scales spanning the jurisdictions of many developed and developing nations. Trends in Pacific leatherback populations illustrate that successful conservation and recovery will be dependent upon cooperation and coordination among diverse peoples throughout the Pacific region. When such coordinated efforts are lacking, recovery can be difficult or impossible, as illustrated by continued population declines despite decades of local conservation efforts (Sarti Martínez et al. 2007; Santidrián Tomillo et al. 2008).

The western Pacific leatherback population is the most robust remaining population with the best chance of survival in the Pacific (Dutton and Squires 2008), but it also experienced a dramatic decline during at least the past three decades, similar to that previously documented in the eastern Pacific and Malaysia (fig. 10.2). The population is still declining at an alarming rate of about 5.9 percent annually (Tapilatu et al. 2013), and effective long-term conservation and recovery actions must be implemented immediately to ensure the survival of this population.

The most critical needs are to (1) increase hatchling production, (2) eliminate killing of nesting females on the beaches, and (3) reduce at-sea bycatch of adults and subadults (Dutton and Squires 2008). Programs to protect nests and increase hatchling production are well established at major beaches in the eastern Pacific and are in varying stages of development at key western Pacific nesting beaches, but positive effects will take decades to realize (e.g., as in St. Croix; Dutton et al. 2005). Although bycatch in pelagic longline fisheries, particularly for swordfish, has received the most attention in recent years, it is clear that bycatch in small-scale coastal fisheries has been a significant contributor to population declines in many regions (Kaplan 2005; Alfaro-Shigueto et al. 2011).

Complicating this picture are the uncertain future effects of climate variability. Saba et al. (2007; 2008) identified a correlation between eastern Pacific leatherback reproductive frequency and El Niño / La Niña events, which could potentially increase this population's vulnerability to anthropogenic impacts compared to other leatherback populations. Longer term, a warming climate and rising sea levels could further affect leatherbacks through changes in beach morphology, increased sand temperatures leading to a greater incidence of lethal incubation temperatures, changes in hatchling sex ratios, and the loss of nests or nesting habitat due to beach erosion (chapter 16). Beach conservation measures must explicitly address these emerging challenges, requiring an even greater commitment of time and resources by biologists, local peoples, and the governments of many nations. Further, it will be important to preserve as many nesting populations, small and large, as possible to maintain the greatest geographic diversity in order to increase the resiliency of leatherback nesting populations in the face of a changing climate (McClenachan et al. 2006). This is particularly important given that current-driven hatchling dispersal patterns connect all western Pacific leatherbacks, such that adverse impacts in one region can have detrimental effects on leatherback persistence in many other regions throughout the Pacific (Gaspar et al. 2012).

Leatherbacks have survived many climate variations spanning millions of years, but at the current, critically low population sizes, past mechanisms of adaptation may no longer be effective, and human intervention is essential to prevent extirpation in the Pacific. The decline of Pacific leatherback populations is a shared international problem that can only be reversed by an immediate, holistic approach that enhances hatchling production through local programs and reduces anthropogenic mortality of adults and subadults wherever it occurs, particularly near nesting beaches and in key foraging areas (Kaplan 2005; Dutton and Squires 2008).

ACKNOWLEDGMENTS

We thank numerous people for sharing their data from multiple nesting sites throughout the Pacific.

These individuals include the coordinators of index beach monitoring projects that are part of Proyecto Laúd: Enrique Ocampo, Alejandro Tavera, Ana Barragán, Ma. del Rosario Juárez (CONANP); Karla López, Carlos Salas and Francesca Vannini (Kutzari, AC); Jim Spotila and Frank Paladino of the Leatherback Trust; Rotney Piedra of Parque Nacional Marino Las Baulas; beach rangers from north Bird's Head villages; students from the State University of Papua (UNIPA); and John Ben (Huon Coast Leatherback Turtle Project). Aerial surveys of leatherback nesting beaches in the western Pacific were aided by John Pita, Peter Ramohia, Joe Horoku, Peter Pikacha, Vagi Rei, Rachel Groom, Barry Kreuger, Betuel Samber, Bas Wurlanty, Karen Frutchey, and Paul Lokani. Funding for the aerial surveys was provided by NOAA-NMFS-Pacific Islands Regional Office (Western Pacific) and NOAA-NMFS-SWFSC, CONABIO, and NFWF (Mexico and Central America). We thank P. Dutton and J. Seminoff for their helpful reviews of an earlier draft of this chapter.

LITERATURE CITED

Alfaro-Shigueto, J., P. H. Dutton, M.-F. Van Bressem, and J. Mangel. 2007. Interactions between leatherback turtles and Peruvian artisanal fisheries. Chelonian Conservation and Biology 6: 129–134.

Alfaro-Shigueto, J., J. Mangel, F. Bernedo, P. H. Dutton, J. A. Seminoff, and B. J. Godley. 2011. Small-scale fisheries of Peru: a major sink for marine turtles in the Pacific. Journal of Applied Ecology 48: 1432–1440.

Alfaro-Shigueto, J., J. C. Mangel, P. H. Dutton, J. A. Seminoff, and B. J. Godley. 2012. Trading information for conservation: a novel use of radio broadcasting to reduce sea turtle bycatch. Oryx 46: 332–339.

Arauz, R., E. López, E. Lyons, B. Wilton, L. Verrier, and W. Reyes. 2003. Sea turtle conservation and research using coastal community organizations as the cornerstone of support. Report, July–December, 2002. Asociación PRETOMA Programa Restauración de Tortugas Marinas, San Jose, Costa Rica.

Araúz-Almengor, M., and R. Morera-Avila. 1994. Status of the marine turtles *Dermochelys coriacea*, *Chelonia agassizii*, and *Lepidochelys olivacea* at Playa Naranjo, Parque Nacional Santa Rosa, Costa Rica. Abstract. In: K. A. Bjorndal, A. B. Bolten, D. A. Johnson, and P. J. Eliazar (compilers), Proceedings of Fourteenth Annual Symposium on Sea Turtle Biology and Conservation. NOAA Technical Memorandum NMFS-SEFSC-351. NOAA, Miami, FL, p. 175.

Bedding, S., and B. Lockhart. 1989. Sea turtle conservation emerging in Papua New Guinea. Marine Turtle Newsletter 47: 13.

Benson, S. R., T. Eguchi, D. G. Foley, K. A. Forney, H. Bailey, C. Hitipeuw, B. P. Samber, et al. 2011. Large-scale movements and high-use areas of western Pacific leatherback turtles, *Dermochelys coriacea*. Ecosphere 2(7): art84. doi: 10.1890/ES11-00053.1.

Benson, S. R., K. M. Kisokau, L. Ambio, V. Rei, P. H. Dutton, and D. Parker. 2007. Beach use, internesting movement, and migration of leatherback turtles, *Dermochelys coriacea*, nesting on the north coast of Papua New Guinea. Chelonian Conservation and Biology 6: 7–14.

Benson, S., V. Rei, C. Hitipeuw, B. Samber, R. Tapilatu, J. Pita, R. Ramohia, et al. 2012. A tri-national aerial survey of leatherback nesting activity in New Guinea and the Solomon Islands. Abstract. In: L. Belskis, M. Frick, A. Panagopoulou, A. F. Rees, and K. Williams (compilers), Proceedings of the Twenty-ninth Annual Symposium on Sea Turtle Biology and Conservation. NOAA Technical Memorandum NMFS-SEFSC-620. NOAA, Miami, FL, pp. 19–20.

Betz, W., and M. Welch. 1992. Once thriving colony of leatherback sea turtles declining at Irian Jaya, Indonesia. Marine Turtle Newsletter 56: 8–9.

Bhaskar, S. 1985. Mass nesting by leatherbacks in Irian Jaya. In: World Wildlife Fund monthly report, January 1985, pp. 15–16. Internal report, World Wildlife Fund-Indonesia, Jakarta, Indonesia.

Bhaskar, S. 1987. Management and research of marine turtle nesting sites on the north Vogelkop coast of Irian Jaya, Indonesia. World Wildlife Fund, Jakarta, Indonesia.

Brown, C. H., and W. M. Brown. 1982. Status of sea turtles in the southeastern Pacific: emphasis on Peru. In: K. A. Bjorndal (ed.), Biology and conservation of sea turtles. Smithsonian Institution Press, Washington, DC, USA, pp. 235–240.

Chan, E. H., and H. C. Liew. 1996. Decline of the leatherback population in Terengganu, Malaysia, 1956–1995. Chelonian Conservation Biology 2: 196–203.

Chan, E. H., H. C. Liew, and A. G. Mazlan. 1988. The incidental capture of sea turtles in fishing gear in Terengganu, Malaysia. Biological Conservation 43: 1–7.

Donoso, M., and P. H. Dutton. 2010. Sea turtle bycatch in the Chilean pelagic longline fishery in the southeastern Pacific: opportunities for conservation. Biological Conservation 143: 2672–2684.

Dutton, D. L., P. H. Dutton, M. Chaloupka, and R. H. Boulon. 2005. Increase of a Caribbean leatherback turtle *Dermochelys coriacea* nesting population linked to long-term nest protection. Biological Conservation 126: 186–194.

Dutton, P. H., A. Frey, R. Leroux, and G. Balazs. 2000. Molecular ecology of leatherbacks in the Pacific. In: N. Pilcher and G. Ismael (eds.), Sea turtles of the Indo-Pacific: Research, management and conservation. ASEAN Academic, London, UK, pp. 248–253.

Dutton, P. H., C. Hitipeuw, M. Zein, S. R. Benson, G. Petro, J. Pita, V. Rei, et al. 2007. Status and genetic structure of nesting populations of leatherback turtles (*Dermochelys coriacea*) in the western Pacific. Chelonian Conservation and Biology 6: 47–53.

Dutton, P. H., and D. Squires. 2008. Reconciling biodiversity with fishing: a holistic strategy for Pacific sea turtle recovery. Ocean Development & International Law 39: 200–222.

Eckert, K. L., B. P. Wallace, J. G. Frazier, S. A. Eckert, and P.C.H. Pritchard. 2012. Synopsis of the biological data on the leatherback sea turtle (*Dermochelys coriacea*). US Fish and Wildlife Service, Biological Technical Publication BTP-R4015-2012. US Fish and Wildlife Service, Washington, DC, USA.

Eckert, S. A., and L. Sarti. 1997. Distant fisheries implicated in the loss of the world's largest leatherback nesting population. Marine Turtle Newsletter 78: 2–6.

Fritts, T. H., M. Stinson, and R. Márquez. 1982. Status of sea turtle nesting in southern Baja California, Mexico. Bulletin of South California Academy of Science 81: 51–60.

Gaspar, P., S. R. Benson, P. H. Dutton, A. Réveillère, G. Jacob, C. Meetoo, and S. Fossette. 2012. Oceanic dispersal of juvenile western pacific leatherback turtles: going beyond passive drift modeling. Marine Ecology Progress Series 457: 265–284.

Guinea, M. 1993. Sea turtles of Fiji. South Pacific Regional Environmental Programme, Reports and Studies No. 65. Apia, Fiji. http://www.sprep.org/att/IRC/eCOPIES/Countries/Fiji/35.pdf.

Hamann, M., C. T. Cuong, N. D. Hong, P. Thuoc, and B. T. Thuhien. 2006. Distribution and abundance of marine turtles in the Socialist Republic of Viet Nam. Biodiversity and Conservation 15: 3703–3720.

Hamann, M., K. Ibrahim, and C Limpus. 2006. Status of leatherback turtles in Malaysia. In: M. Hamann, C. Limpus, G. Hughes, J. Mortimer, and N. Picher (eds.), Assessment of the conservation status of the leatherback turtle in the Indian Ocean and South East Asia. IOSEA Species Assessment: Volume 1. IOSEA Marine Turtle MoU Secretariat, Bangkok, Thailand, pp. 78–82.

Hitipeuw, C., P. H. Dutton, S. R. Benson, J. Thebu, and J. Bakarbessy. 2007. Population status and internesting movement of leatherback turtles, *Dermochelys coriacea*, nesting on the northwest coast of Papua, Indonesia. Chelonian Conservation and Biology 6: 28–36.

Julian, F., and M. Beeson. 1998. Estimates of marine mammal, turtle, and seabird mortality for two California gillnet fisheries: 1990–1995. Fishery Bulletin 96: 271–284.

Kaplan, I. C. 2005. A risk assessment for Pacific leatherback turtles (*Dermochelys coriacea*). Canadian Journal of Fisheries and Aquatic Science 62: 1710–1719.

Kinch, J., S. Benson, P. Anderson, and K. Anana. 2012. Leatherback turtle nesting in the Autonomous Region of Bougainville, Papua New Guinea. Marine Turtle Newsletter 132: 15–17.

Leary, T., and M. Laumani. 1989. Marine turtles of Isabel Province. A report of a survey of nesting beaches (7th–21st of November 1989). Unpublished report. Solomon Islands Fisheries Division, Honiara, Solomon Islands.

Liew, H.-C. 2011. Tragedy of the Malaysian leatherback population: what went wrong. In: P. H. Dutton, D. Squires, and A. Mahfuzuddin (eds.), Conservation and sustainable management of sea turtles in the Pacific Ocean. University of Hawai'i Press, Oahu, HI, USA, pp. 97–107.

Limpus, C. J. 2009. A biological review of Australian marine turtles, 6. Leatherback turtle, *Dermochelys coriacea* (Vandelli). Environmental Protection Agency of the Queensland Government, Brisbane, Australia.

Márquez, R., A. Villanueva, and C. Peñaflores. 1981. Anidación de la tortuga laúd *Dermochelys coriacea schlegelli* en el Pacífico mexicano. Ciencia Pesquera 1: 45–52.

Mast, R. B. (ed.). 2006. State of the world's sea turtles (SWOT) report, Vol. 1. State of the World's Sea Turtles, Arlington, VA, USA.

McClenachan L., J.B.C. Jackson, and M.J.H. Newman. 2006. Conservation implications of historic sea turtle nesting beach loss. Frontiers in Ecology and the Environment 4: 290–296.

McCracken, M. L. 2000. Estimation of sea turtle take and mortality in the Hawaiian longline fisheries. NOAA, National Marine Fisheries Service, Southwest Fisheries Science Center, Administrative Report H-00-06. NOAA. Honolulu, HI, USA.

McKeown, A. 1977. Marine turtles of the Solomon Islands. Ministry of Natural Resources Fisheries Division, Honiara, Solomon Islands.

Molony, B. 2005. Estimates of the mortality of non-target species with an initial focus on seabirds, turtles and sharks. Information Paper WCPFC-SC1-EBWP-1, First Meeting of the Scientific Committee, Noumea, New Caledonia, 8–19 August 2005. Western and Central Pacific Fisheries Commission, Kolonia, Pohnpei State, Federated States of Micronesia.

Morales, V. R., and P. Vargas. 1996. Legislation protecting marine turtles in Peru. Marine Turtle Newsletter 75: 22–23.

Mortimer, J. A. 1988. The pilot project to promote sea turtle conservation in southern Thailand. Unpublished Report to Wildlife Fund Thailand and World Wildlife Fund/USA. http://college.holycross.edu/faculty/kprestwi/chelonia/pubs/9_unpublished/Mortimer_Thailand_1988.pdf.

NMFS and USFWS. 1998. Recovery Plan for U.S. Pacific populations of the leatherback turtle (*Dermochelys coriacea*). National Marine Fisheries Service, Silver Spring, MD, USA.

Petro, G., F. R. Hickey, and K. MacKay. 2007. Leatherback turtles in Vanuatu. Chelonian Conservation and Biology 6: 135–137.

Pilcher, N. 2012. Community-based conservation of leatherback turtles along the Huon coast, Papua New Guinea. Final report under contract 11-turtle-002 to the Western Pacific Regional Fishery Management Council, Honolulu, HI, USA.

Pritchard, P.C.H. 1971. The leatherback or leathery turtle. IUCN Monograph No. 1: Marine Turtle Series. International Union for Conservation of Nature and Natural Resources, Morges, Switzerland.

Pritchard, P.C.H. 1982. Nesting of the leatherback turtle, *Dermochelys coriacea* in Pacific Mexico, with a new estimate of the world population status. Copeia 1982: 741–747.

Quinn, N. J., and B. L. Kojis. 1985. Leatherback turtles under threat in Morobe Province, Papua New Guinea. PLES 85: 79–99.

Reina, R. D., P. A. Mayor, J. R. Spotila, R. Piedra, and F. V. Paladino. 2002. Nesting ecology of leatherback turtles, *Dermochelys coriacea*, at Parque Nacional Marino Las Baulas, Costa Rica: 1988–1989 to 1999–2000. Copeia 2002: 653–664.

Robins, C. M., S. J. Bache, and S. R. Kalish. 2002. Bycatch of sea turtles in pelagic longline fisheries—Australia. Fisheries Research and Development Corporation, Canberra, Australia.

Saba, V. S., P. Santidrián Tomillo, R. D. Reina, J. R. Spotila, J. A. Musick, D. A. Evans, and F. V. Paladino. 2007. The effect of the El Niño Southern Oscillation on the reproductive frequency of eastern Pacific leatherback turtles. Journal of Applied Ecology 44: 395–404.

Saba, V. S., G. L. Shillinger, A. M. Swithenbank, B. A. Block, J. R. Spotila, J. A. Musick, and F. V. Paladino. 2008. An oceanographic context for the foraging ecology of eastern Pacific leatherback turtles: consequences of ENSO. Deep Sea Research Part I: Oceanographic Research Papers 55: 646–660.

Saba, V. S., C. A. Stock, J. R. Spotila, F. V. Paladino, and P. Santidrián Tomillo. 2012. Projected response of an endangered marine turtle population to climate change. Nature Climate Change 2: 814–820. doi: 10.1038/nclimate1582.

Salm, R. V. 1981. Terengganu meets competition: Does Irian Jaya harbor Southeast Asia's densest leatherback nesting beaches? Conservation Indonesia, Newsletter of WWF Indonesia 5: 18–19, as reprinted in Marine Turtle Newsletter 20: 10–11 (1982).

Santidrián Tomillo, P., V. S. Saba, R. Piedra, F. V. Paladino, and J. R. Spotila. 2008. Effects of illegal harvest of eggs on the population decline of leatherback turtles in Las Baulas Marine National Park, Costa Rica. Conservation Biology 22: 1216–1224.

Santidrián Tomillo, P., E. Vélez, R. D. Reina, R. Piedra, F. V. Paladino, and J. R. Spotila. 2007. Reassessment of the leatherback turtle (*Dermochelys coriacea*) nesting population at Parque Nacional Marino Las Baulas, Costa Rica: effects of conservation efforts. Chelonian Conservation and Biology 6: 54–62.

Sarti, A. L., and A. R. Barragán. 2011. Importance of networks for conservation of the Pacific leatherback turtle: the case of "Proyecto Laúd" in Mexico. In: P. H. Dutton, D. Squires, and A. Mahfuzuddin (eds.), Conservation of Pacific sea turtles. University of Hawai'i Press, Oahu, HI, USA, pp. 120–131.

Sarti, L., S. Eckert, P. Dutton, A. Barragán, and N. García. 2000. The current situation of the leatherback population on the Pacific coast of Mexico and Central America, abundance and distribution of the nestings: an update. In: H. J. Kalb and T. Wibbels (compilers), Proceedings of Nineteenth Annual Symposium on Sea Turtle Biology and Conservation, NOAA Technical Memorandum NMFS-SEFSC-443. NOAA, Miami, FL, pp. 85–87.

Sarti Martínez, A. L. 2000. *Dermochelys coriacea*. In: IUCN 2012. IUCN Red List of Threatened Species. Version 2012.2. www.iucnredlist.org. Accessed 11 November 2012.

Sarti Martínez, L., A. R. Barragán, D. Garcia-Muñoz, N. Garcia, P. Huerta, and F. Vargas. 2007. Conservation and biology of the leatherback turtle in the Mexican Pacific. Chelonian Conservation and Biology 6: 70–78.

Sarti Martínez, L., A. Barragán, F. Vargas, P. Huerta, E. Ocampo, A. Tavera, A. Escudero, et al. 2003. The decline of the Eastern Pacific leatherback and its relation to changes in nesting behavior and distribution. In: N. Pilcher (compiler), Proceedings of the Twenty-third International Symposium on Sea Turtle Biology and Conservation. NOAA Technical Memorandum NMFS-SEFSC-536. NOAA, Miami, FL, pp 133–136.

Seminoff, J. A., S. R. Benson, K. E. Arthur, T. Eguchi, P. H. Dutton, R. Tapilatu, and B. N. Popp. 2012. Stable isotope tracking of endangered sea turtles: validation with satellite telemetry and ∂^{15}N analysis of amino acids. PLoS ONE 7(5); e37403. doi: 10.1371/journal.pone.0037403.

Seminoff, J. A., and B. P. Wallace (eds.). 2012. Sea turtles of the eastern Pacific: Advances in research and conservation. University of Arizona Press, Tucson, AZ, USA.

Shillinger, G. L., D. M. Palacios, H. Bailey, S. J. Bograd, A. M. Swithenbank, P. Gaspar, B. P. Wallace, et al. 2008. Persistent leatherback turtle migrations present opportunities for conservation. PLoS Biology 6: e171. doi: 10.1371/journal.pbio.0060171.

Spotila, J. R., A. E. Dunham, A. J. Leslie, A. C. Steyermark, P. T. Plotkin, and F. V. Paladino. 1996. Worldwide population decline of *Dermochelys coriacea*: Are leatherback turtles going extinct? Chelonian Conservation and Biology 2: 209–222.

Spotila, J. R., R. D. Reina, A. C. Steyermark, P. T. Plotkin, and F. V. Paladino. 2000. Pacific leatherback turtles face extinction. Nature 405: 529–530.

Spring, C. S. 1982a. Status of marine turtle populations in Papua New Guinea. In: K. A. Bjorndal (ed.), Biology and conservation of sea turtles. Smithsonian Institution Press, Washington DC, USA, pp. 281–289.

Spring, C. S. 1982b. Subsistence hunting of marine turtles in Papua New Guinea. In: K. A. Bjorndal (ed.), Biology and conservation of sea turtles. Smithsonian Institution Press, Washington, DC, USA, pp. 291–295.

Stewart K., M. Sims, A. Meylan, B. Witherington, B. Brost, and L. B. Crowder. 2011. Leatherback nests increasing significantly in Florida, USA: trends assessed over 30 years using multilevel modeling. Ecological Applications 21: 263–273.

Stobutzki, I., E. Lawrence, N. Bensley, and W. Norris. 2006. Bycatch mitigation approaches in Australia's eastern tuna and billfish fishery: seabirds, turtles, marine mammals, sharks and non-target fish. Information Paper WCPFC-SC2/EBSWG-IP4, Ecosystem and Bycatch Specialist Working Group of the Second Meeting of the Scientific Committee of the WCPFC, Manila, Philippines. Western and Central Pacific Fisheries Commission. Kolonia, Pohnpei State, Federated States of Micronesia.

Suarez, M., and C. Starbird, C. 1995. A traditional fishery of leatherback turtles in Maluku, Indonesia. Marine Turtle Newsletter 68: 15–18.

Tapilatu, R. F., P. H. Dutton, M. Tiwari, T. Wibbels, H. V. Ferdinandus, W. G. Iwanggin, and B. H. Nugroho. 2013.

Long-term decline of the western Pacific leatherback, *Dermochelys coriacea*, a globally important sea turtle population. Ecosphere 4: 25. http://dx.doi.org/10.1890/ES12-00348.1.

Tapilatu, R. F., and M. Tiwari. 2007. Leatherback turtle, *Dermochelys coriacea*, hatching success at Jamursba-Medi and Wermon beaches in Papua, Indonesia. Chelonian Conservation and Biology 6: 154–158.

Tiwari, M., D. L. Dutton, and J. A. Garner. 2011. Nest relocation: a necessary management tool for western Pacific leatherback nesting beaches. In: P. H. Dutton, D. Squires, and A. Mahfuzuddin (eds.), Conservation and sustainable management of sea turtles in the Pacific Ocean. University of Hawai'i Press, Oahu, HI, USA, pp. 87–96.

Turtle Expert Working Group. 2007. An assessment of the leatherback turtle population in the Atlantic Ocean. NOAA Technical Memorandum NOFS-SWFSC-555. NOAA, Miami, FL.116 p.

Urteaga, J., P. Torres, O. Gaitan, G. Rodríguez, and P. Dávila. 2012. Leatherback, *Dermochelys coriacea*, nesting beach conservation on the Pacific coast of Nicaragua, (2002–2010). Abstract. In: T. T. Jones and B. P. Wallace (compilers), Proceedings of the Thirty-first Annual Symposium on Sea Turtle Biology and Conservation. NOAA Technical Memorandum NMFS-SEFSC-631. NOAA, Miami, FL, USA, p. 237.

Vaughan, P. W. 1981. Marine turtles: A review of their status and management in the Solomon Islands. Ministry of Natural Resources Fisheries Division, Honiara, Solomon Islands.

Wallace, B. P., M. Tiwari, and M. Girondot. 2013. *Dermochelys coriacea*. In: IUCN 2013. IUCN Red List of Threatened Species. Version 2013.2. iucn-redlist.org. Accessed 27 February 2014.

Wetherall, J. A., G. H. Balazs, R. A. Tokunga, and M.Y.Y. Yong. 1993. Bycatch of marine turtles in North Pacific high seas drift net fisheries and impacts on the stocks. In: J. Ito, W. Shaw, and R. L. Burgner (eds.), International North Pacific Fisheries Commission (INPFC) Symposium on Biology, Distribution, and Stock Assessment of Species Caught in the High Seas Driftnet Fisheries in the North Pacific Ocean. Bulletin International North Pacific Fisheries Commission 53: 519–538. International North Pacific Fisheries Commission, Vancouver, Canada.

11

Leatherback Turtle Populations in the Indian Ocean

RONEL NEL, KARTIK SHANKER,
AND GEORGE HUGHES

Leatherback turtles (*Dermochelys coriacea*) in the Indian Ocean were first noted around the Cape of Good Hope in 1849 (Smith 1849), but the first detailed studies were undertaken by Deraniyagala (1939) in his systematic descriptions of reptiles of Ceylon (now Sri Lanka). Leatherbacks are now known throughout the Indian Ocean but nesting areas are restricted to a few locations (Bourjea et al. 2008). There is an extended rookery in the southwest Indian Ocean shared by South Africa and Mozambique, and scattered but important sites in the Bay of Bengal. Little, if any, nesting occurs on mainland beaches and islands of the northern and eastern Arabian Sea. Although leatherbacks once nested in small numbers on the Indian mainland coast, there have been no confirmed nests in the last four decades. In the northern Indian Ocean the most consistent nesting sites are on islands in the Bay of Bengal, including Sri Lanka and the Nicobar and Andaman Islands (Nel 2012). Low-level, scattered nesting occurs off the coasts of Thailand and Indonesia (Sumatra and Java; Steering Committee BSTCI 2008). Two regional management units (RMUs) are recognized for the Indian Ocean (Wallace et al. 2010). These are the Southwest Indian Ocean (SWIO), South Africa and Mozambique, and their associated foraging areas, and the Northeast Indian Ocean (NEIO), Sri Lanka, Nicobar and Andaman Islands, and their associated foraging areas.

The Southwest Indian Ocean (SWIO)

Nesting Sites

Nesting of leatherback turtles in the Southwest Indian Ocean RMU is centered along an approximately 300 km stretch of broad (50–100 m wide) silica sand beaches at the shared border of South Africa and Mozambique. The beaches are flanked by the warm, fast-flowing Agulhas Current (Lutjeharms 2001). Leatherback turtles live in the waters off the western Indian Ocean islands, including Madagascar (Rakotonirina and Cooke 1994), but no consistent nesting occurs outside of South Africa or Mozambique. Nesting effectively starts at Cape Vidal, South Africa (28°7'35.50" S 32°33'33.19" E), and continues up to Inhaca Island,

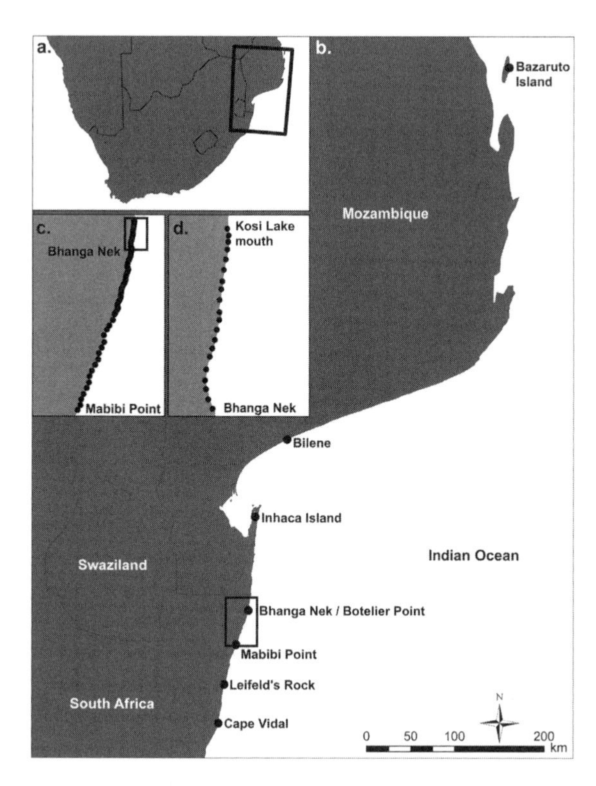

Fig. 11.1. The nesting distribution of leatherback turtles in the western Indian Ocean.

Mozambique (25°58'28.20" S 32°59'35.13" E), with another few nests per annum spread between Bilene (25°19'44.53" S 33°14'27.26" E) and Bazaruto Island (21°31'38.13" S 35°29'28.48" E) (Videira et al. 2011). The bulk of the nesting, however, takes place in South Africa, between Bhanga Nek and Leifeld's Rock (fig. 11.1). This coastal area is characterized by pristine, intact, vegetated dunes rising as high as 100 m above sea level, interspersed with a few mobile dune sheets and fringed, small, primary dunes. Leatherback turtles primarily nest in the areas with little vegetation and open sand (Botha 2010).

Turtle Monitoring

Apart from a brief mention in Rose (1950), no attention was paid to leatherbacks (nor indeed to any other species of sea turtle) in the southwestern Indian Ocean until the first survey carried out by conservation staff of the Natal Parks Board, South Africa, in 1963. Vehicle and foot patrols then took place on the beaches of Maputaland (then known as Tongaland), immediately south of the border with Mozambique. McAllister et al. (1965) reported that not only were loggerhead turtles (*Caretta caretta*) nesting in the area, but there was also a nesting population of leatherback turtles. Since 1963, regular and consistent annual monitoring has taken place, representing one of the longest-running, quantitative monitoring programs of sea turtle nesting activity in the world.

Turtle nesting in this population, which has the most southern distribution in the world, starts in mid-October and lasts until late February each season. Monitoring patrols continue until mid-March to provide protection to emerging hatchlings. There were only six nesting leatherbacks reported in the first surveying season that covered 13 km (McAllister et al. 1965), but with surveys expanded to 56 km starting in 1973–1974 (Nel et al. 2013), numbers grew to approximately 80. In combination with cooperative programs in Mozambique, the southern African nesting population now numbers about 100 females per season (Nel 2010).

Even though nesting takes place between Bhanga Nek and Leifeld's Rock, monitoring takes place only over the northern half of this distance. Nesting distribution is fairly even throughout this region, with some preference for bays fringed by dynamic surf zones and with an absence of low-lying intertidal rock (Hughes 1974; Botha 2010). Nesting females use the rip currents running out from the beach as guides to a safe landing site. In general, leatherbacks show little nest site fidelity with repeat nesting events sometimes as much as 40 km apart, but more commonly about 10 km apart (Botha 2010). A recent quantitative analysis of long-term data confirms that the number of nesting leatherbacks increased after initiation of beach conservation, but that population growth was not sustained over time (Nel et al. 2013). However, despite being small, the population is stable with consistent conservation efforts continuing.

South Africa

Full legal and actual physical protection accompanied by regular monitoring in South Africa started in 1963 (McAllister et al. 1965); it expanded over time and is now effectively taking place over 150 km of the 250 km nesting area as far north as Inhaca Island. The most consistent long-term data set is available for South Africa, especially the 53 km area between Kosi Lake and Mabibi point (fig. 11.1c), which is the "monitoring area." For interannual comparisons, the most robust data with standardized effort are collected for the 13 km "index area" (between Bhanga Nek and Kosi Lake; fig. 11.1d). Analyses of these data sets indicated that the population increased immediately after the onset of protection (fig. 11.2) and grew for about 15 years (Hughes 1996; Nel

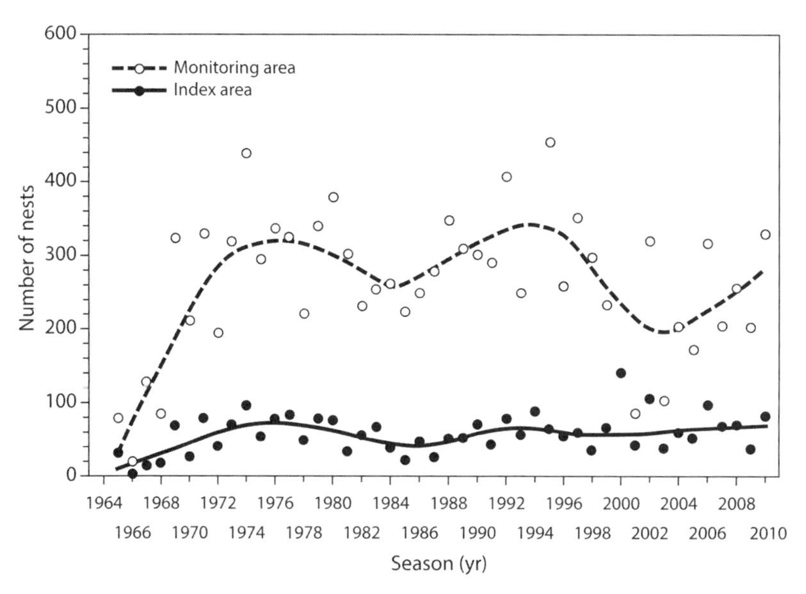

Fig. 11.2. Number of nests over time recorded in a 53 km monitoring area, and in a 13 km index area that is controlled for effort along the Maputaland coast of South Africa. (Both curves use a Lowess spline fit.)

et al. 2013). The population has since stabilized with fluctuations roughly every ten years, and with annual number of nests ranging from 150 to 450 in the monitored area (fig. 11.2). Mean number of nests in the index area ranges between 50 and 80 nests per season. The reason for the lack of long-term population growth is uncertain, but it may be a combination of ineffective monitoring due to the large distance over which turtles nest and lack of nest site fidelity and preference (Thorson et al. 2012), and/or increased threats offshore, especially longline fishing (Petersen et al. 2009; De Wet 2012; Nel et al. 2013).

In the 2010–2011 season (see Nel 2010 for patrol and tagging methods) we recorded 375 emergences of adult turtles and 336 nests in South Africa (Ezemvelo KZN Wildlife, unpublished data), and 61 turtle emergences (number of nests unknown) north of the South Africa / Mozambique border (Videira et al. 2011). Thus, the proportion of turtles nesting in South Africa is about 80% of the total population. This is a shared population because each season a number of turtles nest in both countries and are often identified by carrying both South African (ZA) and Mozambican (MZ) coded flipper tags with unique numbers.

Mozambique

Monitoring and local protection started in 1996 in Mozambique as a partnership between the Natal Parks Board and a local tourist operator at Ponto do Ouro. Although initially supported by the Natal Parks Board, the program became independent, added more partners, and now operates almost continuously from the South African border to Inhaca Island. A total of 102 km was patrolled in 2010, although with varying temporal effort, and there were 61 leatherback emergences (Videira et al. 2011). The number of nesting events and emergences in Mozambique is about 20% of the total in this metapopulation.

This South African / Mozambique population is the only known nesting population in the southern Indian Ocean (Nel 2010). There is no known nesting in Madagascar, other Indian Ocean islands, or Western Australia, although at-sea sightings are not uncommon (Limpus 2009; Waayers 2011).

Hatchling Distributions

At an average of 104 eggs per nest, clutch sizes are higher than for other nesting colonies (Hughes 1974). Most metrics like female size, egg size, incubation time, and hatching success are above average; adult females have a mean size of about 160 cm curved carapace length (Ezemvelo KZN Wildlife, unpublished data), and hatching and emergence success is more than 70% (J. Tucek, unpublished data). The greater number of eggs may be attributed to the marginal, high latitude position of this nesting ground, and the availability of two ocean basins (i.e., the Atlantic and Indian Oceans) to foraging turtles. Larger leatherbacks lay more eggs (chapter 7) and adult females in South Africa forage closer to nesting beaches than Pacific and Atlantic leatherbacks (Luschi et al. 2006; chapters 9 and 10), reducing their energy costs during migration. The beaches are relatively cool and the larger number of eggs may raise the temperature of the clutch and produce more females. Greater

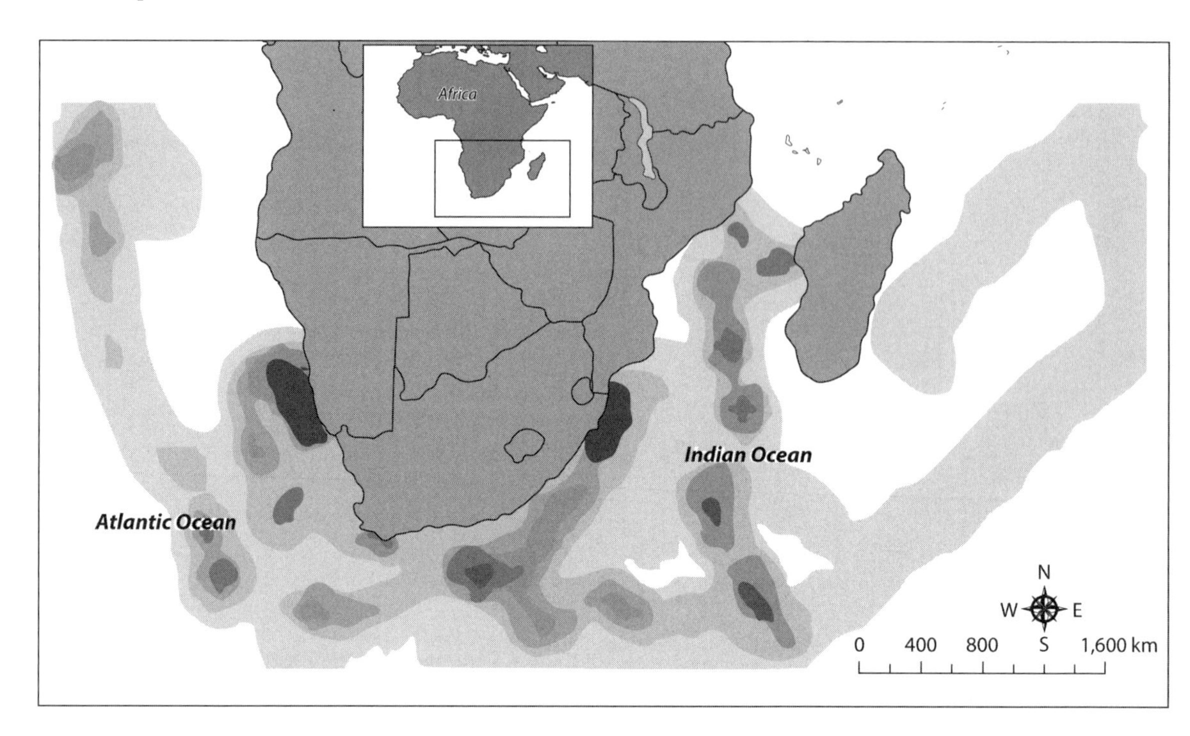

Fig. 11.3. Distribution of post-nesting female leatherback turtles, satellite tagged on the beach in Maputaland, KwaZulu-Natal. Greyscale indicates density of locations from telemetered turtles. Distributions based on unpublished satellite tracking data from Department of Environmental Affairs / Directorate Oceans and Coast, Cape Town, South Africa.

hatchling production may offset losses caused by hatchlings being swept into the very cold waters around the Cape of Good Hope. Hatchlings emerge from early January and enter the Agulhas Current, then reaching its maximum annual velocity, which carries the hatchlings southward toward Cape Agulhas, into the large gyres spinning off into the southern Indian Ocean from the current and probably around Cape Agulhas and into the Atlantic Ocean (Hughes 1978). It is possible that this rapid southward distribution of hatchlings, with concomitant danger of exposure to very cold water, provides an additional reason for the poor response of the nesting population under protection. In contrast to the loggerhead population, which has demonstrated a remarkable increase in nesting females (Nel et al. 2013), the leatherback population appears to be barely holding its own (fig. 11.2). However, since loggerhead hatchlings would experience the same currents, bycatch in fisheries or some other cause may play an important role in the lack of growth in the leatherback population.

Post- and Inter-nesting Movements

Some 30 leatherbacks were satellite tagged in Maputaland, recording both inter-nesting and post-nesting migration movements (fig. 11.3). Hughes and coworkers described post-nesting behavior of leatherbacks as extended feeding journeys (Hughes et al. 1998; Luschi et al. 2003; Luschi et al. 2006; Sale et al. 2006; Lambardi et al. 2008), with females having an unpredictable end destination. These spent females move into the cold Benguela Current, off the Atlantic coast of Africa, or into the warmer Indian Ocean via the Agulhas retroflection and return currents (Luschi et al. 2006); alternatively, they move back along the Mozambique Channel without an obvious feeding ground. This uncertainty is partly due to lack of data as most satellite tags failed in less than 13 months after deployment. Very few hatchling or adult stranding events have been reported along the South African coast (De Wet 2012).

Botha (2007) evaluated inter-nesting behavior of leatherback turtles equipped with backpack-fitted SPOT 5 tags. During the inter-nesting period, excursions of variable distances were undertaken outside the protection of South Africa's iSimangaliso Wetland Park, a UNESCO world heritage site that provides protection from St. Lucia below Cape Vidal in the south, to the border of Mozambique, and 5 km out in the ocean (Botha 2010). The average distance traveled offshore was about 36 km, with swimming speeds varying between 1.6 km h^{-1}–2.2 km h^{-1}. These interesting data confirm that the population is shared between two countries with excursions into Mozambique, outside of the protected area, and on occasion, to the edge of the 200 nm Exclusive Economic Zone (EEZ). No clear hotspots of concentrated leatherback activity were identified, but the overall gravitation was toward Cape Vidal in the southern end of the iSimangaliso Wetland Park.

Threats and Prospects

Threats to southwestern Indian Ocean (SWIO) leatherback turtles are similar to other populations when they are away from the nesting area. With most destructive fisheries like drift nets banned in the SWIO, longlining is the biggest single threat (Petersen et al. 2009; Nel et al. 2013). Swordfish and tuna longline fisheries, in particular, catch leatherback turtles in disproportionately high numbers relative to nesting numbers (Petersen et al. 2009). In South African coastal waters, bather protection nets (Brazier et al. 2012) and boat strikes pose the second biggest suite of threats resulting in mortality of less than ten leatherbacks per annum. The impact of artisanal fisheries along the east African coast has not been quantified but is expected to be numerically small because leatherbacks are most frequently absent in catch reports (Rakotonirina and Cooke 1994; Humber et al. 2011; De Wet 2012).

In South Africa, prior to 1963, substantial numbers of leatherback eggs were harvested from beaches, probably contributing to critically low numbers of nesting females. Since then, every decade has seen an increased monitoring effort and expanded conservation. Leatherback turtles on nesting beaches between St. Lucia in KwaZulu-Natal and Inhaca Island now receive intensive and effective protection, being included in a UNESCO world heritage site (iSimangaliso Wetland Park) and the Ponto Do Ouro–Kosi Transfrontier Conservation Area. Despite these efforts, the population remains small but relatively stable and deserves continuous conservation attention.

The Northeast Indian Ocean (NEIO)

There is a small leatherback nesting population laying a few hundred nests in southern Sri Lanka (Ekanayake et al. 2002), which constitutes the only significant nesting population in the South Asian region. Leatherbacks also nest in reasonable numbers in the Andaman and Nicobar Islands, but the population in the Andamans seems to have declined in the last few decades (Andrews, Krishnan, and Biswas 2006). Nesting activity at West and South Bays of Little Andaman Island totals 100 to 200 nests per annum.

The largest rookeries in the Bay of Bengal are on the Nicobar Islands. There are several nesting beaches on both Great and Little Nicobar Island (Andrews, Krishnan, and Biswas 2006). Beaches like Galathea on the southeast coast of Great Nicobar, and Alexandria and Dagmar on the west coast, each had 400 to 500 nests per season before the December 2004 tsunami (Andrews, Krishnan, and Biswas 2006). Similarly, beaches on Little Nicobar likely received more than 100 nests each (M. Chandi, personal communication). Not far from Great Nicobar Island, Sumatra (Indonesia) also has nesting leatherback turtles, but no quantified data are yet available for these beaches.

In Southeast Asia, there are nesting populations in Indonesia (Sumatra, Java, and Bali), Malaysia, and Thailand; some have declined or been extirpated, the best known of which is the population at Terengganu, Malaysia (Hamann et al. 2006). Few other countries in the region have documented records of leatherback nesting.

Leatherback turtles are seen in offshore waters of most countries in the region. However, it is not certain as to whether these are migratory or foraging animals.

Leatherbacks in India

MAINLAND.

There are numerous sighting and stranding records for leatherback turtles from the Indian mainland coast (Krishna Pillai et al. 2003). Many of these records appear in India's Marine Fisheries Information Service newsletter, a publication of the Central Marine Fisheries Research Institute that initiated a stranding database for marine fauna in the 1970s. The earliest records date to the early 1900s, when a British officer recorded a leatherback turtle on the Kerala coast (Cameron 1923). At the time, local fishermen claimed that leatherback turtles were common on that coast in earlier years, with up to 40 turtles caught each season, but numbers had declined and leatherbacks were seen only in the vicinity of a nearby reef. Cameron had to wait more than a decade to see a specimen, but this may be because most turtles that were caught were quickly slaughtered and sold. In 1956, nesting of a leatherback turtle was observed on the Kerala coast (Jones 1959).

From the 1970s onward, there were periodic strandings and incidental captures in fishing nets (gill nets, trawlnets, and shore seines) in Kerala and Tamil Nadu (the southwest and southeast coasts of the mainland). There were occasional strandings further north on both the west (Maharashtra) and east (Andhra Pradesh) coasts (Krishna Pillai 2003). Older fishermen on the Tamil Nadu coast claim that leatherbacks nested there in earlier decades (K. Shankar, personal observation), but there are no documented records of this activity. Though there are no records of leatherback turtles from Orissa (Odisha) and West Bengal on the northeast coast of India, there is one documented stranding recorded (in 1997) on St. Martins Island in Bangladesh (Rashid and Islam 2006).

Leatherback turtles are not widely consumed, but they are slaughtered for their meat on the southern coast of Kerala. They were frequently captured and eaten in the early 1900s (Cameron 1923). More recently, a leatherback accidentally captured in September 2002 was slaughtered in Vizhinjam, Kerala (Krishna Pillai 2003). Two other turtles captured around the same time (April 2001 and December 2002) were rescued and released by the community, one at the behest of a foreign tourist who was present (Krishna Pillai et al. 2003). Similarly, a leatherback turtle found tied to the pillars of a bridge in Manakudi on the southern Tamil Nadu coast, likely for slaughter, was rescued and released by local biologists (Balachandran et al. 2009).

ANDAMAN AND NICOBAR ISLANDS

Leatherback turtles are known to nest on numerous islands in the Andaman and Nicobar group. The Madras Crocodile Bank Trust conducted the first surveys in the late 1970s (Bhaskar 1979). Satish Bhaskar, who conducted these and subsequent surveys, contributing a vast amount of information about sea turtles of India and its offshore islands (Kar and Bhaskar 1982; Bhaskar 1993), also conducted the first surveys of the now well-known leatherback nesting beaches of West Papua, Indonesia. Namboothri, Swaminathan, and Shanker (2012) compiled a summary of his work in the Andaman and Nicobar Islands.

In the Andaman Islands, nesting occurs on the islands of Middle Andaman (Cuthbert Bay), North Andaman, Little Andaman (West Bay), and Rutland Island (Bhaskar 1979). Nesting at most of these beaches has been sporadic in the decades since leatherbacks were first documented, although there were about 70 nests on Little Andaman (Andrews, Krishnan, and Biswas 2006). Larger numbers of nests occur on Great Nicobar Island (on Alexandria and Dagmar River beaches on the west coast and Galathea on the southeast coast) with over 1,500 nests recorded there. Hence, Great Nicobar and Little Andaman Islands are the prime nesting locations for leatherback turtles in the region.

GREAT NICOBAR

Surveys conducted in the early 1990s revealed that leatherback turtles nested in large numbers at several nesting beaches on Great and Little Nicobar Islands. During the 1991–1992 season, there were over 800 nests at 8 of 9 nesting beaches on Great Nicobar Island (Bhaskar and Tiwari 1992; Bhaskar 1993; Tiwari 2012). The nesting beach at Galathea National Park on the southeastern coast of Great Nicobar was important with over 150 nests reported during that season (Bhaskar 1993). A few leatherback turtles were tagged during that period, and one returned to nest in late 2000 (Andrews, Krishnan, and Biswas 2006). There were 237 nests at Galathea during 1993–1994 (Bhaskar 1994), 282 during 1995–1996, and 124 in 1997–1998. These figures are based on local Forest Department records and may not be comprehensive.

As a significant nesting beach, Galathea was an obvious choice for monitoring because it was accessible by road (until the December 2004 tsunami). Dagmar and Alexandria River beaches were also monitored. The monitoring program began in 2000 and was carried out by the Andaman and Nicobar Monitoring Team, a division of the Madras Crocodile Bank Trust. Intensive daily monitoring of the beaches occurred between November and March each year (Andrews et al. 2001). In 2000–2001, 444 nests were counted on Galathea, and 146 individuals were tagged; the following year, 425 nests were counted and 177 individuals tagged.

During the 2000–2001 surveys, over 1,000 body pits were counted at the Dagmar and Alexandria River beaches (Andrews, Krishnan, and Biswas 2006), with more than 1,500 nests on the three beaches of Great Nicobar. During 2003–2004, there were 575 nests on Galathea Beach.

Along with the nesting on Little Nicobar and Little Andaman Islands, more than 2,000 nests were estimated on Andaman and Nicobar Islands during the 2000–2001 nesting season. Accounting for annual clutch frequency (~5 nests per season) and remigration interval (~2.5 years), an estimated 1,000 leatherback turtles nested on the beaches of these islands during the 2000–2002 period (Andrews and Shanker 2002).

At the beginning of the 2004–2005 season, there were about 100 nests on Galathea Beach (Andrews, Tripathy, et al. 2006), but the tsunami on 24 December destroyed the nesting beaches. The island's infrastructure was destroyed, several researchers died, and the monitoring program was terminated (Andrews, Chandi, et al. 2006). Subsequent surveys showed that beaches were reestablished and that there were over 100 leatherback nests near the Galathea River mouth in 2006 (Andrews, Chandi, et al. 2006; N. Namboothri, personal communication).

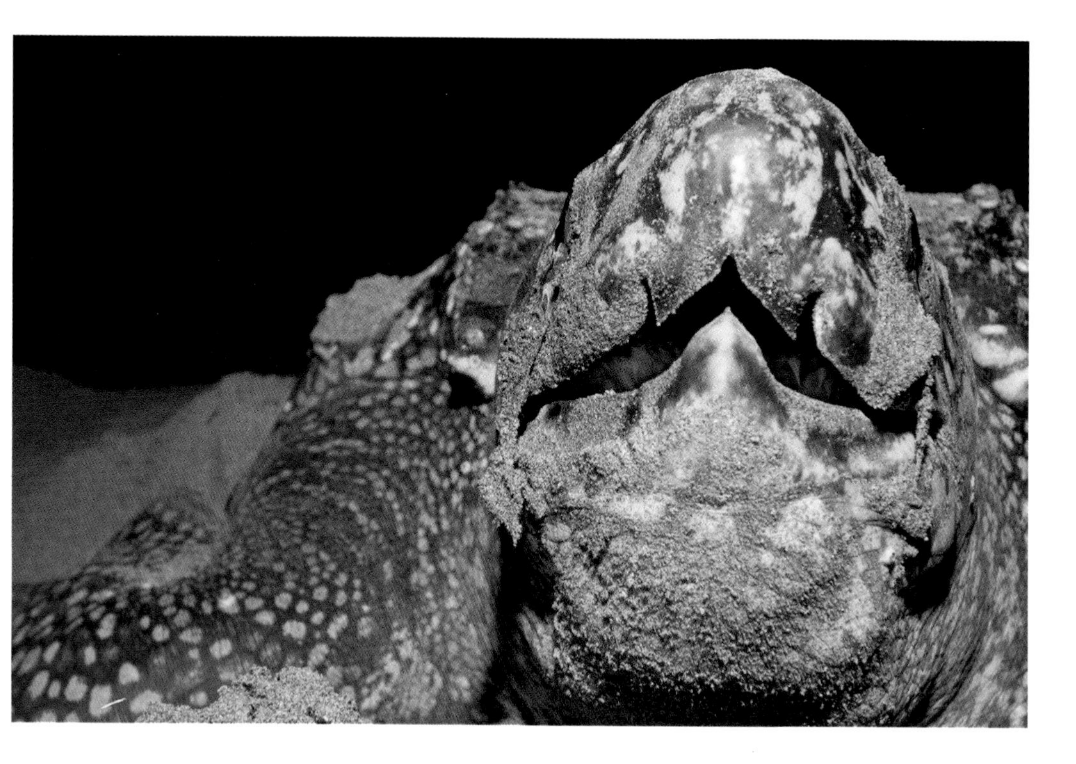

Plate 9. *Top*, Detail of the mouth of a leatherback turtle. Photo: © Jason Bradley | BradleyPhotographic.com. *Bottom*, Bibi Santidrián excavates leatherback nest on Playa Grande while Johnna Holding opens eggs to determine stage of death of embryos. Photo: Barbara J. Bergwerf.

Plate 10. *Top,* Leatherback turtle hatchlings remain inactive for a short period of time after emerging from the nest. Photo: Erin Keene. *Bottom,* Eastern Pacific leatherback turtle hatchlings on the beach at dusk heading to the water. Photo: © Jason Bradley | BradleyPhotographic.com.

Plate 11. *Top,* Leatherback turtle nesting while research team and tourists are present. Photo: Kip Evans. *Bottom,* Leatherback turtle emerging to nest at Playa Grande, Costa Rica. East Pacific leatherback turtles are among the most endangered populations in the world. Photo: James R. Spotila and Frank Paladino.

Plate 12. *Top,* Leatherback turtle on the boat with Scott Benson's research team. The turtle was scooped aboard the boat with a hoop net. Photo: Scott Hansen. NMFS-ESA Permit No. 1596. *Bottom,* Detail of the head of a leatherback turtle after surfacing to breathe. Pink or pineal spot is visible on top of the head. Photo: Scott Hansen. NMFS-ESA Permit No. 1596.

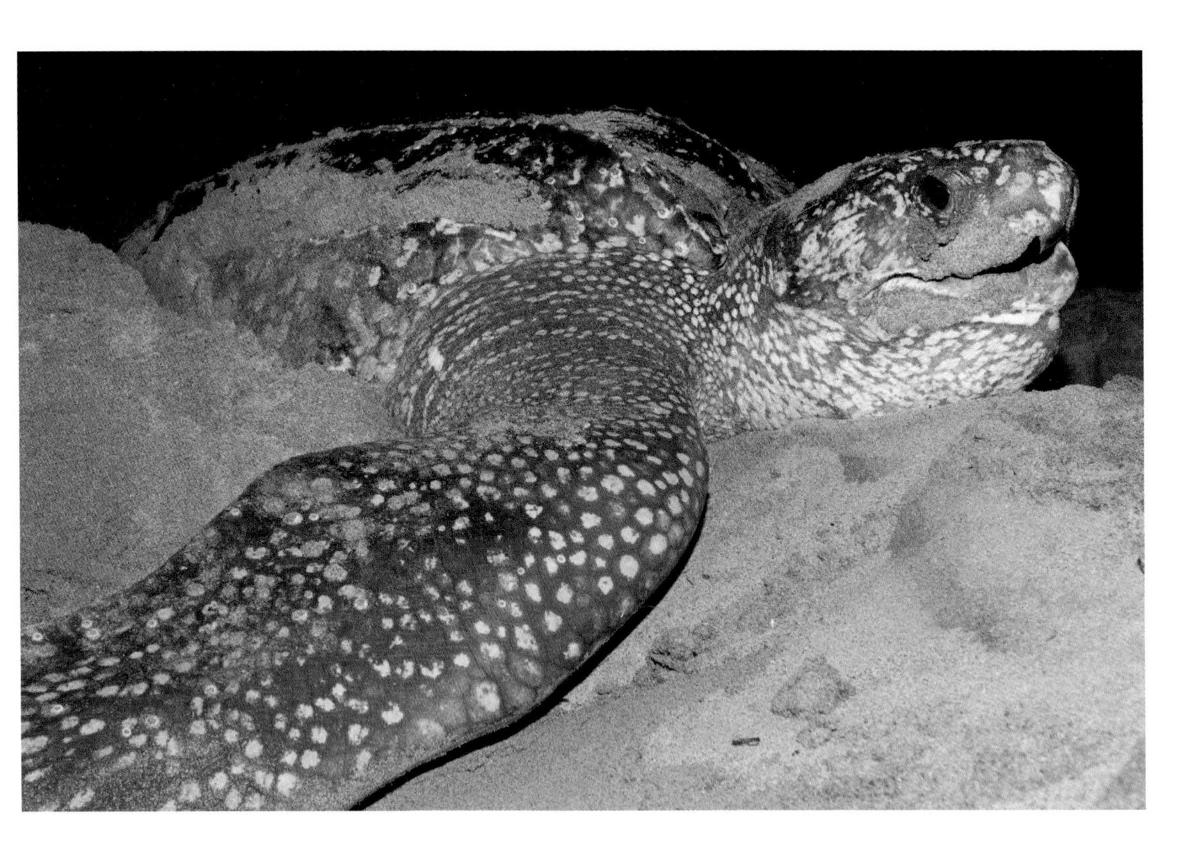

Plate 13. *Top,* Female leatherback turtle on the beach after nesting at Playa Grande, Costa Rica. Photo: © Jason Bradley | BradleyPhotographic.com. *Bottom,* Scott Benson and female leatherback turtle on a nesting beach in Solomon Islands. Photo: Karin Forney.

Plate 14. *Top,* Leatherback turtle hatchlings are effective swimmers. Photo: Sam Friederichs. *Bottom,* Leatherback turtle hatchlings do not feed during the first few days of their life in the ocean. Photo: Nathan Robinson.

Plate 15. *Top,* Dr. T. Todd Jones holding a two-year-old juvenile leatherback turtle that he raised in captivity. Photo: Vancouver Sun. *Bottom,* Dr. T. Todd Jones measuring a two-year-old juvenile leatherback turtle raised in captivity at the University of British Columbia. Photo: Michael Carey.

Plate 16. *Top,* Leatherback turtle swimming in the ocean. Photo: Brian J. Skerry / National Geographic Creative. *Bottom,* Leatherback turtles feed on jellyfish or other soft body prey like this salp. Photo: Brian J. Skerry / National Geographic Creative.

LITTLE ANDAMAN

The beaches at Little Andaman Island have a smaller nesting population than those at the Nicobars. First surveyed in the 1970s, these beaches received little attention until recently, apart from a brief study on nest placement and predation in the 1990s (Sivasundar and Devi Prasad 1996). Following the December 2004 tsunami, the beaches were surveyed with the prospect of establishing a monitoring program in 2007 (M. Chandi, personal communication). A long-term program began at South Bay in 2007–2008, and at West Bay in 2010–2011. At South Bay, there were 25 nests during the first season, followed by 39 nests, 7 nests, 58 nests, and 36 nests in 2008–2009, 2009–2010, 2010–2011, and 2011–12, respectively. There were 91 nests at West Bay during the 2010–2011 season and 148 nests during the 2011–2012 season (Swaminathan et al. 2011; Swaminathan et al. 2012). In January 2011 and 2012, seven leatherback turtles were fitted with satellite transmitters. Two turtles transmitted for about six months and four turtles transmitted for two to three months; one turtle transmitted only a single data point. Of these, four turtles traveled southeast of the Andaman and Nicobar Islands, two along the coast of Sumatra, and two beyond Cocos (Keeling) Island, toward Western Australia. Two turtles moved southwest of the islands, one of which traveled south of the British Ocean Territory (Namboothri, Swaminathan, et al. 2012).

Leatherbacks in Sri Lanka

Sri Lanka has a special place in the history of leatherback biology. Deraniyagala, a pioneering Sri Lankan zoologist, wrote extensively about sea turtle taxonomy, morphology, embryology, and behavior. The publications included his account of the "nesting habit" of leatherback turtles and superb illustrations (Deraniyagala 1936, 1939).

While leatherback nesting is known to occur at many beaches including Yala, Rekawa, Kosgoda, Bundala, and others, the highest numbers nest on Godavaya; between March and November 2001, 333 nests were reported on this beach (Ekanayake et al. 2002). However, only two nests were recorded at the beach after the 2004 tsunami (Hamann et al. 2006); the reasons for the decline are not known, and it is unclear if the population has recovered.

Leatherbacks in Southeast Asia

There are several nesting sites for leatherback turtles in Indonesia that have not been quantified. These include beaches in the Sumatran provinces of Aceh, West Sumatra, North Sumatra, Bengkulu, and Lampung; several beaches in West Java and East Java; and Pecatu and Lepang beaches in Bali (Adnyana 2006, cited in Hamann et al. 2006). Only the beaches in Meru Betiri and Alas Purwo National Parks in East Java have been monitored and together currently receive less than 50 nests per year.

Currently, a small number of nests (< 20) are recorded each year at a few beaches on the west coast of Phanga and Phuket Provinces in Thailand. Historical nesting may have been greater.

Threats and Prospects

There are a number of current (fisheries, depredation) and potential threats (development and climate change) to leatherback populations in the region. In the Andaman and Nicobar Islands, the main threats to leatherback turtles are from depredation of eggs by wild pigs, monitor lizards, and feral dogs. The threat from feral dogs has increased over the years, and may have led to the decline and extirpation of some populations in the Andaman Islands. The threat from both fisheries and development activities on the islands have been increasing.

In order to assess the impact of these threats, important or index beaches need to be monitored regularly. However, many or most beaches have only been infrequently surveyed, if surveyed at all. Few nesting beaches have been monitored, and only Little Andaman Island has been annually monitored in recent years. Additionally, there are important nesting beaches in the Nicobar Islands that need to be monitored on a regular basis to assess population trends. Research is also required on various aspects of leatherback biology, such as the impacts of climate change on sex ratios, turtle migratory routes and foraging areas, and other factors determining hatchling and adult survival.

Population Genetics

Very little has been published on the genetic composition of either the SWIO or NEIO leatherbacks. To date, only the SWIO rookeries, particularly in South Africa, have contributed samples to the global phylogeographic analysis (Dutton et al. 1999). Nothing is currently known about the detailed population structures for either RMU although studies are currently underway to evaluate multiple paternity for SWIO leatherbacks (K. Naish, personal communication).

LITERATURE CITED

Andrews, H. V., M. Chandi, A. Vaughan, J. Aungthong, S. Aghue, S. Johnny, S. John, and S. Naveen. 2006. Marine turtle status and distribution in the Andaman and Nicobar Islands after the 2004 M 9 quake and tsunami. Indian Ocean Turtle Newsletter 4: 3–11.

Andrews, H. V., S. Krishnan, and P. Biswas. 2001. The status and distribution of marine turtles around the Andaman and Nicobar Archipelago. A GOI-UNDP National Sea Turtle Project Report (unpublished), IND/97/964. Centre for Herpetology / Madras Crocodile Bank Trust, Tamil Nadu, India.

Andrews, H. V., S. Krishnan, and P. Biswas. 2006. Distribution and status of marine turtles in the Andaman and Nicobar Islands. In: K. Shankar and B. C. Choudhury (eds.), Marine turtles of the Indian subcontinent. Universities Press, Hyderabad, India, pp. 33–57.

Andrews, H. V., and K. Shanker. 2002. A significant population of leatherback turtles in the Indian Ocean. Kachhapa 6: 17.

Andrews, H. V., A. Tripathy, S. Aghue, S. Glen, S. John, and K. Naveen. 2006. The status of sea turtles in the Andaman and Nicobar Islands. In: K. Shankar and H. Andrews (eds.), Monitoring and networking for sea turtle conservation in India. A UNEP CMS (unpublished) Project Report. Centre for Herpetology / Madras Crocodile Bank Trust, Tamil Nadu, India.

Balachandran, S., P. Sathiyaselvam, and P. Dhakshinamoorthy. 2009. Rescue of a leatherback turtle (*Dermochelys coriacea*) at Manakudi beach, Kanniyakumari District, Tamil Nadu and the need for an awareness campaign. Indian Ocean Turtle Newsletter 10: 19–20.

Bhaskar, S. 1979. Sea turtle survey in the Andamans and Nicobars. Hamadryad 4: 2–26.

Bhaskar, S. 1993. The status and ecology of sea turtles in the Andaman and Nicobar Islands. Unpublished Report, ST 1/93. Centre for Herpetology, Madras Crocodile Bank Trust, Tamil Nadu, India.

Bhaskar, S. 1994. Andaman and Nicobar sea turtle project. Phase-V. Unpublished Report. Centre for Herpetology / Madras Crocodile Bank Trust, Tamil Nadu, India.

Bhaskar, S., and M. Tiwari. 1992. Andaman and Nicobar Sea Turtle Project. Phase-I: Great Nicobar Island. Unpublished Report. Centre for Herpetology / Madras Crocodile Bank Trust, Tamil Nadu, India.

Botha M. 2007. The importance of the Greater St. Lucia Wetland Park in conserving the leatherback turtle *Dermochelys coriacea* during interesting periods. BSc (honors) thesis, Nelson Mandela Metropolitan University, Port Elizabeth, South Africa.

Botha M. 2010. Nest site fidelity and nest selection of loggerhead, *Caretta caretta* and leatherback *Dermochelys coriacea*, turtles in KwaZulu-Natal, South Africa. MSc thesis, Nelson Mandela Metropolitan University, Port Elizabeth, South Africa.

Bourjea. J., R. Nel, N. S. Jiddawi, M. S. Koonjul, and G. Bianchi. 2008. Sea turtle bycatch in the Western Indian Ocean: review, recommendations and research priorities. Western Indian Ocean Journal of Marine Science 7: 137–150.

Brazier, W., R. Nel, G. Cliff, and S. Dudley. 2012. Impact of protective shark nets on sea turtles in KwaZulu-Natal, South Africa, 1981–2008. African Journal of Marine Science 34: 249–257. doi: 10.2989/1814232x.2012.709967.

Cameron, T. H. 1923. Notes on turtles. Journal of the Bombay Natural History Society 29: 299–300.

Deraniyagala, P.E.P. 1936. The nesting habit of leathery turtle, *Dermochelys coriacea*. Ceylon Journal of Science B 19: 331–336.

Deraniyagala, P.E.P. 1939. The tetrapod reptiles of Ceylon, Volume I. Testudinates and crocodilians. Dulau, London, UK.

De Wet, A. 2012. Factors affecting survivorship of loggerhead (*Caretta caretta*) and leatherback (*Dermochelys coriacea*) sea turtles of South Africa. MSc thesis, Nelson Mandela Metropolitan University, Port Elizabeth, South Africa.

Dutton, P. H., B. W. Bowen, D. W. Owens, A. Barrigan, and S. K. Davis. 1999. Global phylogeography of the leatherback turtle (*Dermochelys coriacea*). Journal of Zoology 248: 397–409.

Ekanayake, E.M.L., T. Kapurasinghe, M. M. Samanand, and M.G.C. Premakumara. 2002. Estimation of the number of leatherbacks (*Dermochelys coriacea*) nesting at the Godavaya rookery in southern Sri Lanka during the nesting season in 2001. Kachhapa 6: 11–12.

Hamann, M., C. Limpus, G. Hughes, J. Mortimer, and N. J. Pilcher. 2006. Assessment of the conservation status of the leatherback turtle in the Indian Ocean and South East Asia. IOSEA Marine Turtle MoU Secretariat, Bangkok, Thailand.

Hughes, G. R. 1974. The sea turtles of South East Africa. PhD diss., University of Natal, Pietermaritzburg, South Africa.

Hughes, G. R. 1978. Marine turtles. In: A.E.F. Heydorn (ed.), Ecology of the Agulhas Current region. Proceedings of the Royal Society of South Africa, 43: 151–190.

Hughes, G. R. 1996. Nesting of the leatherback turtle (*Dermochelys coriacea)* in Tongaland, KwaZulu-Natal, South Africa, 1963–1995. Chelonian Conservation and Biology 2: 153–158.

Hughes, G. R., P. Luschi, R. Mencacci, and F. Papi. 1998. The 7000-km oceanic journey of a leatherback turtle tracked by satellite. Journal of Experimental Marine Biology and Ecology 229: 209–217.

Humber, F., B. J. Godley, V. Ramahery, and A. C. Broderick. 2011. Using community members to assess artisanal fisheries: the marine turtle fishery in Madagascar. Animal Conservation 14: 175–85.

Jones, S. 1959. A leathery turtle *Dermochelys coriacea* (Linnaeus) coming ashore for laying eggs during the day. Journal of the Bombay Natural History Society 56: 137–139.

Kar, C. S., and S. Bhaskar. 1982. Status of sea turtles in the eastern Indian Ocean. In: K. Bjorndal (ed.), Biology and conservation of sea turtles. Smithsonian Institution Press, Washington, DC, USA, pp. 365–372.

Krishna Pillai, S. 2003. Instance of meat of leatherback turtle *Dermochelys coriacea* used as food. Fishing Chimes 23: 46–47. (Reprinted in Kachhapa 9: 23 [2003]).

Krishna Pillai, S., K. K. Suresh, and P. Kannan. 2003. Leatherback turtle released into the sea at Vizhinjam in Kerala, India. Kachhapa 9: 5–6.

Lambardi, P., J.R.E. Lutjeharms, R. Mencacci, G. C. Hays, and P. Luschi. 2008. Influence of ocean currents on long-distance movement of leatherback sea turtles in the Southwest Indian Ocean. Marine Ecology Progress Series 353: 289–301.

Limpus, C. J. 2009. A biological review of Australian marine turtles, 6. Leatherback turtles, *Dermochelys coriacea* (Vandelli). In: L. Fien (ed.), A biological review of Australian marine turtles. Queensland Environmental Protection Agency, Brisbane, Queensland, Australia.

Luschi, P., J.R.E. Lutjeharm, R. Lambardi, R. Mencacci, G. R. Hughes, and G. C. Hays. 2006. A review of migratory behaviour of sea turtles off Southeastern Africa. South African Journal of Science 102: 51–58.

Luschi, P., A. Sale, R. Mencacci, G. R. Hughes, J. R. Lutjeharms, and F. Papi. 2003. Current transport of leatherback sea turtles (*Dermochelys coriacea*) in the ocean. Proceedings of the Royal Society B 270 (Suppl. 2): S129-S132. doi: 10.1098/rsbl.2003.0036.

Lutjeharms, J.R.E. 2001. Agulhas Current. In: J. H. Steele, K. K. Turekian, and S. A. Thorpe (eds.), Ocean currents: Encyclopaedia of ocean sciences. Academic Press, London, UK, pp. 104–113.

McAllister, H. J., A. J. Bass, and H. J. Van Schoor. 1965. Marine turtles on the coast of Tongaland, Natal. The Lammergeyer 3: 12–40.

Namboothri, N., A. Swaminathan, B. C. Choudhury, and K. Shanker. 2012. Post-nesting migratory routes of leatherback turtles from Little Andaman Island. Indian Ocean Turtle Newsletter 16: 421–123.

Namboothri, N., A. Swaminathan, and K. Shanker. 2012. A compilation of data from Satish Bhaskar's sea turtle surveys of the Andaman and Nicobar Islands. Indian Ocean Turtle Newsletter 16: 21–23.

Nel, R. 2010. Sea turtles of KwaZulu-Natal: data report for the 2009/10 season. Unpublished Report. Ezemvelo KZN Wildlife, Pietermaritzburg, South Africa.

Nel, R. 2012. Assessment of the conservation status of the leatherback turtle in the Indian Ocean and South-East Asia. IOSEA Marine Turtle MoU Secretariat, Bangkok, Thailand.

Nel, R., A. E. Punt, and G. R. Hughes. 2013. Coastal protection for nesting sea turtles: Are MPAs always successful to achieve population recovery? PLoS ONE 8(5): e63525. doi: 10.1371.

Petersen, S. L., M. B. Honig, P. G. Ryan, R. Nel, and L. G. Underhill. 2009. Turtle bycatch in the pelagic longline fishery off southern Africa. African Journal of Marine Science 31: 87–96.

Rakotonirina, B., and A. Cooke. 1994. Sea turtles of Madagascar—their status, exploitation and conservation. Oryx 28: 51–61.

Rashid, S.M.A., and M. Z. Islam. 2006. Status and conservation of marine turtles in Bangladesh. In: K. Shanker and B. C. Choudhury (eds.), Marine turtles of the Indian subcontinent. Universities Press, Hyderabad, India, pp. 200–216.

Rose, W. 1950. The reptiles and amphibians of southern Africa. Maskew Miller, Cape Town, South Africa.

Sale, A., P. Luschi, R. Mencacci, P. Lambardi, G. Hughes, G. Hays, S. Benvenuti, and F. Papi. 2006. Long-term monitoring of leatherback turtle diving behaviour during oceanic movements. Journal of Experimental Marine Biology and Ecology 328: 197–210.

Sivasundar, A., and K. V. Devi Prasad. 1996. Placement and predation of nests of leatherback sea turtles in the Andaman Islands, India. Hamadryad 21: 36–42.

Smith, A. 1849. Appendix to illustrations of the zoology of South Africa: Reptiles. Smith, Elder, London, UK.

Steering Committee BSTCI. 2008. Strategic planning for long-term financing of Pacific leatherback conservation and recovery. Proceedings of the Bellagio Sea Turtle Conservation Initiative, Terengganu, Malaysia. WorldFish Center, Penang, Malaysia.

Swaminathan, A., N. Namboothri, M. Chandi, and K. Shanker. 2012. Leatherback turtles at South Bay and West Bay, Little Andaman (2011–12). Unpublished Report to Forest Department, Andaman and Nicobar Islands. Madras Crocodile Bank Trust, Tamil Nadu, India.

Swaminathan, A., N. Namboothri, and K. Shanker. 2011. Post-tsunami status of leatherback turtles on Little Andaman Island. Indian Ocean Turtle Newsletter 14: 5–9.

Thorson, J. T., A. E. Punt, and R. Nel. 2012. Evaluating population recovery for sea turtles under nesting beach protection while accounting for nesting behaviours and changes in availability. Journal of Applied Ecology 49: 601–610.

Tiwari, M. 2012. Sea turtles in the southern Nicobar Islands: results of surveys from February–May 1991. Indian Ocean Turtle Newsletter 16: 14–18.

Videira, F.J.S., M.A.M. Pereira, and C.M.M. Louro. 2011. Monitoring, tagging and conservation of marine turtles in Mozambique: Annual Report (unpublished) 2010/11. AICM/GTT, Maputo, Mozambique.

Waayers, D. 2011. Overview of marine turtle research in Western Australia. IOSEA Newsletter, July 2011.

Wallace, B. P., A. D. DiMatteo, B. J. Hurley, E. M. Finkbeiner, A. B. Bolten, M. Y. Chaloupka, B. J. Hutchinson, et al. 2010. Regional management units for marine turtles: a novel framework for prioritizing conservation and research across multiple scales. PLoS ONE 5: e15465. doi: 10.1371/journal.pone.0015465.

Part IV FROM EGG TO ADULTHOOD

12

Leatherback Turtle Eggs and Nests, and Their Effects on Embryonic Development

PAUL R. SOTHERLAND,
BRYAN P. WALLACE, AND
JAMES R. SPOTILA

Before a leatherback turtle (*Dermochelys coriacea*) hatchling emerges from its nest, dashes down the beach, and begins its life at sea, it must develop successfully within an egg surrounded by 60 or 80 other eggs containing developing embryos. These are all buried beneath more than a half meter of beach sand, consuming oxygen obtained through the clutch and surrounding sand from the above-ground air. This amazing feat, accomplished while living on nutrients and water provisioned in its egg by a hatchling's mother, has piqued the curiosity of biologists for decades and generated fascinating questions that remain unanswered. Why do leatherback eggs contain so much albumin? Do shelled albumin gobs (SAGs, aka "false eggs" or "yolkless eggs") laid by female leatherbacks at the end of every oviposition bout play a role in development? Why is hatching success so low? And, how do interactions between leatherback embryos and their surrounding environment affect embryo development and hatching success?

Leatherback turtle eggs and their nests are a study in extremes. When compared with other sea turtle species (table 12.1), female leatherbacks lay the largest eggs (approximately 60–80 g, 50–60 mm in diameter) that produce the largest hatchlings (>40 g), deposit the most numerous (6–8 per season) and most massive (>3 kg) clutches of eggs, and construct the widest and deepest nests (>70 cm from the sand surface to the bottom of the nest) for their eggs (Miller 1997; Eckert et al. 2012). These characteristics can affect developmental trajectories and survivorship of embryos. A better understanding of the biological and physical factors that affect leatherback embryos can improve management of their eggs and nests, as well as nesting beaches, to enhance hatchling production and contribute to population recovery.

Eggs, SAGs, and Maternal Investment

Composition of Leatherback Eggs

Leatherback eggs are simple and self-sufficient packages of maternal investment in offspring survival. Egg yolk, at the center of a freshly laid egg and enclosed initially by a vitelline membrane, provides energy and

Table 12.1. Egg characteristics of sea turtles. Numbers in parentheses are the number of studies upon which data are based.

Species	Clutch size (n)	Egg mass (g)	Egg diameter (mm)	Hatchling mass (g)
Dermochelys coriacea	82 (38)	79.3 (4)	53.4 (9)	44.4 (5)
Chelonia mydas	113 (24)	46.1 (10)	44.9 (17)	24.6 (11)
Natator depressus	53 (6)	76.0 (3)	51.5 (6)	42.0 (3)
Lepidochelys kempii	110 (1)	30.0 (1)	38.9 (1)	17.3 (1)
Lepidochelys olivacea	110 (11)	35.7 (1)	39.3 (6)	17.0 (1)
Eretmochelys imbricata	130 (17)	26.6 (5)	37.8 (1)	14.8 (5)
Caretta caretta	112 (19)	32.7 (7)	40.9 (14)	19.9 (7)

Source: Based on tables in Miller (1997) and Eckert et al. (2012).

building materials that support development of the embryo (Vleck and Hoyt 1991). Together, the yolk and vitelline membrane comprise the ovum (i.e., "egg") that is ovulated from the mother's ovary, located dorsally and posterior to the lungs in her body cavity (Spotila 2004, pp. 36–37). Albumin, layered onto the yolk as it passes through the oviduct, contributes water and proteins with antimicrobial, water storage, and nutritive properties to the developing embryo (Palmer and Guillette 1991). A pliable eggshell surrounds the albumin and provides a protective, yet gas-permeable, barrier between the developing embryo and the external environment. Shell membranes are about 250 μm thick and the flexible shell is about 125 to 150 μm thick (Simkiss 1962; Sahoo et al. 2010).

Sea turtle eggshells are composed of two layers, an outer inorganic layer and an inner organic membrane composed of multiple layers of fibers. The fibrous membrane forms a barrier and increases diffusion distance, slowing down water movement. It also helps to maintain egg shape, support egg contents, and protect the embryo from mechanical injury. Fibers form a dense crisscrossing mat and act as a barrier to bacteria and fungi (Palmer and Guillette 1991). The outer layer is a calcareous matrix that is primarily formed of aragonite crystals with some calcite present, especially in leatherback eggshells (Solomon and Watt 1985; Packard and Demarco 1991; Sahoo et al. 2010). Aragonite crystals with a knob-like or conical shape sit on top of the organic membrane, and gas molecules diffuse between the crystals and not through pores as they do in bird eggs.

Respiratory gases (oxygen and carbon dioxide) and water vapor diffuse quite easily through the shell as its porosity (47.9 ± 6.0 mg H_2O day^{-1} kPa^{-1} cm^{-2}) is nearly 30 times that of the eggshell of a similarly sized bird egg (Ar et al. 1974; Rahn and Paganelli 1990; Wallace, Sotherland, et al. 2006). Porosity of the leatherback eggshell is similar to that of the eggshells of the loggerhead turtle, *Caretta caretta* (49.9 mg H_2O day^{-1} kPa^{-1} cm^{-2}), painted turtle, *Chrysemys picta* (66.1 mg H_2O day^{-1} kPa^{-1} cm^{-2}), and snapping turtle, *Chelydra serpentina* (54.7 and 86.4 mg H_2O day^{-1} kPa^{-1} cm^{-2}) (Deeming and Thompson 1991). Inside the shell, the embryo develops a vascularized yolk sac that envelopes the yolk and delivers nutrients from the yolk to the growing embryo. Other extra-embryonic membranes that surround the embryo (Spotila 2004, p. 15) support development into a hatchling.

Leatherback eggs contain far more albumin than yolk (Wallace, Sotherland, et al. 2006) and, as a result, are more like bird eggs in composition (Sotherland and Rahn 1987) than those of other nonavian reptiles (Finkler and Claussen 1997), including other sea turtle species (Hewavisenthi and Parmenter 2002). Albumin comprises roughly two-thirds of leatherback egg mass, causing these eggs to have a fraction of yolk in the contents (FYC = 0.33) unlike that of most nonavian reptile eggs, which exhibit a higher FYC than bird eggs—about 49% of egg mass in eggs of other sea turtles and about 90% in eggs of other nonavian reptiles (table 12.2). As such, leatherback eggs could be classified as "endohydric" eggs (e.g., bird and crocodilian eggs), into which females allocate sufficient water to support embryogenesis, rather than "ectohydric" eggs (e.g., those of most nonarchosaur reptiles), in which embryos typically depend, in part, upon water from the environment outside the eggshell (Tracy and Snell 1985). However, the thin and somewhat pliable shells of leatherback eggs, like eggshells of many nonarchosaur reptiles (Packard and DeMarco 1991), allow them to absorb water from the nest during incubation, making them "ectohydric" during development (Ackerman 1991; Packard 1991). When incubated on moist sand in the laboratory they increase in mass (Wallace, Sotherland, et al. 2006). Therefore, leatherback eggs have characteristics of both types of eggs.

Variation in leatherback egg size correlates positively with variation in albumin quantity. As total egg mass

Table 12.2. Comparison of egg components, fraction of mean hatchling mass to mean egg wet mass, and proportional contribution of each egg component and water content to whole egg wet mass across turtle taxa. From Wallace, Sotherland, et al. (2006).

Species	Egg wet mass (g)	Hatchling wet mass (g)	Hatchling mass (g) / Egg Mass (g)	Yolk mass (%)	Albumin mass (%)	Shell mass (%)	Water (%)
Chelydra serpentina	9.5	7.8	0.82	–	–	–	72.6
Chelydra serpentina	11.6	9.2	0.79	88.1*	–	11.0	70.7
Lepidochelys kempii	30.0	17.3	0.58	–	–	–	–
Lepidochelys olivacea	35.7	17.0	0.48	–	–	–	–
Eretmochelys imbricata	26.2	14.8	0.56	–	–	–	59.4
Caretta caretta	32.7	20.0	0.61	49.2	45.5	4.8	–
Chelonia mydas	46.1	24.6	0.53	–	–	–	66.7
Natator depressus	71.7	39.4	0.55	49.6	45.2	5.2	78.8
D. coriacea Malaysia	76.0	44.4	0.58	46.8	48.9	4.3	67.5
D. coriacea Playa Grande	80.9	40.1	0.50	33.0	63.0	4.0	64.0

* Combination of yolk and albumin.

varies over 30 g, albumin mass varies 20 g, while yolk mass varies 8 g or less. Therefore most of the variation in total egg mass is due to differences in the mass of albumin. These characteristics agree with observations from ultrasonography, which show little variation in ova size within and among females (Rostal et al. 1996; chapter 5). Oddly, the correlation between egg mass and albumin mass is, again, similar to the pattern in bird eggs, in which albumin contributes the most to variation in egg size of many species (Sotherland et al. 1990; Hill 1995). Though leatherback hatchling size increases with egg size (Wallace et al. 2007), it correlates most highly with yolk size (Wallace, Sotherland, et al. 2006), as in other nonavian oviparous animals (Roff 1992). However, hatchling mass is 50–100% greater than yolk mass (Wallace, Sotherland, et al. 2006), indicating that albumin water, possibly water absorbed from the nest, and albumin solids contribute to growth and development of leatherback embryos. Do leatherback embryos ingest albumin (which migrates into the amnion) like bird embryos (Deeming 1991; Nelson et al. 2010), or do they rely on other, yet to be discovered, means of incorporating water and solids from the albumin into their bodies during development?

SAGs

Unique to leatherbacks is the consistent presence of shelled albumin gobs (SAGs; Sotherland et al. 2003; Wallace et al. 2004) in every clutch, although other species sometimes produce small numbers of SAGs in some clutches. The SAGs are masses of albumin enclosed in shell that resemble little eggs. In the past, authors have referred to them as "yolkless eggs" or "false eggs," although SAGs do not contain an ovum and, therefore, are not eggs. The number (~10–100), shape and size (~10–50 mm diameter, ranging from spheres to teardrops), and mass (~1 g to nearly the mass of an egg) of SAGs vary widely within and among clutches (table 12.3). They are laid toward the end of oviposition (chapter 5) and deposited primarily on top of the eggs, although some are laid earlier when one oviduct finishes expelling eggs before the other. The SAGs have a total mass of nearly 1 kg and comprise about 20% of material deposited in every clutch (Wallace et al. 2007). Researchers have hypothesized possible functions of SAGs, such as (1) predator deterrence or satiation (Caut et al. 2006), (2) gas exchange enhancement and/or the buffering of temperature fluctuations within the nest environment (Wallace et al. 2004), or (3) facilitation of synchronized hatchling emergence by allowing neonates to group together at the top of the nest chamber in a space remaining after SAGs "deflate" through water loss (Patiño-Martínez et al. 2010). Area-specific water vapor conductance of eggshells and SAG shells are not statistically different, indicating that shell material and functional structure are similar in both (Wallace, Sotherland, et al. 2006). Total SAG mass varies negatively with number of eggs and positively with egg mass. These observations suggest that SAGs are "production over-runs" of oviducts producing copious albumin for eggs (Wallace et al. 2007). This hypothesis stipulates that when production of albumin exceeds that needed for the number of yolks (ova) in a given clutch (i.e., smaller clutches), excess albumin is covered with shell material and deposited to a greater degree into SAGs, whereas in larger clutches there is less albumin left over and total SAG mass is lower. Regardless of whether

Table 12.3. Regional variation in clutch size of eggs and SAGs of leatherback turtles. Number in parentheses is the number of studies upon which averages are based.

Region	Clutch size (*n*)	SAGs (*n*)
Western Atlantic	82 (18)	30 (17)
Eastern Atlantic	75 (1)	26 (1)
Eastern Pacific	64 (17)	39 (2)
Western Pacific	86 (11)	52 (3)
Indian Ocean	102 (2)	–

Source: Averages in table are averages of data presented in table 9 in Eckert et al. (2012).

SAGs serve a particular function for development, their contents and shells represent important investments of fresh water, protein, and calcium (in the shell). It remains for someone to quantify the investment that leatherbacks place into SAGs. In addition, investigators should report the total mass of SAGs rather than number, because SAGs vary widely in size, whereas mass provides information on maternal investment in reproduction (Wallace et al. 2007).

Maternal Investment in Eggs and Clutches

Leatherbacks lay large and numerous eggs in each of several clutches per season every two to four years, depending upon productivity of the different oceans (Eckert et al. 2012; chapter 15). On the Caribbean coast of Costa Rica they lay about 85 eggs in a clutch (Leslie et al. 1996), and in the eastern Pacific they lay about 64 (Saba et al. 2007; Santidrián Tomillo et al. 2009). Leatherbacks in the Indian Ocean lay the greatest number of eggs per clutch, at more than 100 (table 12.3). Egg size varies less than clutch size and frequency (number of clutches), supporting the hypothesis that leatherbacks increase reproductive output by increasing number, rather than size, of eggs (Wallace et al. 2007). Because hatching success and emergence success (i.e., the proportion of hatchlings that successfully emerge from the nest) are not related to clutch size (Wallace et al. 2004; Wallace et al. 2007; Santidrián Tomillo et al. 2009), leatherbacks increase seasonal fecundity and hatchling production primarily through increased clutch frequency and secondarily through increased clutch size (because number of hatchlings produced is affected by clutch size).

Leatherbacks exhibit patterns of maternal investment in eggs and clutches that reflect the evolutionary history of the species as well as environmental conditions in nests. Gaining fuller knowledge of these biological factors is essential for understanding reproduction in proximate and ultimate ecological and evolutionary contexts. For more details about patterns of reproductive output see chapters 6 and 7.

Embryonic Development

Embryonic development is similar in all sea turtles and begins immediately after fertilization (Miller 1997). The fertilized ovum moves into the oviduct where it is covered in albumin, and the inner shell membrane begins to form as secretions from cells in the shell-forming segment of the oviduct. Aragonite crystals then form to complete the outer part of the shell. The shell takes at least seven days to form. Cleavage begins within hours of fertilization but development stops at the midgastrulation stage while the egg is in the oviduct. Once the egg is laid development resumes within about four to eight hours. If eggs are moved after that time, particularly if they are rotated along their horizontal axis, the developing embryos can die if this movement causes them to detach from membranes in the upper part of the egg. Development may begin sooner in leatherback eggs than in other sea turtles. Movement of eggs even immediately after they are laid may reduce hatching success because success in a hatchery is often lower than in natural nests (Sieg et al. 2011).

The first sign that development is proceeding is formation of a white spot on the upper egg surface. This is where the embryo attaches to the shell. In turtle eggs, this attachment is essential to successful development, unlike in bird eggs in which the embryo does not attach and the adult turns the egg periodically during development (Deeming 1991). As the white spot grows to cover the entire shell, embryonic membranes (amnion, allantois, and chorion) form (see image in Spotila 2004, p. 15). It was once thought eggs that appeared undeveloped at the end of incubation (stage 0 eggs; Leslie et al. 1996) were unfertilized. However, nearly all of those eggs were fertilized but did not develop beyond the gastrula stage (Bell et al. 2003).

Miller (1985) described embryological development and provided sources for detailed anatomical descriptions of embryonic and hatchling sea turtles. Development is similar in all sea turtles until formation of the carapace begins, at which time differences between hard-shelled sea turtles and leatherbacks becomes increasingly apparent. Deraniyagala (1939) provided a detailed account of embryonic development in leatherback embryos and defined 12 postovipositional stages. His account included plates picturing each stage and descriptions of skull ossification. Pehrson (1945) provided a detailed description of the embryonic skull, and Renous et al. (1989) identified 22 embryonic stages between formation of the first somites and hatching. Development takes about 55 to 60 days and depends upon incubation temperature (Eckert et al. 2012).

Hatching Success

The low (around 50%) and highly variable (within and among individual females) hatching success of leatherback clutches has been widely documented, but identifying the specific cause(s) of this "success" has proven elusive (chapter 7; Eckert et al. 2012). Hatching success is lower than in other sea turtles (Miller 1997). Since improvements in hatching success could contribute to recovery of leatherback populations, solving the mystery of low hatching success should be a very high priority.

Intrinsic and Extrinsic Influences on Hatching Success

Hatching success is influenced by intrinsic factors (e.g., quantity or quality of maternal investment, fertility, or genetics) and extrinsic factors (e.g., nest environment conditions). Early stage embryonic death is primarily due to intrinsic factors because development during the first half of incubation is not dependent on respiratory gases, normal temperatures, or other external variables (Ackerman 1997; Wallace et al. 2004), although high tides and waves that inundate nests can kill embryos at any developmental stage. In contrast, late stage embryonic death is probably due primarily to extrinsic factors, such as limitations in or lethal levels of respiratory gas concentrations, high temperatures, or moisture within nests, because accelerated somatic growth of embryos during the second half of incubation requires adequate levels of O_2, CO_2, water, and heat (Ackerman 1997).

Intrinsic Factors: Fertility and Maternal Identity

In early studies, frequent field observations of eggs with no apparent development (i.e., no visible embryo) led researchers to infer that these eggs were not fertilized and that low hatching success of leatherback eggs was due to infertility of either male or female adults (Chan et al. 1985; Whitmore and Dutton 1985). If female-skewed sex ratios in hatchlings (Sieg et al. 2011) persist to adult stages, then the relatively few adult males in a population could negatively affect mating and fertilization of eggs. Although leatherbacks, like other sea turtle species, exhibit both polygyny and polyandry, single paternity occurs more frequently than multiple paternity (e.g., St. Croix, US Virgin Islands [Caribbean Sea], Stewart and Dutton 2011; Playa Grande, Costa Rica [Pacific Ocean], Crim et al. 2002). In addition, males make migrations to breeding areas more frequently than females, perhaps annually (James et al. 2005), which could balance sex ratios of breeding adults at breeding areas in a given season (Stewart and Dutton 2011). Finally, careful microscopic examination of leatherback embryos throughout development confirmed that low hatching success is not due to infertility of eggs (Bell et al. 2003). Thus, mating opportunities and related fertility issues do not appear to be the cause of low hatching success.

Emergence success, which is frequently lower than hatching success because it does not include hatchlings that die in the nest, is not related to either clutch size or frequency, and reproductive traits vary widely within and among leatherback females (Wallace et al. 2007). Although more than 90% of eggs are fertilized, the fraction of eggs containing embryos declines in later clutches laid by the same female during the nesting season. Since mean fertility differs significantly among females there may be inter-individual variation in female "quality" (Bell et al. 2003). Clutches produced by "remigrant" females (i.e., turtles that had nested in previous seasons) exhibited consistently higher hatching success than clutches produced by "new" females—that is, turtles presumed to be neophytes (Rafferty et al. 2011). This illustrates the critical role that maternal quality plays not just in maternal investment patterns (e.g., provision of egg components, number and size of eggs per clutch, number of clutches per season), but also in successful embryonic development. Specific causes of early embryonic death in "low-quality" females remain to be determined. Insights gained through long-term monitoring of inter- and intra-individual reproduction patterns can shed light on ways in which maternal identity might affect hatching success.

Nest Environment

Successful development of turtle embryos occurs within optimal ranges of O_2, CO_2, temperature, and water in the nest environment (Ackerman 1997). Exposure to mild chronic hypoxia during incubation results in decreased metabolism and modified cardiovascular physiology in embryos of freshwater turtles (Kam 1993; Kam and Lillywhite 1994), alligators, *Alligator mississipiensis* (Warburton et al. 1995), and crocodiles, *Crocodylus porosus* (Booth 2000), as well as in hatchling chickens (Dzialowski et al. 2002). Limited O_2 in the oviduct arrests embryonic development of turtle embryos (Rafferty et al. 2013). Thus, the biotic (e.g., large number of large embryos in a nest consuming O_2 and generating heat and carbon dioxide) and abiotic (e.g., deep nests separated by sand from the atmospheric source of O_2) characteristics of leatherback nests create challenging developmental environments for embryos. By measuring, describing, and correlating these characteristics, their effects on the nest environment, and ultimately their effects on emergence success, we can improve our understanding of factors that affect development of sea turtle embryos, while informing decisions about nesting beach management directed toward enhanced hatchling production.

Oxygen that supports embryonic growth is ultimately obtained from the atmosphere and must move through surrounding beach sand to reach eggs within a nest. Thus, a clutch of sea turtle embryos and the sand in which they are buried can affect respiratory gas concentrations in the nest, much like a porous eggshell affects the gas concentrations observed in developing bird embryos (Ackerman 1977). Levels of O_2 in a nest are dependent on sand characteristics that affect gas conductance (e.g., sand grain size and moisture content) and on the collective embryonic O_2 consumption (Ackerman 1977, 1980). During the second half of development, when most somatic growth occurs and O_2 demands are highest, the rate at which a clutch of growing embryos consumes O_2 outpaces the rate at which it can be replaced by gas diffusion. This causes levels of O_2 in the nest to decline, creating a progressively greater O_2 gradient from the surrounding sand toward the interior of the nest (i.e., center of the egg clutch) as development continues (Ackerman 1977; Wallace et al. 2004).

Oxygen levels in leatherback nests should be lower than in other sea turtle nests because (1) peak rates of O_2 consumption by individual leatherback embryos (Thompson 1993) are two to three times those of green turtle (*Chelonia mydas*) and loggerhead (*Caretta caretta*) embryos (Ackerman 1981), (2) surface area to volume ratios of leatherback clutches are lower, and (3) leatherback embryos develop farther from the atmospheric source of O_2. However, at Playa Grande, on the Pacific coast of Costa Rica, O_2 levels in nests were actually higher than predicted (Wallace et al. 2004).

Carbon dioxide levels also affect development. High CO_2 concentrations (hypercapnia) have strong physiological effects in mammals, and changes in atmospheric CO_2 concentration lead to significant physiological homeostatic responses. Hypercapnia causes an increase in the rate of ventilation in leatherback and olive ridley turtle, *Lepidochelys olivacea*, hatchlings (Price et al. 2007); this also occurs in painted turtle hatchlings (Silver and Jackson 1985) at CO_2 levels similar to those seen in sea turtle nests (Ackerman 1977; Clusella Trullas and Paladino 2007). Moderate hypercapnia (6.6% CO_2) for several days was not lethal for the broad-shelled river turtle, *Chelodina expansa* (Booth 1998). However, there are no data on the effect of hypercapnia on sea turtle embryonic development.

Temperature also affects embryonic development. The temperature at which a clutch develops determines hatchling sex and incubation duration (Standora and Spotila 1985; Mrosovsky 1994; chapter 8), and hatching success decreases significantly as nest temperatures increase above 30°C (Santidrián Tomillo et al. 2009). Sand moisture content can affect temperature in nests and indirectly affect hatching success. A leatherback clutch's hatching success is higher with increased precipitation (as long as eggs are not submerged) and this is coupled with lower nest temperatures (Houghton et al. 2007; Santidrián Tomillo et al. 2009). This relationship between sand moisture and temperature is correlated with interannual patterns of hatching success and inferred sex ratios of hatchlings, with increased production of male hatchlings associated with increased precipitation and decreased incubation temperatures (Sieg et al. 2011).

Empirical Studies of Leatherback Nest Conditions and Embryonic Development

Most research to date on effects of nest environmental conditions on leatherback development and hatching success has been conducted at Playa Grande, the primary nesting beach in Parque Nacional Marino Las Baulas (PNMB), on the northern Pacific coast of Costa Rica. Wallace et al. (2004) described nest conditions at PNMB and how they might influence embryonic development of leatherbacks. As for sea turtles generally (Ackerman 1997), oxygen partial pressures (PO_2) decrease and temperatures increase inside leatherback nests during the second half of incubation, reaching minimum and maximum levels, respectively, just before hatchling emergence. Although minimum PO_2 decreases and maximum temperatures increases with larger clutch sizes, these nest parameters are more strongly related to number of metabolically active embryos (i.e. embryos alive and contributing to O_2 consumption and heat generation) calculated as the sum of all hatchlings and late stage embryos (Wallace et al. 2004). However, there was no statistically significant relationship between hatching success of clutches and any nest environment parameter.

Might there be localized conditions within nests that cause intraclutch variation in hatching success? Just as respiratory gas gradients are created across bird eggshells during development as embryonic O_2 consumption and CO_2 production outpace diffusion into and out of the egg, gas concentration gradients occur from the center of egg clutches, extending into the surrounding sand (Ackerman 1977). Measurements of PO_2 and temperatures in different locations within leatherback clutches—top, bottom, sides, and center—followed this pattern; PO_2 increases and temperature decreases from the center to periphery of clutches (Wallace et al. 2004). Spatial gradients of respiratory gas partial pressures and temperatures occurred within nests and hatching success was significantly lower for embryos that developed in eggs at the centers of clutches (Ralph et al. 2005). However, there are no significant relationships between spatial patterns of hatching success and respiratory gases or temperatures, and late stage embryonic death is not significantly related to low PO_2, high PCO_2, or high temperature. Thus, although there appears to be a relationship between where an embryo develops relative to other embryos in a clutch and the embryo's hatching success, the mechanism explaining this relationship remains unclear.

Effects of Tidal Pumping on Leatherback Nest Environments

Though minimum PO_2 in leatherback nests decreases as the number of metabolizing embryos and concomitant demand for O_2 in the nests increases, as predicted from Fick's Law, PO_2 minima reported from studies at PNMB are much higher than expected. Streams of water flowing continuously from the sand along the beach at low tide suggested that twice-daily vertical excursions of the subterranean water table driven by large tidal fluctuations were effectively ventilating leatherback nests by pulling fresh air from above the sand down into nests as the tide and water table fell (Wallace et al. 2004). This idea was proposed in earlier studies (Prange and Ackerman 1974), but was rejected because differences between high and low tides, and related vertical movements of the water table, were inadequate to ventilate sea turtle nests on nesting beaches monitored in the Caribbean and elsewhere (Prange and Ackerman 1974; Maloney et al. 1990; Leslie et al. 1996). In contrast, daily tidal fluctuations at Playa Grande can be in excess of 4 m (Reina et al. 2002), so ventilation of leatherback nests by tidal pumping is plausible. Testing the tidal pumping hypothesis required measuring vertical excursions of the water table below the sand surface and demonstrating that fluctuations in nest PO_2 were correlated with these vertical excursions.

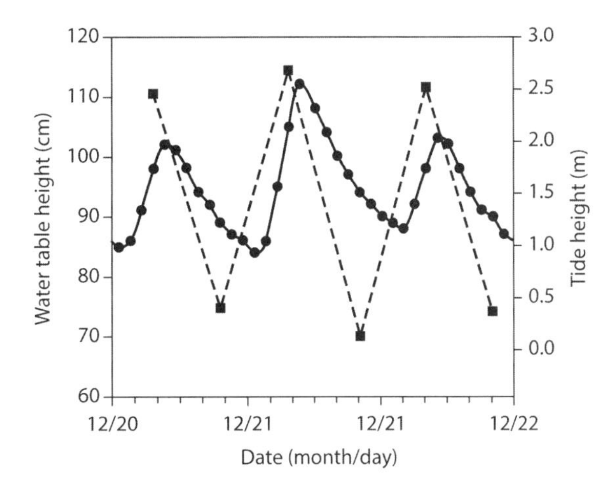

Fig. 12.1. Vertical excursions of the water table (circles) and tidal cycles (squares) at Parque Nacional Marino Las Baulas (PNMB), Costa Rica, recorded over three days in December 2002 using the manometer system shown in figure 12.2. Similar measurements were made, hourly over a period of four weeks, in 2003; these were located close to the PNMB beach hatchery where the water table was about 4 m below the sand surface.

The water table at PNMB was about 1.5 m below the sand surface, at a location about 0.5 m beyond the high tide line, when the tide was at its lowest daily level. It underwent twice-daily vertical excursions of approximately 20 cm that corresponded to and lagged slightly behind twice-daily tidal excursions (fig. 12.1). A water table device (fig. 12.2) measured a rapid rise in water table as the tide rose, followed by a more gradual fall as the tide fell.

Twice-daily vertical excursions of the water table recorded over a time span of four weeks (i.e., approximately 56 tidal periods) in 2003 followed patterns similar to those found in 2002 and caused minimal variation in nest pressure. Differences in pressure between the above-ground air and the gas in the "nest" during 48 falling tides (-0.3 ± 1.4 mm H_2O) and during 40 rising tides (-0.1 ± 1.6 mm H_2O) were not significantly different from zero. However, pressure measurements made during a falling tide after a heavy rain (~4 cm in one hour) on 15 December 2003 revealed nest pressures 14 mm H_2O below that of the above-ground air. The rain-soaked sand apparently formed a barrier to air movement, and (according to Boyle's Law) the expanding volume of gas in the "closed" space between the wet sand at the surface and the water table "piston" caused pressures in the "nest" to drop precipitously, corroborating

Fig. 12.2. Device constructed to measure vertical excursions of the water table during tidal cycles at PNMB, Costa Rica. A. Sections of PVC pipe in a pit dug in the beach with a shovel handle as size reference. B. Top of 2002 water table device, showing vertical graduations on rigid plastic "dipstick" and location of high tide line in the background. A toilet float ball, with rigid plastic tubing attached and inserted into the PVC pipe, floated on the surface of the water table and changed height with the tide. C. Water manometer attached to 2003 water table device located close to the PNMB beach hatchery where the water table was about 4 m below the sand surface. This device was connected via Tygon™ tubing to a Wiffle™ ball gas-sampling port (described in Wallace et al. 2004) buried in a "clutch" of perforated plastic golf balls in a "nest" 75 cm below the sand surface. The manometer was used to monitor the difference in pressure between the above-ground air and the air in the "nest."

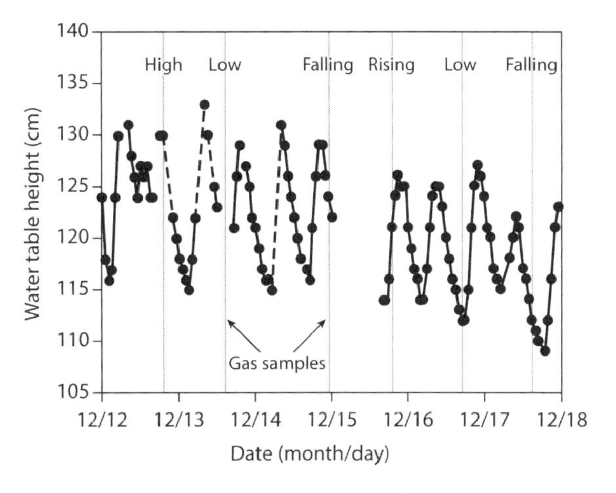

Fig. 12.3. Water table heights, timing of gas sampling from "bionic egg clutches," and status of the water table during tidal cycles over six days in December 2003 at PNMB, Costa Rica.

Table 12.4. Water table status and ΔPO_2, oxygen partial pressure (kPa) difference, between above-ground air, gas in a control clutch of plastic golf balls, and gas in a "bionic" clutch of zinc-air "eggs" during the five-day period shown in figure 12.3. Status of the water table corresponds to those periods shown in figure 12.3 when nest gases were sampled. The bionic clutch contained 40 zinc-air "eggs" at the beginning of the experiment, and 39 functional "eggs" by the end. Both control and bionic clutches were buried in natural beach conditions, exposing the clutches to ventilation induced by vertical excursions of the water table that were driven by the tide.

Water table status	Control nest ΔPO_2 (kPa)	Bionic clutch ΔPO_2 (kPa)
High	0.13	0.99
Low	0.20	0.57
Falling*	0.23*	0.61*
Rising	0.22	0.55
Low	0.18	0.50
Falling	0.08	0.12

* Heavy rain (~4 cm in one hour) drenched the beach sand six hours before recording these data on 15 December 2003 (see fig. 12.3).

similar observations made in previous studies (Prange and Ackerman 1974).

Gas flowing vertically through the sand, pulled from the above-ground air by the falling water table and pushed from below as the water table rose, caused fluctuations in the PO_2 of a nest chamber filled with "bionic eggs" consuming O_2 at a constant rate (table 12.4). The "bionic eggs" were used to simulate leatherback clutch metabolism. Each "egg," based on the design for a Columbus Instruments Micro-Oxymax O_2 standard test cell, was made from a zinc-air battery (~9 V) connected to a constant-current (~5.8 mA) circuit that consumed oxygen at a constant rate. A "clutch" (initially 40) of these "eggs," having a metabolic rate similar to that of a very small (six egg) clutch of leatherback eggs, was buried in a "nest" 75 cm below the sand surface, along with a thermocouple and gas sampling port (used to measure PO_2 as described in Wallace et al. 2004).

The ΔPO_2, oxygen partial pressure (kPa) difference, between the above-ground air and nest was consistently greater for the bionic clutch than for a control nest of plastic golf balls (table 12.4) during a tidal cycle (fig. 12.3), reflecting depletion of O_2 in the nest cavity caused by "egg" oxygen consumption. Note, however, the difference between the two "falling" tide measurements of ΔPO_2. During a normal falling tide, when the above-nest sand is relatively dry and above-ground air can be pulled down through the sand by the retreating water table, the ΔPO_2 of the bionic clutch is essentially the same as that of the control clutch (shown in the last row of table 12.4). In contrast, after the heavy rain, the ΔPO_2 in the bionic clutch is considerably greater than that in the control clutch, presumably because above-ground air is prevented from being pulled into the nest.

Preventing a nest from being ventilated by tidal pumping caused ΔPO_2 to rise and blocked effects of the falling tide (table 12.5). A second clutch of bionic eggs was isolated from tide-induced ventilation by burying it in sand within a large (~210 liter) plastic barrel buried in the beach near the control clutch of golf balls. The barrel blocked the effects of the tide but allowed diffusion of gases to and from the sand around the nest and to and from the surface. At the beginning of the experiment, when all 40 bionic eggs were functioning, the ΔPO_2 between the above-ground air and the isolated bionic clutch (table 12.5) was greater than that measured with the bionic clutch exposed to tidal pumping (table 12.4) and considerably greater than the ΔPO_2 for the control clutch (table 12.5). Though 40% of the batteries stopped functioning by the end of the experiment, the ΔPO_2 for the isolated bionic clutch remained greater than for the control clutch and for the bionic clutch exposed to tidal pumping (table 12.4), especially during a falling tide. The ΔPO_2 for the isolated bionic clutch did not vary during the second cycle of rising and falling tides at the end of the experiment (table 12.5), suggesting that O_2 in the nest remained depleted by clutch metabolism because diffusion of gas from above-ground air was not sufficient to replenish the O_2 in the nest.

Interactions between characteristics of a nesting beach and those of a clutch of developing leatherback

Table 12.5. Water table status and ΔPO_2, oxygen partial pressure (kPa) difference, between above-ground air and gas in a control clutch of plastic golf balls and gas in a "bionic" clutch of zinc-air "eggs" during two tidal cycles. The bionic clutch contained 40 zinc-air "eggs" at the beginning of the experiment and 24 functional "eggs" by the end. While the control nest was buried 75 cm in natural beach conditions, the bionic clutch was buried 75 cm in beach sand within a large (210 liter) plastic barrel, blocking the clutch from potential effects of ventilation by tidal pumping.

Water table status	Control nest ΔPO_2 (kPa)	Bionic clutch ΔPO_2 (kPa)
Rising	0.09	2.91
Falling	0.11	1.91
Rising	0.11	1.07
Falling	0.06	1.08

embryos can affect environmental conditions, respiratory gases in particular, in nests containing large numbers of metabolizing eggs. During the last weeks of incubation, when metabolic demand for O_2 peaks, a clutch of eggs can affect its environment by depleting O_2 and reducing PO_2 in the nest cavity. This leads to the following tidal-pumping hypothesis: If daily tidal fluctuations are large enough to cause tide-driven vertical excursions of the water table to ventilate a nest of metabolizing eggs, as at PNMB, then PO_2 in the nest cavity will be greater than expected, as described by Wallace et al. (2004), and it will likely fluctuate throughout the day (as illustrated above). A corollary to this hypothesis is as follows: If nests are isolated from tide-induced flow of air (as described above), or if nests are located in a beach where differences between high and low tides are minimal (as in the Caribbean), then the nests will not be ventilated and the clutches of eggs will have a more profound effect on their own environments (Prange and Ackerman 1974). Therefore, regional differences in tides could produce noticeably different environments in leatherback nests that could, in turn, affect hatching success of eggs in those nests.

Testing the Tidal-Pumping Hypothesis: Leatherback Nests at St. Croix, US Virgin Islands

Relatively minor (< 1 m) vertical excursions of the tide in the Caribbean Sea (Kjerfve 1981) create a natural experiment to test the tidal-pumping hypothesis. Garrett et al. (2010) measured conditions in leatherback nests and recorded developmental consequences associated with those conditions at Sandy Point National Wildlife Refuge (SPNWR; St. Croix, US Virgin Islands) that, in effect, tested (and supported) the tidal-pumping hypothesis. Environmental conditions in SPNWR nests, and responses of embryos to those conditions, were notably different from those at PNMB. Although hatching success at SPNWR is similar to that reported at PNMB (Eckert and Eckert 1990), the majority of embryos in SPNWR clutches die during later stages of development and not during early stages typical of PNMB clutches (Bell et al. 2003; Rafferty et al. 2011). The timing of embryo death at SPNWR suggests that instead of intrinsic factors influencing hatching success, as at PNMB, extrinsic factors (i.e., nest environment conditions influenced to a large extent by the embryos in the absence of nest ventilation) influence hatching success of leatherback eggs.

There were significant curvilinear relationships between three nest parameters—PO_2, PCO_2, and temperature—and hatching success at SPNMR, whereas no such relationships existed in PNMB clutches. Consistent with measurements made in PNMB nests, minimum PO_2 decreased, maximum PCO_2 increased, and temperature increased as the number of metabolizing embryos increased. The SPNMR embryos collectively affected their nest environments until the gaseous and thermal conditions within the nest chamber reached possibly lethal levels at excessively low PO_2 and/or high PCO_2 and high temperatures (Garrett et al. 2010). Thus, although leatherback embryos at both PNMB and SPNMR collectively affected nest environment conditions, those altered nest conditions significantly affected development and hatching success at SPNMR, but not at PNMB. This disparity indicates that tide-induced nest ventilation at PNMB ameliorates effects of clutch metabolism on the nest environment.

Values for minimum PO_2 in nests at SPNWR aligned more closely with expected values than PO_2 minima observed in PNMB nests (fig. 12.4), indicating that tides of sufficient magnitude can cause ventilation of nests. We plotted observed minimum nest PO_2 as a function of the number of metabolizing embryos in each nest at SPNMR and PNMB and then superimposed expected nest PO_2 values, calculated using Fick's Law as described in Wallace et al. (2004). Results indicate stark differences in developmental environments between the two sites. In contrast to the significant divergence between observed and expected declines in minimum nest PO_2 with increasing numbers of metabolizing embryos at PNMB, observed nest PO_2 values generally agreed with expected PO_2 values as metabolic clutch sizes increased in SPNWR nests (fig. 12.4). The greater than expected PO_2 observed in nests at PNMB is likely

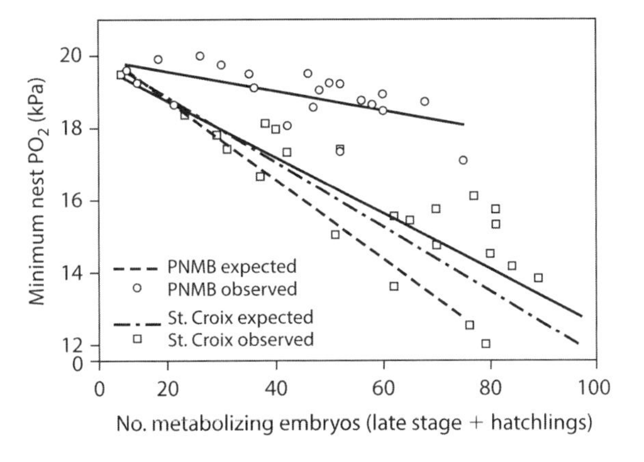

Fig. 12.4. Observed and expected minimum oxygen partial pressures in leatherback nests at PNMB, Costa Rica (solid line: observed values; dashed line: expected values), and Sandy Point, St. Croix, US Virgin Islands (solid line: observed values; dash-dot line: expected values). See Wallace et al. (2004) for full description of how expected values were calculated.

attributable to tidal pumping. In contrast, in the absence of external convective mechanisms ventilating nests at SPNWR, nest environment conditions were changed by the progression of embryonic development, which can influence hatching success of leatherback clutches (Garrett et al. 2010).

What We Have Learned and Have Yet to Learn

We now have a better understanding of the intrinsic and extrinsic factors that affect embryonic development and hatching success of leatherback eggs. Most of the research discussed here was conducted on the nesting population at PNMB on the Pacific coast of Costa Rica. The PNMB nesting population and other eastern Pacific leatherbacks have the smallest adult female body sizes, smallest clutch sizes, and longest remigration intervals of any leatherback population in the world (Saba et al. 2008; Eckert et al. 2012). Resource availability accounts for some of the differences in egg composition and reproductive output among regional populations (Wallace, Kilham, et al. 2006; Saba et al. 2008; Wallace and Saba 2009), and could account for intra- and interseasonal variation in reproductive investment in eggs and clutches. For example, yolks in PNMB eggs are smaller, both relative to egg mass and in absolute mass, than those in eggs from the Malaysian population (Simkiss 1962), and the fraction of yolk in contents of PNMB eggs is lower than in eggs of other sea turtle species (e.g., Hewavisenthi and Parmenter 2002). In addition, the paradoxical nature of albumin allocation and use observed at PNMB should be noted—that is, a high fraction of albumin in eggs, SAGs as possible "production overruns" of albumin and eggshell material outpacing yolk production and provision to eggs, and incomplete albumin and water uptake by embryos (Wallace, Sotherland, et al. 2006); this may indicate limitation of resources available to PNMB females when producing ova (yolks). Future research at leatherback nesting sites around the world could determine whether these reproductive patterns represent consequences of resource limitation on foraging grounds (Wallace, Kilham, et al. 2006), where vitellogenesis occurs (Rostal et al. 1996), or are adaptations to unique characteristics of leatherback nest environments.

Tidal pumping appears to be important in the gas exchange of leatherback eggs on Playa Grande. It probably plays a smaller or minimal role at beaches such as Tortuguero and St. Croix that have a lower tidal amplitude (Prange and Ackerman 1974; Leslie et al. 1996; Garrett et al. 2010).

In addition to investigating possible environmental effectors of intra- and interpopulation variation in reproductive investment, the influence of interfemale genetic variation on reproductive phenotypes also warrants investigation. Coordinated longitudinal analyses of both intra- and inter-individual reproductive investment (e.g., egg and hatchling production, hatching and emergence success) in several leatherback populations would help characterize broader species and population specific life history patterns while revealing whether individual leatherbacks are consistently more "successful" or "productive" (e.g., Rafferty et al. 2011) than other individuals.

In conclusion, despite recent advances in research on leatherback eggs and nests, significant gaps remain in our understanding of how maternal investment and environmental conditions affect embryonic development. In light of what has been learned, we recommend that future research focuses on the following: (1) possible effects of resource limitation on maternal investment in eggs; (2) albumin allocation to eggs and SAGS, and the extent to which albumin is utilized (or not) by developing embryos; (3) how physical characteristics of nesting beaches might affect nest conditions; and (4) mechanisms by which maternal identity can influence embryonic development and hatching success. Ultimately, answering these research questions will help explain patterns of maternal investment in reproduction by leatherbacks, other sea turtles, and oviparous amniotes generally, while informing and augmenting conservation efforts on leatherback nesting beaches so as to increase hatchling production and help restore populations.

ACKNOWLEDGMENTS

We thank the many biologists at PNMB who contributed to research reviewed here; we especially thank P. Santidrián Tomillo, B. Franks, E. Price, V. Saba, S. Bouchard, and particularly PNMB Director R. Piedra, for their hard work, dedication, and good cheer. Many investigators were instrumental in design and execution of studies of leatherback nest environments at PNMB; C. Ralph, T. Muir, J. Spotila, R. Reina, and F. Paladino were especially helpful. We acknowledge the important contributions of K. Garrett, J. Garner, and S. Garner at Sandy Point, St. Croix, USVI. Research presented here was supported by the Earthwatch Institute, the Betz Chair of Environmental Science at Drexel University, and the Shrey Chair of Biology at Purdue University. The support of the Goldring-Gund Marine Biology Station and the Leatherback Trust were essential to successful completion of this research.

LITERATURE CITED

Ackerman, R. A. 1977. The respiratory gas exchange of sea turtle nests (Chelonia, Caretta). Respiration Physiology 31: 19–38.

Ackerman, R. A. 1980. Physiological and ecological aspects of gas exchange by sea turtle eggs. American Zoologist 20: 575–583.

Ackerman, R. A. 1981. Growth and gas exchange of embryonic sea turtles (Chelonia, Caretta). Copeia 1981: 757–765.

Ackerman, R. A. 1991. Physical factors affecting the water exchange of buried reptile eggs. In: D. C. Deeming and M.W.J. Ferguson (eds.), Egg incubation: Its effects on embryonic development in birds and reptiles. Cambridge University Press, Cambridge, UK, pp. 193–210.

Ackerman, R. A. 1997. The nest environment and the embryonic development of sea turtles. In: P. L. Lutz and J. A. Musick (eds.), The biology of sea turtles. CRC Press, Boca Raton, FL, USA, pp. 83–106.

Ar, A., C. V. Paganelli, R. B. Reeves, D. G. Greene, and H. Rahn. 1974. The avian egg: water vapor conductance, shell thickness, and functional pore area. Condor 76: 153–158.

Bell, B. A., J. R. Spotila, F. V. Paladino, and R. D. Reina. 2003. Low reproductive success of leatherback turtles, *Dermochelys coriacea*, is due to high embryonic mortality. Biological Conservation 115: 131–138.

Booth, D. T. 1998. Nest temperature and respiratory gases during natural incubation in the broad-shelled river turtle, *Chelodina expansa* (Testudinata: Chelidae). Australian Journal of Zoology 46: 183–191.

Booth, D. T. 2000. The effect of hypoxia on oxygen consumption of embryonic estuarine crocodiles (*Crocodylus porosus*). Journal of Herpetology 34: 478–481.

Caut, S., E. Guirlet, P. Jouquet, and M. Girondot. 2006. Influence of nest location and yolkless eggs on the hatching success of leatherback turtle clutches in French Guiana. Canadian Journal of Zoology 84: 908–915.

Chan, E. H., H. U. Salleh, and H. C. Liew. 1985. Effects of handling on hatchability of eggs of the leatherback turtle, *Dermochelys coriacea*, (L.). Pertanika 8: 265–271.

Clusella Trullas, S., and F. V. Paladino. 2007. Micro-environment of olive ridley turtle nests deposited during an aggregated nesting event. Journal of Zoology 272: 367–376.

Crim, J. L., L. D. Spotila, J. R. Spotila, M. P. O'Connor, R. D. Reina, C. J. Williams, and F. V. Paladino. 2002. The leatherback turtle, *Dermochelys coriacea*, exhibits both polyandry and polygyny. Molecular Ecology 11: 2097–2106.

Deeming, D. C. 1991. Reasons for the dichotomy in egg turning in birds and reptiles. In: D. C. Deeming and M.W.J. Ferguson (eds.), Egg incubation: Its effects on embryonic development in birds and reptiles. Cambridge University Press, Cambridge, UK, pp. 307–323.

Deeming, D. C., and M. B. Thompson. 1991. Gas exchange across reptilian eggshells. In: D. C. Deeming and M.W.J. Ferguson (eds.), Egg incubation: Its effects on embryonic development in birds and reptiles. Cambridge University Press, Cambridge, UK, pp. 277–284.

Deraniyagala, P.E.P. 1939. The tetrapod reptiles of Ceylon, Vol. I. Testudinates and crocodilians. Dulau, London, UK.

Dzialowski, E. M., D. von Plettenberg, N. A. Elmonoufy, and W. W. Burggren. 2002. Chronic hypoxia alters the physiological and morphological trajectories of developing chicken embryos. Comparative Biochemistry and Physiology 131A: 713–724.

Eckert, K. L., and S. A. Eckert. 1990. Embryo mortality and hatch success in in situ and translocated leatherback sea turtle, *Dermochelys coriacea*, eggs. Biological Conservation 53: 37–46.

Eckert, K. L., B. P. Wallace, J. G. Frazier, S. A. Eckert, and P.C.H. Pritchard. 2012. Synopsis of the biological data on the leatherback sea turtle (*Dermochelys coriacea*). U.S. Fish and Wildlife Service, Biological Technical Publication BTP-R4015-2012. US Department of Interior, U.S. Fish and Wildlife Service. Washington, DC, USA.

Finkler, M. S., and D. L. Claussen. 1997. Within and among clutch variation in composition of *Chelydra serpentina* eggs with initial egg mass. Journal of Herpetology 31: 620–624.

Garrett, K., B. P. Wallace, J. Alexander-Garner, and F. V. Paladino. 2010. Variation in leatherback turtle nest environments and their consequences for hatching success. Endangered Species Research 11: 147–155.

Hewavisenthi, S., and C. J. Parmenter. 2002. Egg components and utilization of yolk lipids during development of the flatback turtle *Natator depressus*. Journal of Herpetology 36: 43–50.

Hill, W. L. 1995. Intraspecific variation in egg composition. Wilson Bulletin 107: 382–387.

Houghton, J.D.R., A. E. Myers, C. Lloyd, R. S. King, C. Isaacs, and G. C. Hays. 2007. Protracted rainfall decreases temperature within leatherback (*Dermochelys coriacea*) clutches in Grenada, West Indies: ecological implications for a species displaying temperature dependent sex determination. Journal of Experimental Marine Biology and Ecology 345: 71–77.

James, M. C., R. A. Myers, and C. A. Ottensmeyer. 2005. Behaviour of leatherback sea turtles, *Dermochelys coriacea*, during the migratory cycle. Proceedings of the Royal Society B 272: 1547–1555.

Kam, Y. C. 1993. Physiological effects of hypoxia on metabolism and growth of turtle embryos. Respiration Physiology 92: 127–138.

Kam, Y. C., and H. B. Lillywhite. 1994. Effects of temperature and water on critical oxygen tension of turtle embryos. Journal of Experimental Zoology 268: 1–8.

Kjerfve, B. 1981. Tides of the Caribbean Sea. Journal of Geophysical Research 86: 4243–4247.

Leslie, A. J., D. N. Penick, J. R. Spotila, and F. V. Paladino. 1996. Leatherback turtle, *Dermochelys coriacea*, nesting and nest success at Tortuguero, Costa Rica, in 1990–1991. Chelonian Conservation and Biology 2: 159–168.

Maloney, J. E., C. Darian-Smith, Y. Takahashi, and C. J. Limpus. 1990. The environment of the embryonic loggerhead turtle (*Caretta caretta*) in Queensland. Copeia 1990: 378–387.

Miller, J. D. 1985. Embryology of marine turtles. In: C. Gans, F. Billett, and P.F.A. Maderson (eds.), Biology of the Reptilia, Vol. 14, Development A. John Wiley and Sons, New York, NY, USA, pp. 269–238.

Miller, J. D. 1997. Reproduction in sea turtles. In: P. L. Lutz and J. A. Musick (eds.), The biology of sea turtles. CRC Press, Boca Raton, FL, USA, pp. 51–82.

Mrosovsky, N. 1994. Sex ratios of sea turtles. Journal of Experimental Zoology 270: 16–27.

Nelson, T. C., K. D. Groth, and P. R. Sotherland. 2010. Maternal investment and nutrient use affect phenotype of American alligator and domestic chicken hatchlings. Comparative Biochemistry and Physiology A Molecular and Integrative Physiology 157: 19–27.

Packard, G. C. 1991. Physiological and ecological importance of water to embryos of oviparous reptiles. In: D. C. Deeming and M.W.J. Ferguson (eds.), Egg incubation: Its effects on embryonic development in birds and reptiles. Cambridge University Press, Cambridge, UK, pp. 213–228.

Packard, M. J., and V. G. Demarco. 1991. Eggshell structure and formation of eggs of oviparous reptiles. In: D. C. Deeming and M.W.J. Ferguson (eds.), Egg incubation: Its effects on embryonic development in birds and reptiles. Cambridge University Press, Cambridge, UK, pp. 53–69.

Palmer, B., and L. J. Guillette. 1991. Oviductal proteins and their influence on embryonic development in birds and reptiles. In: D. C. Deeming and M.W.J. Ferguson (eds.), Egg Incubation: Its effects on embryonic development in birds and reptiles. Cambridge University Press, Cambridge, UK, pp. 29–46.

Patiño-Martínez, J., A. Marco, L. Quiñones, and C. P. Calabuig. 2010. Los huevos falsos (SAGs) facilitan el compartamiento social de emergencia en las crías de la tortuga laúd *Dermochelys coriacea* (Testudines: Dermochelyidae). Revista de Biología Tropical 58: 943–954.

Pehrson, T. 1945. Some problems concerning the development of the skull of turtles. Acta Zoologica 26: 157–184.

Prange, H. D., and R. A. Ackerman. 1974. Oxygen consumption and mechanisms of gas exchange of green turtle (*Chelonia mydas*) eggs and hatchlings. Copeia 1974: 758–763.

Price, E. R., F. V. Paladino, K. P. Strohl, P. Santidrián Tomillo, K. Klann, and J. R. Spotila. 2007. Respiration in neonate sea turtles. Comparative Biochemistry and Physiology 146A: 422–428.

Rafferty, A. R., R. G. Evans, T. F. Scheelings, and R. D. Reina. 2013. Limited oxygen availability in utero may constrain the evolution of live birth in reptiles. American Naturalist 181: 245–253.

Rafferty, A. R., P. Santidrián Tomillo, J. R. Spotila, F. V. Paladino, and R. D. Reina. 2011. Embryonic death is linked to maternal identity in the leatherback turtle (*Dermochelys coriacea*). PLoS ONE 6(6): e21038. doi: 10.1371/journal.pone.0021038.

Rahn, H., and C. V. Paganelli. 1990. Gas fluxes in avian eggs: driving forces and the pathway for exchange. Comparative Biochemistry and Physiology 95A: 1–15.

Ralph, C. R., R. D. Reina, B. P. Wallace, P. R. Sotherland, J. R. Spotila, and F. V. Paladino. 2005. Effect of egg location and respiratory gas concentrations on developmental success in nests of the leatherback turtle, *Dermochelys coriacea*. Australian Journal of Zoology 53: 289–294.

Reina, R. D., P. A. Mayor, J. R. Spotila, R. Piedra, and F. V. Paladino. 2002. Nesting ecology of the leatherback turtle, *Dermochelys coriacea*, at Parque Nacional Marino Las Baulas, Costa Rica: 1988–1989 to 1999–2000. Copeia 2002: 653–664.

Renous, S., F. Rimblot-Baly, J. Fretey, and C. Pieau. 1989. Caractéristiques du développement embryonnaire de la tortue luth, Dermochelys coriacea (Vandelli, 1761). Annales des Sciences Naturelles. Zoologie et Biologie Animale 10: 197–229.

Roff, D. A. 1992. The evolution of life histories: Theory and analysis. Chapman and Hall, London, UK.

Rostal, D. C., F. V. Paladino, R. M. Patterson, and J. R. Spotila. 1996. Reproductive physiology of nesting leatherback turtles (*Dermochelys coriacea*) at Las Baulas National Park, Costa Rica. Chelonian Conservation and Biology 2: 230–236.

Saba, V. S., P. Santidrián Tomillo, R. D. Reina, J. R. Spotila, J. A. Musick, D. A. Evans, and F. V. Paladino. 2007. The effect of the El Niño Southern Oscillation on the reproductive frequency of eastern Pacific leatherback turtles. Journal of Applied Ecology 44: 395–404.

Saba, V. S., J. R. Spotila, F. P. Chavez, and J. A. Musick. 2008. Bottom-up and climatic forcing on the worldwide population of leatherback turtles. Ecology 89: 1414–1427.

Sahoo, G., B. K. Mohapatra, and S. K. Dutta. 2010. Chemical composition and ultrastructure of shells of unfertilized eggs of olive ridley turtles, *Lepidochelys olivacea*. Current Herpetology 29: 37–43.

Santidrián Tomillo, P., J. S. Suss, B. P. Wallace, K. D. Magrini, G. Blanco, F. V. Paladino, and J. R. Spotila. 2009. Influence of emergence success on the annual reproductive output of leatherback turtles. Marine Biology 156: 2021–2031.

Sieg, A. E., C. Binckley, B. P. Wallace, P. Santidrián Tomillo, R. D. Reina, F. V. Paladino, and J. R. Spotila. 2011. Sex ratios of leatherback turtles: hatchery translocation decreases metabolic heating and female bias. Endangered Species Research 15: 195–204.

Silver, R. B., and D. C. Jackson. 1985. Ventilatory and acid-base responses to long-term hypercapnia in the freshwater turtle, *Chrysemys picta bellii*. Journal of Experimental Biology 114: 661–672.

Simkiss, K. 1962. The source of calcium for the ossification of the embryos of the giant leathery turtle. Comparative Biochemistry and Physiology 7: 71–79.

Solomon, S. E., and J. M. Watt. 1985. The structure of the egg shell of the leatherback turtle (*Dermochelys coriacea*). Animal Technology 36: 19–27.

Sotherland, P. R., and H. Rahn. 1987. On the composition of bird eggs. Condor 89: 48–64.

Sotherland, P. R., R. D. Reina, S. Bouchard, B. P. Wallace, B. F. Franks, and J. R. Spotila. 2003. Egg mass, egg composition, clutch mass, and hatchling mass of leatherback turtles (*Dermochelys coriacea*) nesting at Parque Nacional Las Baulas, Costa Rica. Abstract. In: J. A. Seminoff (compiler), Proceedings of the Twenty-second International Symposium on Sea Turtle Biology and Conservation. NOAA Technical Memorandum NMFS SEFSC-503. NOAA, Miami, FL, USA, p. 31.

Sotherland, P. R., J. A. Wilson, and K. M. Carney. 1990. Naturally occurring allometric engineering experiments in avian eggs. American Zoologist 30: 86A.

Spotila, J. 2004. Sea turtles: A complete guide to their biology, behavior, and conservation. Johns Hopkins University Press, Baltimore, MD, USA.

Standora, E. A., and J. R. Spotila. 1985. Temperature dependent sex determination in sea turtles. Copeia 1985: 711–722.

Stewart, K. R., and P. H. Dutton. 2011. Paternal genotype reconstruction reveals multiple paternity and sex ratios in a breeding population of leatherback turtles (*Dermochelys coriacea*). Conservation Genetics 12: 1101–1113.

Thompson, M. B. 1993. Oxygen consumption and energetics of development in eggs of the leatherback turtle, *Dermochelys coriacea*. Comparative Biochemistry and Physiology 104A: 449–453.

Tracy, C. R., and H. L. Snell. 1985. Interrelations among water and energy relations of reptilian eggs, embryos, and hatchlings. American Zoologist 25: 999–108.

Vleck, C. M., and D. F. Hoyt. 1991. Metabolism and energetics of reptilian and avian embryos. In: D. C. Deeming and M.W.J. Ferguson (eds.), Egg incubation: Its effects in embryonic development in birds and reptiles. Cambridge University Press, Cambridge, UK, pp. 285–306.

Wallace, B. P., S. S. Kilham, F. V. Paladino, and J. R. Spotila. 2006. Energy budget calculations indicate resource limitation for eastern Pacific leatherback turtles. Marine Ecology Progress Series 318: 263–270.

Wallace, B. P., and V. S. Saba. 2009. Environmental and anthropogenic impacts on intra-specific variation in leatherback turtles: opportunities for targeted research and conservation. Endangered Species Research 7: 11–21.

Wallace, B. P., P. R. Sotherland, P. Santidrián Tomillo, S. S. Bouchard, R. D. Reina, J. R. Spotila, and F. V. Paladino. 2006. Egg components, egg size, and hatchling size in leatherback turtles. Comparative Biochemistry and Physiology 145A: 524–532.

Wallace, B. P., P. Sotherland, P. Santidrián-Tomillo, R. Reina, J. Spotila, and F. Paladino. 2007. Maternal investment in reproduction and its consequences in leatherback turtles. Oecologia 152: 37–47.

Wallace, B. P., P. R. Sotherland, J. R. Spotila, R. D. Reina, B. R. Franks, and F. V. Paladino. 2004. Biotic and abiotic factors affect the nest environment of embryonic leatherback turtles, *Dermochelys coriacea*. Physiological and Biochemical Zoology 77: 423–432.

Warburton, S. J., D. Hastings, and T. Wang. 1995. Responses to chronic hypoxia in embryonic alligators. Journal of Experimental Zoology 273: 44–50.

Whitmore, C. P., and P. H. Dutton. 1985. Infertility, embryonic mortality and nest-site selection in leatherback and green sea turtles in Suriname. Biological Conservation 34: 251–272.

13

Leatherback Turtle Physiological Ecology

Implications for Bioenergetics and Population Dynamics

BRYAN P. WALLACE AND
T. TODD JONES

Leatherback turtles, *Dermochelys coriacea*, are superlative in nearly every way. They plumb dark ocean depths where few air breathers ever venture. They thrive in waters cold enough to kill a human, but also frequent areas prized by people for warm-as-bathwater seas. They feed on stinging sacs of jelly as salty as seawater, and convert this low-energy, low-nutrient food into hulking bodies that undergo epic migrations and lay massive egg clutches. They cross ocean basins and forge across churning currents to transit between feeding and breeding areas. This species' unique suite of anatomical and physiological traits sets it apart in the animal kingdom, and ensures that it plies across the world's oceans today.

Because leatherbacks are air-breathing divers and egg-laying reptiles, their distributions extend from the ocean's surface to hundreds of meters depth, and from open ocean to sandy shores. This incredibly diverse set of habitats means that environmental conditions influencing leatherback energetics and life history, such as water temperatures, salinity of prey and water, and patchy distribution of prey in three dimensions, vary tremendously. While understanding how leatherbacks meet each of these challenges is important, it is through an integrative, holistic perspective of leatherback physiological ecology that we gain insight into the species' evolutionary history and into how leatherbacks respond to environmental and anthropogenic factors that influence their future survival.

In the past few decades, scientists have used equipment ranging from weather balloons and turtle masks to isotopic tracers and electronic data loggers to reveal many fascinating secrets of leatherback physiological ecology that all seem to start with the simple question: "How do they do that?" In this chapter, we synthesize research on leatherback physiological ecology. In the first section, we focus on specific regulatory processes, including osmoregulation and thermoregulation (physiological and anatomical adaptations for an aquatic, diving lifestyle); the integrated processes of feeding, digestion, assimilation; and growth. In the second section, we present a revised model of leatherback bioenergetics to examine how integration of these systems and processes influences variation in leatherback life history traits. Finally, we conclude with recommendations for applications of leatherback ecophysiology research to the emerging field of "conservation physiology."

How Leatherbacks Work

Osmoregulation

Although they pass virtually their entire lives in water, leatherback turtles live in a desiccating environment; their plasma is one-third the salinity of sea water, and they feed on immense quantities of gelatinous zooplankton (~1000 metric tons in their lifetimes; Jones et al. 2012) that are high in water content, isosmotic to seawater, and low in energy content. Therefore, leatherbacks experience tremendous osmotic stress throughout their lives, starting from the first moments outside of their eggs. As moments-old hatchlings, they crawl to and enter the ocean, and then drink seawater to increase mass and regulate plasma sodium (Na^+) concentrations (Reina et al. 2002). For their entire lives at sea, they eat sea jellies (hyperosmotic to leatherback body fluids), sustaining rapid growth rates as juveniles, and maintaining large body mass as adults; leatherbacks also embark on long-distance migrations, and have both a high reproductive output and significant thermoregulatory requirement.

Major routes of water and ion loss are through respiration, tear production, excretion (urine and feces), and across integument (transcutaneous), while major routes of water and ion gain are across the integument and through ingestion (i.e., across the gut). Leatherbacks can reduce water loss and ion gain through reduction in membrane permeability and by actively pumping ions. Respiratory water loss (RWL) is low in sea turtles because they inhale air at the water's surface that is nearly saturated with water vapor, and since most species are ectothermic (i.e., body temperatures similar to ambient), inspired and expired air have similar percentages of humidity (Lutz 1997). However, RWL may play a more important role in water balance for adult leatherbacks that venture into temperate foraging areas where they experience differences between body and ambient temperatures of 8°C or more (James and Mrosovsky 2004). Davenport et al. (2009a) identified specialized tracheal structures that might serve to minimize heat and water loss during respiration; these would represent a likely adaptation for life in temperate waters. In addition, the high efficiency of the sea turtle lung means that there is a greater amount of oxygen taken up with each breath and, therefore, a lower ratio of respiratory water loss for a given amount of oxygen uptake (Lutz 1997). Transcutaneous water loss is probably negligible because sea turtles have keratinized skin with an embedded adipose (i.e., hydrophobic) layer (see Lutz 1997 for review). Ion regulation is primarily controlled by tear-producing extrarenal glands located in the eye orbit, called the lachrymal salt glands (Schmidt-Nielsen and Fange 1958). These glands handle nearly 90% of all sodium flux in sea turtles (Kooistra and Evans 1976). Sea turtle kidneys produce fluid that is only isosmotic to their plasma (Prange and Greenwald 1980), so urine production functions to maintain body water volume and rid the body of waste (Prange 1985).

Decades ago, researchers considered sea turtles to be water conservationists; that is, they were thought to avoid taking in saltwater, thereby minimizing water turnover (Minnich 1982). In contrast to these early hypotheses, Wallace et al. (2005) measured water turnover rates of 16–30% total body water (TBW) per day for adult female leatherbacks during the inter-nesting period (i.e., the days between consecutive nesting events during a nesting season). Because turtles only forage opportunistically, if at all, during this period, seawater intake may come strictly from drinking (Reina et al. 2005; Southwood et al. 2005). These results, and those from studies of other sea turtle species (Ortiz et al. 2000; Clusella Trullas et al. 2006; Jones et al. 2009), demonstrate conclusively that sea turtles, especially leatherbacks, are active consumers of seawater; this suggests that seawater drinking is important in maintaining osmotic homeostasis. In addition, leatherbacks handle large quantities of salt due to their gelatinous zooplankton (i.e., sea jelly) diet. Sea jellies are isosmotic with seawater (Mills and Vogt 1984), composed of up to 96% water by mass, and are low in energy density (Doyle et al. 2007), requiring leatherbacks to consume them in extremely high quantities.

When seawater is ingested through drinking or food consumption, Na^+ and Cl^- are actively absorbed across the gut, and water passively follows, leading to a net absorption of water and electrolytes (Sullivan and Field 1991). Sea turtles must then deal with the salt load by subsequent excretion of ions via salt glands and reduction of increased extracellular fluid volume via the kidneys. Thus, dual osmoregulatory costs include pumping ions in the gut to gain water, and again across salt glands to secrete hyperosmotic tears to rid the body of excess salt, leaving osmotically free water.

How the paired lachrymal salt glands handle large salt loads and at what energetic cost are interesting physiological questions, especially since leatherbacks have the highest water flux rates for sea turtles (Jones et al. 2009; Ortiz et al. 2000; Wallace et al. 2005; Southwood et al. 2006) and consume enormous quantities of hyperosmotic prey (Jones et al. 2012). Lachrymal salt glands of sea turtles produce tears up to six times more concentrated than sea turtle blood plasma and twice as concentrated as seawater (Hudson and Lutz 1986; Nicoloson and Lutz 1989; Reina et al. 2002). However, tear production rates of leatherback hatchlings (Reina et al. 2002) are not sufficient to rid them of their consumed salt load if they actually consumed prey quantities equivalent to their body mass daily, as hypothesized by Lutcavage and Lutz (1986). Furthermore, water turnover rates of juvenile and adult sea turtles are 20–30% of TBW per day, regardless of feeding, suggesting that salt glands need to work well above their reported capacity (Wallace et al. 2005; Jones et al. 2009). Leatherbacks could eliminate much of the saltwater and bound salts by trapping food in the esophagus using esophageal papillae and expelling fluids, although this would not eliminate the water and ions bound in prey tissues (Lutz 1997; Reina et al. 2002). Another possibility is that leatherbacks could temporarily allow their extracellular fluid to concentrate, as occurs in pipping hatchings (Reina et al. 2002), and then remove much of the concentrated hypervolume through the kidneys. Turtles could then remove the remaining portion of ions via lachrymal glands to restore homeostasis. Despite advances in understanding osmoregulatory adaptations, more detailed experiments are needed to answer specific questions about how leatherback turtles maintain ionic/water balance in a desiccating environment.

Thermoregulation

No aspect of leatherback physiology has received as much attention as thermoregulation. Leatherbacks swim among ice floes at high latitudes, but also inhabit warm waters in the tropics during each reproductive cycle. Thriving in both of these thermal extremes means that leatherbacks possess a highly specialized, yet flexible, array of adaptations for maintaining large thermal gradients (i.e., difference between body temperatures [T_b] and water temperatures [T_w]) in cold water and small thermal gradients in warm water. They have body temperatures more than 8°C above water temperature in cold water (James and Mrosovsky 2004) and body temperatures less than 4°C above water temperature in warm water (Southwood et al. 2005); they can then avoid overheating in tropical seas. Leatherbacks can thus remain active according to the ecological demands of a range of environments. Wallace and Jones (2008) provided a detailed synthesis of research on leatherback thermoregulation; here we provide an updated overview.

Wallace et al. (2005) proposed that leatherback thermoregulatory adaptations likely allowed them to exploit a unique ecological niche that was unavailable to other marine animals (Davenport 1997), similar to the "thermal niche expansion theory" developed by Block et al. (1993) to explain the multiple and diverse origins of endothermy in the suborder Scombroidei (tunas, billfish). The unique suite of leatherback adaptations includes anatomical features, such as (1) countercurrent heat exchangers in their flippers that conserve heat while in cold water and dissipate heat in warm water (Greer et al. 1973; Bostrom et al. 2010); (2) thick peripheral insulation around the body (Goff and Lien 1988; Davenport et al. 1990), head, and neck (Davenport et al. 2009b); (3) brown adipose tissue (Goff and Stenson 1988); (4) specialized tracheal structures that minimize heat loss during respiration (Davenport et al. 2009a); and (5) large body size (Mrosovsky and Pritchard 1971; Frair et al. 1972; Paladino et al. 1990). Physiological features include (1) thermal independence of muscle tissue metabolism (Penick et al. 1998), (2) elevated metabolic rate (Paladino et al. 1990), (3) internal heat generation (Southwood et al. 2005; Bostrom and Jones 2007; Bostrom et al. 2010), and (4) blood flow patterns that distribute internally generated heat (Penick et al. 1998; Bostrom et al. 2010).

In recent years, researchers used data loggers and biophysical modeling to study responses of leatherback T_b to diving behavior, T_w, and other variables (Southwood et al. 2005; Wallace et al. 2005; Bostrom and Jones 2007). Integrating the existing empirical and theoretical research, Wallace and Jones (2008) concluded that leatherback thermoregulation depends on adaptations for heat production (e.g., metabolic and behavioral adjustments) and heat retention (e.g., thermal inertia through large body size, blood flow adjustments, insulation, and behavioral modifications) to both reach and maintain set point differentials between T_b and T_w in varied thermal environments.

In the first experiment that empirically demonstrated leatherback thermoregulation in response to variation in T_w, Bostrom et al. (2010) showed that juvenile leatherbacks (16 and 37 kg) simultaneously adjusted swimming activity to generate heat and reduce heat loss through the carapace, plastron, and flippers while exposed to cold water; and reversed these responses in warm water. A biophysical model demonstrated that an adult leatherback could maintain a large gradient between T_b and T_w in cold water using these mechanisms. Leatherbacks are endotherms that exhibit behavioral control over endogenous heat production similar to lamnid sharks, tuna, and billfish (Block et al. 1993). This cohesive picture of coupled physiological and behavioral responses to varied thermal environments provides an improved comprehension of leatherback thermoregulation and sheds light on the evolutionary history and ecological consequences of leatherback niche expansion. It provides a solid foundation for understanding how leatherbacks interact with and exploit biophysical features of their environments.

Feeding, Ingestion, Assimilation

Leatherbacks find, handle, and consume large quantities of prey, efficiently assimilating scarce energy and nutrients from low-quality gelatinous zooplankton (Davenport and Balazs 1991; Doyle et al. 2007). At the same time, they confront the osmoregulatory challenge of a diet of hyperosmotic prey that is more than 90% water by mass.

Duron (1978) estimated that an adult leatherback could ingest roughly 200 kg per day. More recent studies estimate daily prey intake rates between 20–40% (or more) of total body mass (Lutcavage and Lutz 1986; Wallace et al. 2006; Jones et al. 2012). Using growth rate and feeding efficiency data in a population model Jones et al. (2012) estimated prey biomass required for leatherbacks of different age classes, and reported that juvenile leatherbacks (i.e., two to seven years old) account for more than half of prey biomass consumption of the entire population, whereas adults only account for less than 9% of the total population consumption. Direct observations illustrate that leatherbacks use different optimal foraging strategies, depending on types and availability of resources. Several studies described possible leatherback feeding behavior (e.g., Myers and Hays 2006; Fossette et al. 2008; Casey et al. 2010) during inter-nesting periods in the Caribbean, whereas inter-nesting leatherbacks in the eastern Pacific apparently do not ingest prey but only drink seawater (Reina et al. 2005; Southwood et al. 2005; Wallace et al. 2005).

Two studies documented leatherback feeding during nonbreeding periods. Heaslip et al. (2012) used turtle-borne cameras on leatherbacks in the Atlantic Ocean and reported that leatherbacks consume around 73% (and up to 184%) of their body mass daily in relatively large *Cyanea* jellies. In contrast, Fossette et al. (2012) used video footage of leatherbacks in Solomon Islands to show that leatherbacks could meet their energy requirements by feeding on large numbers of small jellies, as long as the jellies are available in extremely high densities and intake rates are sufficiently rapid. However, estimates in both papers were derived from extremely short observation periods (i.e., minutes to hours) compared to total durations of foraging periods (i.e., months), so should be applied with caution. Nonetheless, replicating these efforts in other important leatherback feeding areas around the world, especially additional locations in the Pacific Ocean, would be particularly important for understanding how leatherback energy acquisition is constrained by physiological and environmental factors.

Virtually nothing is known about leatherback assimilation efficiency. Previous research (Wallace et al. 2006; Jones et al. 2012) assumed a conservative estimate of 80% efficiency based on proxies from studies of digestion and assimilation of gelatinous prey. However, no empirical measures of this parameter in leatherbacks exist. This is an important data gap because leatherbacks almost certainly have high assimilation efficiency given their specialized diet of low-energy prey that fuels enormous energy requirements imposed by their unique life history.

Growth and Age at Maturity

Leatherbacks are the largest and fastest growing turtle species; from hatching to adulthood they undergo a greater than 10,000 fold increase in mass. Jones et al. (2012) estimated the required prey biomass for a leatherback hatchling to reach reproductive maturity to be about 300 metric tons. Therefore, juvenile leatherbacks are under tremendous selective pressure to acquire adequate resources and habitat to support high growth rates. Somatic growth and its life history consequences for leatherbacks, however, remain less studied than in hard-shelled sea turtles.

The paucity of studies on leatherback growth rates is most likely due to difficulties resulting from (1) its oceanic-pelagic lifestyle throughout ontogeny (Bolten 2003), and (2) problems in maintaining leatherbacks in captivity (Birkenmeier 1971; Jones et al. 2000). Therefore, studying leatherback growth rate through mark-recapture studies has not been possible, with the exception of nesting females, and direct observations of growth rates in captivity are limited (Jones et al. 2011).

EARLY GROWTH

Growth rate estimates of juvenile leatherbacks are based entirely on inferences from captive animals. Early research on captive leatherbacks up to one year of age documented growth rates between 22 and more than 50 cm per year (Deraniyagala 1939; Birkenmeier 1971), which led to speculation that leatherbacks reach age at first reproduction (AFR) within 2 to 6 years. Jones et al. (2011) maintained 20 leatherback turtles in captivity, with four turtles surviving for more than 1 year, and growth rates of these animals averaged 32 cm each year over two years. If growth rates are linear until maturation, these turtles would reach a mean nesting length

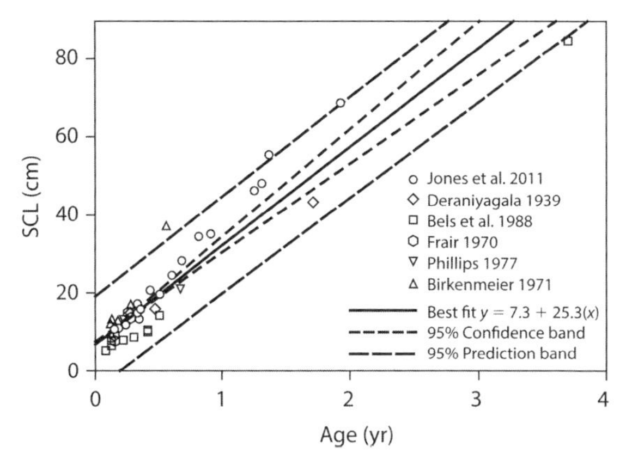

Fig. 13.1. Linear regression of early age-at-length data for captive leatherbacks. Best fit relationship of $y = 7.3 + 25.3(x)$; slope = 25.3 ± 1.33; $p < 0.0001$. Solid line: best fit curve using linear least squares (LLS) regression. Short-dash lines: 95% confidence bands around the regression. Long-dash lines: 95% prediction bands for the scatter of data points.

of 147 cm straight carapace length (SCL) (Stewart et al. 2007) within 5 years. However, it is inappropriate to extrapolate linear growth rate estimates during their first few months of life to the entire life span, because they do not capture exponential decay of growth rate over time (fig. 13.1; Jones et al. 2011).

ADULT GROWTH AND AGE AT MATURITY

Early attainment of sexual maturity was suggested by studies on chondro-osseous (i.e., cartilage and bone) morphology of leatherbacks. Rhodin et al. (1981) and Rhodin et al. (1996) reported that chondro-osseous morphology of hard-shelled sea turtles resembled that of terrestrial and freshwater turtles, whereas chondro-osseous morphology of leatherback turtles was more similar to that of rapidly growing marine mammals. Based on rapid growth in captivity and specialized chondro-osseous morphology, Rhodin (1985) suggested that leatherbacks may reach AFR as early as two to three years, with a more conservative estimate of six years owing to discrepancy in growth for free-living animals versus captive.

Zug and Parham (1996) applied skeletochronological analyses of scleral ossicles of hatchlings, juveniles, and adults to a von Bertalanffy growth function (VBGF), thereby estimating that leatherbacks could reach AFR within 13 to 14 years. Avens et al. (2009) used skeletochronological analyses of scleral ossicles of hatchlings, juveniles, and adults that stranded on the US Atlantic and Gulf of Mexico coasts, and from captive animals, to conclude that North Atlantic leatherbacks reach sexual maturity within 24–29 years, and perhaps as early as 16–23 years. The AFR discrepancy between these studies may be due to differences in the populations and habitats (e.g., genetics, resource availability) or, as suggested by Avens et al. (2009), in the method of counting growth marks; that is, Zug and Parham (1996) counted the lateral edges where rings may be compressed or absorbed, rather than the scleral ossicle tips, thereby resulting in an underestimation of age. Jones et al. (2011) combined data from 46 leatherbacks of known age, ranging in length from 5–85 cm SCL and in age from 0.08–3.7 years (Deraniyagala 1939; Frair 1970; Birkenmeier 1971; Phillips 1977; Bels et al. 1988), with adult data (Zug and Parham 1996; Avens et al. 2009); they then modeled growth rate and age at maturity with several different growth functions. The best-fit age at maturity estimate was 13–20 years using the VBGF.

Caution must be applied when using skeletochronology to provide a growth rate or age at maturity estimation. Skeletochronology matches carapace lengths with age estimates. An adult leatherback turtle may have a 155 cm curved carapace length (CCL) and be an estimated 30 years old, but this does not indicate how long the animal has been mature. For example, Price et al. (2004) reported that adult female growth rate slows to 0.2 cm CCL yr^{-1} after first nesting. Therefore, a 155 cm CCL / 30-year-old turtle may have been mature for 10+ years and only grown 2 cm in carapace length. Furthermore, how well growth rates in controlled laboratory conditions reflect growth rates of wild turtles subject to competing energy demands and environmental variability is unknown, and requires further evaluation to be fully validated for leatherbacks. For all of these reasons, attention should be given to assumptions made and use of growth functions to model known age and length data.

Regardless of how leatherback growth rates are measured, early growth rate is high (20–50+ cm SCL yr^{-1}; fig. 13.1) and adult growth rate is low (0.2 cm CCL yr^{-1}). Therefore, juvenile growth rates either slowly decline with age or juvenile growth rate is linear until plateauing when adulthood is reached. An important data gap remains for growth rates of leatherbacks from 80–120 cm CCL, making modeling growth rate with growth functions the only available option to date. Despite application of several different modeling approaches, all studies show that the VBGF, an asymptotic exponential decay function, best models leatherback growth rate and age at maturity. Hard-shelled sea turtles undergo changes in habitat and diet throughout their ontogeny (Bolten 2003), and these may cause polyphasic growth patterns, which makes VBGF a potentially poor selection for modeling growth rates in these species (Chaloupka and Musick 1997). In contrast, leatherbacks are oceanic-pelagic and feed on gelatinous zooplankton throughout their life (Bjorndal 1997; Salmon et al. 2004), which could explain the better fit of the VBGF to leatherback length-at-age data (Jones et al. 2011).

Several technological and methodological advances show promise for filling critical data gaps in growth rates from 80 to 120 cm CCL. A few examples include (1) time-lapsed hatchling drift models (Gaspar et al. 2012) to determine growth rates of juvenile leatherbacks captured at sea; (2) bio-loggers that can be attached to smaller turtles (Mansfield et al. 2012) to understand time spent in juvenile habitats and temperature regimes; (3) improvements in captive care (Jones et al. 2011) to allow studies of effects of temperature and food availability on growth rates of posthatchling and juvenile leatherbacks; and (4) advancements in skeletochronolgy and radiocarbon dating (Avens et al. 2009) to better determine known age, as well as to back calculate growth rates from multiple lines of arrested growth. Until new data become available, leatherback growth rates and AFR remain the least understood and studied traits among sea turtle species.

Leatherback Bioenergetics: How Physiology Underpins Life History

All animals balance competing life history demands on allocations of time and energy in the context of their physiological state and limitations; biophysical environment; and resource availability, acquisition, and assimilation (Congdon 1989). Put simply, they need energy to fuel essential functions like foraging for food, growth, reproduction, and predator evasion, as well as for regulatory processes. How well, or how poorly, animals strike this balance has important consequences for reproduction and survival.

Previous research on leatherback energy budgets provided estimates of feeding rates to meet energetic demands, and also made inferences about how resource availability might influence differences in life history and population demography among leatherback populations. Wallace et al. (2006) estimated energy expenditure of a typical reproductive cycle for adult female leatherbacks in the East Pacific (EP) and North Atlantic (NA) Oceans, and then added costs incurred during foraging to estimate daily feeding requirements (averaged over varying remigration intervals—that is, years between consecutive nesting seasons) necessary for total energy demands. They hypothesized that resource availability differences in respective feeding areas in the two ocean basins result in leatherbacks in the EP that are smaller, have lower reproductive output, and declining population trajectories compared with those in the NA. Subsequent oceanographic modeling confirms that persistent and substantial differences in environmental conditions drive these observed population differences (Saba et al. 2008).

Important advances toward refining several input parameters present an opportunity to improve the Wallace et al. (2006) energy budget model. First, the original model used the only available data for sea jelly energy content (i.e., a colonial tunicate, *Pyrosomas atlantica*; Davenport and Balazs 1991). However, Doyle et al. (2007) reported energy and nutrient content values for three other species that are known leatherback prey (*Cyanea capillata, Rhizostoma octopus,* and *Chrysaora hysoscella*). Second, new estimates of migratory travel rates (km d^{-1}) are available for the EP (Shillinger et al. 2008) and NA (Fossette et al. 2010) populations. Third, there are new estimates of feeding rates, handling times, ingestion rates, and prey sizes (Fossette et al. 2012; Heaslip et al. 2012). Fourth, biophysical modeling and empirical measurements of leatherback activity and thermoregulation (Bostrom and Jones 2007; Bostrom et al. 2010) provide improved estimates of swimming activity patterns and thermal gradients that leatherbacks can maintain given different water temperatures, turtle body sizes, and activity levels.

We incorporated these new data and updated the Wallace et al. (2006) bioenergetics model to provide revised estimates of the energy intake and feeding rates required for female leatherbacks with different body sizes and reproductive outputs to meet reproductive energy requirements. This approach is intended to bracket possible energy budget scenarios given different life history traits (e.g., seasonal reproductive outputs, body sizes) and energy costs (e.g., different metabolic rates [MRs] for different phases). For specific model details, see Wallace et al. (2006); we briefly describe only the updated parameters here.

To calculate reproductive energy (RE) costs, we used the equation

$$RE = N + E + I + M$$

where N = nesting activity, E = egg clutches, I = internesting periods, and M = migrations to and from nesting areas. We then added foraging and maintenance costs to RE and calculated daily feeding rates necessary to accrue sufficient energy for reproduction at different remigration intervals (RIs). We estimated foraging and maintenance costs during the interval beginning at the end of nesting season 1 and ending at the beginning of nesting season 2, but not during the round-trip migration (~230–250 d), or the breeding season (~60 d) when turtles would be near nesting beaches and not actively foraging, because migration and inter-nesting period costs were included in RE. We used new values for prey energy content (202 kJ kg^{-1} at assumed 80% assimilation rate; Doyle et al. 2007). We also used observed prey intake rates for different feeding strategies; these included many small jellies (approximately 150 kg d^{-1}; Fossette et al. 2012) or fewer large jellies (330 kg d^{-1}; Heaslip et al. 2012), and used minimum estimates of migration distances and travel rates for the EP (4,000 km each way, ~40 km d^{-1}; Shillinger et al. 2008) and NA (5,000 km each way, ~40km d^{-1}; Fossette et al. 2010) populations. Based on improved estimates of leatherback thermoregulation in different thermal environments (Bostrom and Jones 2007; Bostrom et al. 2010), we adjusted MR values for both migration and foraging phases to 0.9 W kg^{-1}. This corresponds to biophysical modeling predictions for the MR required for a leatherback to maintain the approximate 8.0°C thermal gradient observed in temperate feeding areas (James and Mrosovsky 2004) and the metabolic rate required for a leatherback to swim nearly constantly, as reported by Eckert (2002). However, to reflect the lower thermal gradient, and corresponding lower MRs, necessary for the tropical feeding strategy exhibited by some leatherback populations (Benson et al. 2011; Fossette et al. 2012), we also considered an alternative scenario using a slightly lower value (i.e., maximum inter-nesting field metabolic rate (FMR) = 0.74 W kg^{-1}; Wallace et al. 2005) during the foraging phase. These input values for observed swimming speeds and travel rates of leatherbacks during these phases generally agree with the corresponding predicted thermal gradients (Bostrom and Jones 2007).

As Wallace et al. (2006) reported, migration costs dominate the reproductive energy budget (table 13.1). Foraging and maintenance costs differ depending on which foraging MR is used, resulting in differences in required feeding rates of roughly 20 kg d^{-1} between the high and low MR scenarios. Overall, estimated feeding rates range from 86–135 kg d^{-1}, in agreement with estimated daily feeding rates predicted by growth rate and biomass intake population modeling (Jones et al. 2012). Total energy costs (reproduction + maintenance + foraging) increase and daily feeding rates decrease with prolonged remigration intervals (fig. 13.2), because daily feeding rates required to meet the fixed cost of RE are divided over a longer time period, while costs of foraging and movements in nonbreeding years continue to accumulate over time. This corroborates findings of Price et al. (2004) who reported that delayed remigration confers no reproductive output benefit to female leatherbacks in either enhanced growth rates or reproductive output. Furthermore, our analyses show that delayed remigration is, in fact, more energetically costly than minimizing the RI. These results indicate that leatherbacks should acquire sufficient energy resources to remigrate and reproduce as soon as possible to maximize their lifetime reproductive output.

The low-cost (in EP), low-foraging MR scenario was the only one under which a feeding rate of 100 kg d^{-1} was sufficient to meet both foraging (plus maintenance) and reproductive costs, but this would result in remigration intervals of 8 years (fig. 13.2). At minimum observed feeding rates of 150 kg d^{-1} (Fossette et al. 2012), leatherbacks under the EP scenario could achieve remigration intervals of 3 years with either high- or low-foraging MR, while leatherbacks in the high-cost (in NA) scenario would have remigration intervals of 4 years (low-foraging MR) or more than 8 years (high-foraging MR). At feeding rates of 330 kg d^{-1}, leatherbacks in any scenario could achieve remigration intervals of 2 years (fig. 13.2).

This holistic energy budget approach illustrates the many trade-offs leatherbacks can make to reduce the length of remigration intervals, thereby increasing lifetime reproductive output. For example, leatherbacks can reduce energetic costs of thermoregulating in colder water temperatures by foraging in tropical areas, which could also have the benefit of shortening migrations, the largest component of total energy costs (Wallace et al. 2006; table 13.1). Furthermore, turtles could maintain lower daily feeding rates to meet lower energy costs, but would have to find sufficiently high densities of prey to support this strategy (Fossette et al. 2012). However, the "more and smaller prey" strategy observed by Fossette et al. (2012) would still impose significant costs of ingesting excess water in the process of ingesting small prey items. Leatherbacks would still need to deal with thermoregulatory costs of heating ingested water, and osmoregulatory costs under this strategy are likely very high. Alternatively, turtles could expend more energy to migrate longer distances to cold-water feeding areas, but forage for large, higher-energy prey items in adequate patches to meet energy demands (James et al. 2005; Heaslip et al. 2012). Feeding opportunistically during the breeding season (Casey et al. 2010) or during post-nesting migrations (James et al. 2005) would also reduce required feeding times on foraging grounds.

Our updated bioenergetics model also highlights the discrepancy between estimated reproductive en-

Table 13.1. Summary of energy and prey biomass requirements for leatherback turtles under lower cost (i.e., fewer eggs, smaller body sizes; "East Pacific") and higher costs (i.e., more eggs, larger body sizes; "North Atlantic"), and lower and higher foraging, metabolic rate scenarios. See text for details.

Component	Lower costs (small size, fewer eggs, shorter migrations) "Pacific"		Higher costs (larger size, more eggs, longer migrations) "Atlantic"	
Nesting events	14		18	
Egg clutches	438		567	
Inter-nesting periods	327		397	
Round-trip migration	4,829		6,804	
Total reproductive energy costs (RE) (kJ)	5,607		7,786	
Prey biomass required (kg)	27,758		38,545	
	Higher foraging MR	Lower foraging MR	Higher foraging MR	Lower foraging MR
Daily foraging + maintenance costs (kJ d^{-1})	20,995	17,286	27,216	22,408
Prey biomass required (kg d^{-1})	104	86	135	111
Prey biomass remaining @ 100 kg d^{-1}	–4	14	–35	–11
Foraging days to acquire RE	NA	1,623	NA	NA
Prey biomass remaining @ 150 kg d^{-1}	46	64	15	39
Foraging days to acquire RE	508	363	591	433
Prey biomass remaining @ 330 kg d^{-1}	196	214	165	189
Foraging days to acquire RE	119	109	233	204
Required feeding rate (for 2 yr RI)	180	160	246	219
Required feeding rate (for 4 yr RI)	133	114	175	150
Required feeding rate (for 8 yr RI)	117	98	153	128

ergy costs and predicted remigration intervals based on required feeding rates for EP and NA leatherback populations. Consistent with the Wallace et al. (2006) model, we found that smaller, less fecund turtles should meet energy demands faster than larger, more fecund turtles if resources are equally available to both populations. For example, EP leatherbacks could achieve remigration intervals of two years by feeding close to the rates observed by Fossette et al. (2012) (table 13.1; fig. 13.2), and could halve their observed remigration intervals from four to two years through plausible increases in feeding rates (20–30%) that would still be half the rates observed for leatherbacks feeding in Nova Scotia, Canada (Heaslip et al. 2012). In contrast to model predictions, this pattern is the opposite of that reported by long-term monitoring data at index beaches for both nesting populations (see Wallace and Saba 2009 for review). Saba et al. (2008) demonstrated that the feeding areas used by EP leatherbacks showed the highest variability among feeding areas globally, which has resulted in the least fecund, both in number of eggs and length of remigration intervals, leatherbacks in the world. Furthermore, this resource limitation could make EP leatherbacks less resilient to anthropogenic threats (Wallace et al. 2006; Wallace and Saba 2009). Interpreting actual foraging strategies used by leatherbacks in southeastern Pacific feeding areas (Shillinger et al. 2008; Shillinger et al. 2011) in a bioenergetics context could clarify the spatiotemporal nature of resource limitation and how energetic trade-offs affect life history and population demographics of EP leatherbacks.

Despite our attempts to introduce greater detail to our bioenergetics model, it is likely that we have drastically oversimplified, and thus underestimated, energetic costs of simultaneous thermoregulation and feeding in cold waters. For example, Bostrom et al. (2010) estimated that adult leatherbacks foraging in cold temperate waters must maintain MRs that are two to four times greater than resting MRs in order to generate and retain sufficient heat to maintain a thermal gradient of up to 20°C. Furthermore, measurements of swimming speed of adult leatherbacks showed wide variations between being stationary (Southwood et al. 2005) to maximum swim speeds of nearly 3.0 m s^{-1} for short periods of about 20 sec (Eckert 2002). Thus, swimming activity and associated MRs during active feeding in cold water are probably higher than the swim speeds and MRs we used, and vary greatly as turtles search for prey, dive, travel, and move through different water temperatures. More sophisticated approaches to

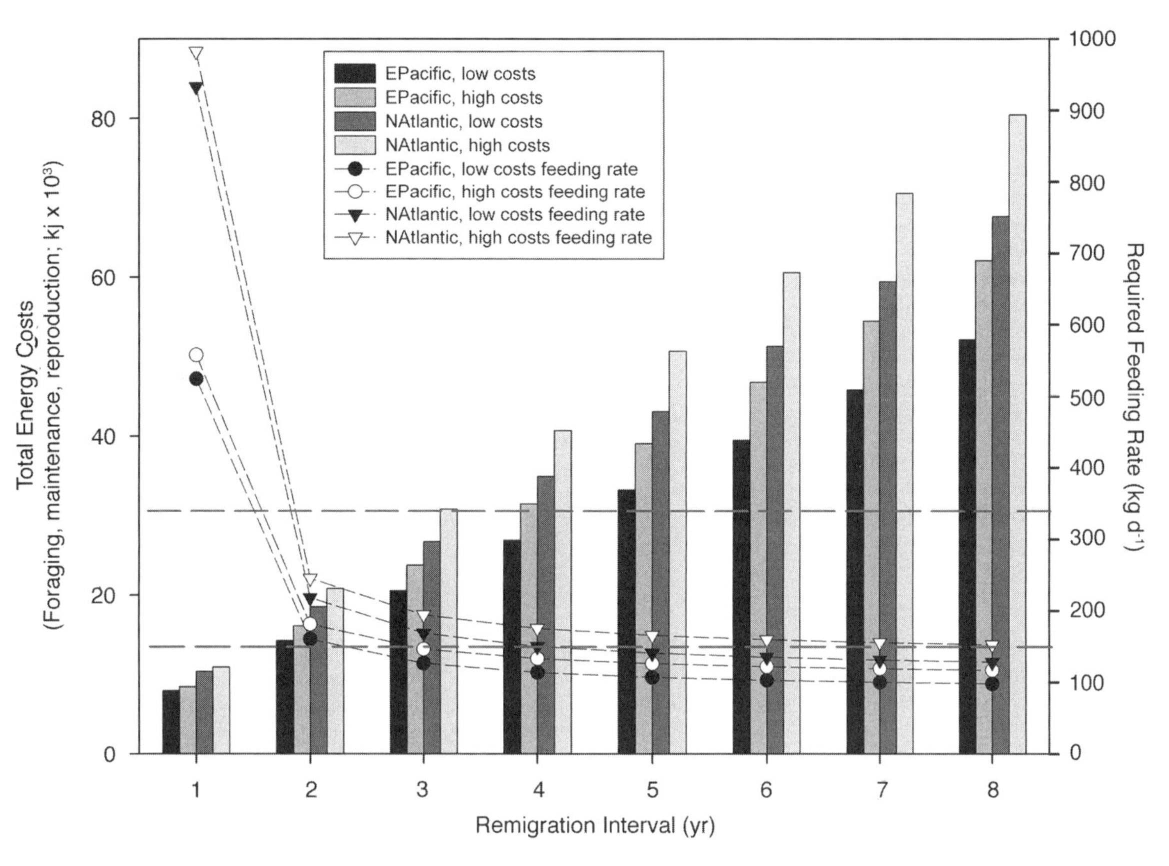

Fig. 13.2. Estimates of energy costs and feeding rate requirements for various remigration intervals for leatherback populations from the East Pacific and North Atlantic. Bars represent total energetic costs (reproductive energy requirements + foraging and maintenance costs, kJ × 10^3) for a given remigration interval per reproductive season. Short dashed lines represent feeding rates required to meet reproductive energy cost for each remigration interval. Long dashed lines show the minimum (150 kg d^{-1}; Fossette et al. 2012) and maximum (330 kg d^{-1}; Heaslip et al. 2012) feeding rates observed for leatherbacks in the wild. See table 13.1 and text for details on model inputs and results.

modeling energy demands and trade-offs for leatherbacks during migration and foraging phases (i.e., using time-activity budgets that account for variation in swimming activity and MRs related to different thermal regimes and diving/swimming patterns) would provide worthwhile refinements to the energy budget model presented here. Nonetheless, our model emphasizes the importance of considering distinct physiological adaptations and processes within a holistic context of bioenergetics—that is, it emphasizes how leatherbacks physiologically and behaviorally respond to energetic and biophysical demands on their life history strategies.

New Applications: Leatherback Conservation Physiology

Given recent declines in leatherback populations in some parts of the world, especially in the East Pacific Ocean (Wallace et al. 2011), integrating insights from leatherback physiological ecology into management strategies would have useful and pertinent applications to leatherback conservation. For example, new information on leatherback physiology can be used to improve input parameters for demographic modeling of leatherback populations. Specifically, improved estimates of each component of leatherback energy budgets and MRs during different behavioral phases would enhance understanding of energy requirements, patterns of food intake and assimilation, and growth dynamics, which could all inform estimates of vital rates such as life-stage durations and AFR. Estimated growth rates and AFR values obtained for posthatchling leatherbacks in controlled temperature and resource conditions (Jones et al. 2011) could be integrated with natural variations in water temperatures, competing energy demands, and resource variability to generate more nuanced estimates of growth rates and age at maturity, perhaps across distinct populations. Improving these models and refining their predictions would inform the development of appropriate targets for recovery by incorporating information about leatherback energy requirements in the context of environmentally driven resource availability (Wallace and Saba 2009).

Another application of insights from leatherback physiological ecology to management could be to develop or improve methods for reducing leatherback bycatch (chapter 17). Bycatch is the primary threat to sea turtles, including leatherbacks, worldwide (Wallace et al. 2011), so characterizing how horizontal and vertical habitat use is constrained by leatherbacks' physiological parameters (e.g., temperature tolerance, oxygen-limited dive durations) could influence distribution, timing, and methods used by fishing fleets in which bycatch occurs. For example, upper latitudinal boundaries have been defined based on temporal shifts in leatherback distribution related to sea surface temperature isotherms in the Atlantic (James et al. 2005; McMahon and Hays 2006; Sherrill-Mix 2007) and Pacific Oceans (Benson et al. 2011; Shillinger et al. 2011). Taken together, integrating physiological constraints on leatherback behavior and distributions could provide valuable guidance for fishing activities to minimize leatherback bycatch, as has occurred with loggerhead turtles (*Caretta caretta*) in the Hawaii shallow longline fishery (Howell et al. 2008), and could also be incorporated into broader marine spatial planning efforts (Shillinger et al. 2011).

Conservation physiology is a largely untapped, but potentially a highly valuable pursuit that not only connects processes, adaptations, and systems from the individual to the population, but also can inform how those populations are managed. We look forward to marine turtle conservation efforts inspiring renewed interest in answering interesting and important questions in marine turtle physiological ecology.

ACKNOWLEDGMENTS

We thank the editors, J. R. Spotila and P. Santidrián Tomillo, for the invitation to write this chapter. Support from the Goldring Family Foundation and the Offield Family Foundation is gratefully acknowledged by B. P. Wallace. Keith Bigelow is thanked by T. T. Jones for allowing him the time to work on this chapter. On a personal note, we gratefully recognize the inspiration and guidance we received from our mentors, J. R. Spotila, F. V. Paladino, D. R. Jones, and P. L. Lutz, who have been true trailblazers in research of sea turtle physiological ecology.

LITERATURE CITED

Avens, L., J. C. Taylor, L. R. Goshe, T. T. Jones, and M. Hastings. 2009. Use of skeletochronological analysis to estimate the age of leatherback sea turtles *Dermochelys coriacea* in the western North Atlantic. Endangered Species Research 8: 165–177.

Bels, V., F. Rimblot-Baly, and J. Lescure. 1988. Croissance et maintien en captivité de la tortue luth Dermochelys coriacea (Vandelli, 1761). Revue française d'aquariologie 15: 59–64.

Benson, S. R., T. Eguchi, D. G. Foley, K. A. Forney, H. Bailey, C. Hitipeuw, B. P. Samber, et al. 2011. Large-scale movements and high-use areas of western Pacific leatherback turtles, *Dermochelys coriacea*. Ecosphere 2(7): art84. doi: 10.1890/ES11-00053.1.

Birkenmeier, E. 1971. Juvenile leatherback turtles, *Dermochelys coriacea* (Linnaeus), in captivity. Brunei Museum Journal 3: 160–172.

Bjorndal, K. A. 1997. Foraging ecology and nutrition of sea turtles. In: P. L. Lutz and J. A. Musick (eds.), The biology of sea turtles. CRC Press, Boca Raton, FL, USA, pp. 199–231.

Block, B. A., J. R. Finnerty, A. F. Stewart, and J. Kidd. 1993. Evolution of endothermy in fish: mapping physiological traits on a molecular phylogeny. Science 260: 210–214.

Bolten, A. B. 2003. Variation in sea turtle life history patterns: neritic vs. oceanic developmental stages. In: P. L. Lutz, J. W. Wyneken, and J. A. Musick (eds.), The biology of sea turtles. CRC Press, Boca Raton, FL, USA, pp. 243–257.

Bostrom, B. L., and D. R. Jones. 2007. Exercise warms adult leatherback turtles. Comparative Biochemistry and Physiology, Part A: Molecular & Integrative Physiology 147: 323–331.

Bostrom, B. L., T. T. Jones, M. Hastings, and D. R. Jones. 2010. Behaviour and physiology: the thermal strategy of leatherback turtles. PloS ONE 5: e13925. doi: 10.1371.

Casey, J., J. Garner, S. Garner, and A. S. Williard. 2010. Diel foraging behavior of gravid leatherback sea turtles in deep waters of the Caribbean Sea. Journal of Experimental Biology 213: 3961–3971.

Chaloupka, M., and J. A. Musick. 1997. Age, growth, and population dynamics. In: P. L. Lutz and J. A. Musick (eds.), The biology of sea turtles. CRC Press, Boca Raton, FL, USA, pp. 233–276.

Clusella Trullas, S., J. R. Spotila, and F. V. Paladino. 2006. Energetics during hatchling dispersal of the olive ridley turtle *Lepidochelys olivacea* using doubly labeled water. Physiological and Biochemical Zoology 79: 389–399.

Congdon, J. D. 1989. Proximate and evolutionary constraints on energy relations of reptiles. Physiological Zoology 62: 356–373.

Davenport, J. 1997. Temperature and the life-history strategies of sea turtles. Journal of Thermal Biology 22: 479–488.

Davenport, J., and G. H. Balazs. 1991. 'Fiery bodies'—Are Pyrosomas an important component of the diet of leatherback turtles? British Herpetological Society Bulletin 37: 33–38.

Davenport, J., J. Fraher, E. Fitzgerald, P. McLaughlin, T. Doyle, L. Harman, and T. Cuffe. 2009a. Fat head: an analysis of head and neck insulation in the leatherback turtle (*Dermochelys coriacea*). Journal of Experimental Biology 212: 2753–2759.

Davenport, J., J. Fraher, E. Fitzgerald, P. McLaughlin, T. Doyle, L. Harman, T. Cuffe, et al. 2009b. Ontogenetic changes in tracheal structure facilitate deep dives and cold water foraging in adult leatherback sea turtles. Journal of Experimental Biology 212: 3440–3447.

Davenport, J., D. Holland, and J. East. 1990. Thermal and biochemical characteristics of the lipids of the leatherback turtle *Dermochelys coriacea*: evidence of endothermy. Journal of the Marine Biological Association of the United Kingdom 70: 33–41.

Deraniyagala, P.E.P. 1939. The tetrapod reptiles of Ceylon, Vol. I. Testudinates and crocodilians. Dulau, London, UK.

Doyle, T. K., J. D. Houghton, R. McDevitt, J. Davenport, and G. C. Hays. 2007. The energy density of jellyfish: estimates from bomb-calorimetry and proximate-composition. Journal of Experimental Marine Biology and Ecology 343: 239–252.

Duron, M. 1978. Contribution a l'étude de la biologie de *Dermochelys coriacea* (Linne) dans les Pertuis Charentais. PhD these, Universite Bordeaux, France.

Eckert, S. A. 2002. Swim speed and movement patterns of gravid leatherback sea turtles (*Dermochelys coriacea*) at St Croix, US Virgin Islands. Journal of Experimental Biology 205: 3689–3697.

Fossette, S., P. Gaspar, Y. Handrich, Y. L. Maho, and J. Y. Georges. 2008. Dive and beak movement patterns in leatherback turtles *Dermochelys coriacea* during internesting intervals in French Guiana. Journal of Animal Ecology 77: 236–246.

Fossette, S., C. Girard, M. López-Mendilaharsu, P. Miller, A. Domingo, D. Evans, L. Kelle, et al. 2010. Atlantic leatherback migratory paths and temporary residence areas. PloS ONE 5: e13908. doi: 10.1371.

Fossette, S., A. C. Gleiss, J. P. Casey, A. R. Lewis, and G. C. Hays. 2012. Does prey size matter? Novel observations of feeding in the leatherback turtle (*Dermochelys coriacea*) allow a test of predator–prey size relationships. Biology Letters 8: 351–354.

Frair, W. 1970. The world's largest living turtle. Salt Water Aquarium 5: 235–241.

Frair, W., R. Ackman, and N. Mrosovsky. 1972. Body temperature of *Dermochelys coriacea*: warm turtle from cold water. Science 177: 791–793.

Gaspar, P., S. R. Benson, P. H. Dutton, A. Réveillère, G. Jacob, C. Meetoo, A. Dehecq, and S. Fossette. 2012. Oceanic dispersal of juvenile leatherback turtles: going beyond passive drift modeling. Marine Ecology Progress Series 457: 265–284.

Goff, G., and J. Lien. 1988. Atlantic leatherback turtles, *Dermochelys coriacea*, in cold water off Newfoundland and Labrador. Canadian Field-Naturalist 102: 1–5.

Goff, G. P., and G. B. Stenson. 1988. Brown adipose tissue in leatherback sea turtles: A thermogenic organ in an endothermic reptile? Copeia 1988: 1071–1075.

Greer, A. E., J. D. Lazell, and R. M. Wright. 1973. Anatomical evidence for a counter-current heat exchanger in the leatherback turtle (*Dermochelys coriacea*). Nature 244: 181.

Heaslip, S. G., S. J. Iverson, W. D. Bowen, and M. C. James. 2012. Jellyfish support high energy intake of leatherback sea turtles (*Dermochelys coriacea*): video evidence from animal-borne cameras. PloS ONE 7: e33259. doi: 10.1371.

Howell, E. A., D. R. Kobayashi, D. M. Parker, G. H. Balazs, and J. J. Polovina. 2008. TurtleWatch: a tool to aid in the bycatch reduction of loggerhead turtles *Caretta caretta* in the Hawaii-based pelagic longline fishery. Endangered Species Research 5: 267–278.

Hudson, D. M., and P. L. Lutz. 1986. Salt gland function in the leatherback sea turtle, *Dermochelys coriacea*. Copeia 1986: 247–249.

James, M. C., and N. Mrosovsky. 2004. Body temperatures of leatherback turtles (*Dermochelys coriacea*) in temperate waters off Nova Scotia, Canada. Canadian Journal of Zoology 82: 1302–1306.

James, M. C., R. A. Myers, and C. A. Ottensmeyer. 2005. Behaviour of leatherback sea turtles, *Dermochelys coriacea*, during the migratory cycle. Proceedings of the Royal Society B 272: 1547–1555.

Jones, T. T., B. L. Bostrom, M. D. Hastings, K. S. Van Houtan, D. Pauly, and D. R. Jones. 2012. Resource requirements of the Pacific leatherback turtle population. PloS ONE 7: e45447. doi: 10.1371.

Jones, T. T., M. D. Hastings, B. L. Bostrom, R. D. Andrews, and D. R. Jones. 2009. Validation of the use of doubly labeled water for estimating metabolic rate in the green turtle (*Chelonia mydas* L.): a word of caution. Journal of Experimental Biology 212: 2635–2644.

Jones, T. T., M. D. Hastings, B. L. Bostrom, D. Pauly, and D. R. Jones. 2011. Growth of captive leatherback turtles, *Dermochelys coriacea*, with inferences on growth in the wild: implications for population decline and recovery. Journal of Experimental Marine Biology and Ecology 399: 84–92.

Jones, T. T., M. Salmon, J. Wyneken, and C. Johnson. 2000. Rearing leatherback hatchlings: protocols, growth and survival. Marine Turtle Newsletter 90: 3–6.

Kooistra, T. A., and D. H. Evans. 1976. Sodium balance in the green turtle, *Chelonia mydas*, in seawater and freshwater. Journal of Comparative Physiology B: Biochemical, Systemic, and Environmental Physiology 107: 229–240.

Lutcavage, M., and P. L. Lutz. 1986. Metabolic rate and food energy requirements of the leatherback sea turtle, *Dermochelys coriacea*. Copeia 1986: 796–798.

Lutz, P. L. 1997. Salt, water and pH balance in the sea turtle. In: P. L. Lutz and J. A. Musick (eds.), The biology of sea turtles. CRC Press, Boca Raton, FL, USA, pp. 343–362.

Mansfield, K. L., J. Wyneken, D. Rittschof, M. Walsh, C. W. Lim, and P. M. Richards. 2012. Satellite tag attachment methods for tracking neonate sea turtles. Marine Ecology Progress Series 457: 181–192.

McMahon, C. R., and G. C. Hays. 2006. Thermal niche, large-scale movements and implications of climate change for a critically endangered marine vertebrate. Global Change Biology 12: 1330–1338.

Mills, C. E., and R. G. Vogt. 1984. Evidence that ion regulation in hydromedusae and ctenophores does not facilitate vertical migration. Biological Bulletin 166: 216–227.

Minnich, J. E. 1982. The use of water. In: C. Gans and F. H. Pough (eds.), Biology of the Reptilia, Vol 12, Physiology C, Physiological ecology. Academic Press, New York, NY, USA, pp. 325–395.

Mrosovsky, N., and P. Pritchard. 1971. Body temperatures of *Dermochelys coriacea* and other sea turtles. Copeia 1971: 624–631.

Myers, A. E., and G. C. Hays. 2006. Do leatherback turtles *Dermochelys coriacea* forage during the breeding season? A combination of data-logging devices provide new insights. Marine Ecology Progress Series 322: 259–267.

Nicoloson, S. W., and P. L. Lutz. 1989. Salt gland function in the green sea turtle *Chelonia mydas*. Journal of Experimental Biology 144: 171–184.

Ortiz, R. M., R. M. Patterson, C. E. Wade, and F. M. Byers. 2000. Effects of acute fresh water exposure on water flux rates and osmotic responses in Kemp's ridley sea turtles (*Lepidochelys kempii*). Comparative Biochemistry and Physiology, Part A: Molecular & Integrative Physiology 127: 81–87.

Paladino, F. V., M. P. O'Connor, and J. R. Spotila. 1990. Metabolism of leatherback turtles, gigantothermy, and thermoregulation of dinosaurs. Nature 344: 858–860.

Penick, D. N., J. R. Spotila, M. P. O'Connor, A. C. Steyermark, R. H. George, C. J. Salice, and F. V. Paladino. 1998. Thermal independence of muscle tissue metabolism in the leatherback turtle, *Dermochelys coriacea*. Comparative Biochemistry and Physiology, Part A: Molecular & Integrative Physiology 120: 399–403.

Phillips, E. 1977. Raising hatchlings of the leatherback turtle, *Dermochelys coriacea*. British Journal of Herpetology 5: 677–678.

Prange, H. D. 1985. Renal and extra-renal mechanisms of salt and water regulation of sea turtles: a speculative review. Copeia 1985: 771–776.

Prange, H. D., and L. Greenwald. 1980. Effects of dehydration on the urine concentration and salt gland secretion of the green sea turtle. Comparative Biochemistry and Physiology, Part A: Physiology 66: 133–136.

Price, E. R., B. P. Wallace, R. D. Reina, J. R. Spotila, F. V. Paladino, R. Piedra, and E. Vélez. 2004. Size, growth, and reproductive output of adult female leatherback turtles *Dermochelys coriacea*. Endangered Species Research 5: 1–8.

Reina, R. D., K. J. Abernathy, G. J. Marshall, and J. R. Spotila. 2005. Respiratory frequency, dive behaviour and social interactions of leatherback turtles, *Dermochelys coriacea* during the inter-nesting interval. Journal of Experimental Marine Biology and Ecology 316: 1–16.

Reina, R. D., T. T. Jones, and J. R. Spotila. 2002. Salt and water regulation by the leatherback sea turtle *Dermochelys coriacea*. Journal of Experimental Biology 205: 1853–1860.

Rhodin, A. G. 1985. Comparative chondro-osseous development and growth of marine turtles. Copeia 1985: 752–771.

Rhodin, A. G., J. A. Ogden, and G. J. Conlogue. 1981. Chondro-osseous morphology of *Dermochelys coriacea*, a marine reptile with mammalian skeletal features. Nature 290: 244–246.

Rhodin, J.A.G., A. G. Rhodin, and J. R. Spotila. 1996. Electron microscopic analysis of vascular cartilage canals in the humeral epiphysis of hatchling leatherback turtles, *Dermochelys coriacea*. Chelonian Conservation and Biology 2: 250–260.

Saba, V. S., J. R. Spotila, F. P. Chavez, and J. A. Musick. 2008. Bottom-up and climatic forcing on the worldwide population of leatherback turtles. Ecology 89: 1414–1427.

Salmon, M., T. T. Jones, and K. W. Horch. 2004. Ontogeny of diving and feeding behavior in juvenile seaturtles: leatherback sea turtles (*Dermochelys coriacea* L) and green sea turtles (*Chelonia mydas* L) in the Florida Current. Journal of Herpetology 38: 36–43.

Schmidt-Nielsen, K., and R. Fange. 1958. Salt glands in marine reptiles. Nature 182: 783–785.

Sherrill-Mix, S. A. 2007. An analysis of migratory cues and potential tagging effects in the leatherback sea turtle. MS thesis, Dalhousie University, Halifax, Nova Scotia, Canada.

Shillinger, G. L., D. M. Palacios, H. Bailey, S. J. Bograd, A. M. Swithenbank, P. Gaspar, B. P. Wallace, et al. 2008. Persistent leatherback turtle migrations present opportunities for conservation. PLoS Biology 6: e171. doi: 10.1371.

Shillinger, G. L., A. M. Swithenbank, H. Bailey, S. J. Bograd, M. R. Castelton, B. P. Wallace, J. R. Spotila, et al. 2011. Vertical and horizontal habitat preferences of post-nesting leatherback turtles in the South Pacific Ocean. Marine Ecology Progress Series 422: 275–289.

Southwood, A., R. Andrews, F. V. Paladino, and D. Jones. 2005. Effects of diving and swimming behavior on body temperatures of Pacific leatherback turtles in tropical seas. Physiological and Biochemical Zoology 78: 285–297.

Southwood, A., R. Reina, V. Jones, J. Speakman, and D. Jones. 2006. Seasonal metabolism of juvenile green turtles (*Chelonia mydas*) at Heron Island, Australia. Canadian Journal of Zoology 84: 125–135.

Stewart, K., C. Johnson, and M. H. Godfrey. 2007. The minimum size of leatherbacks at reproductive maturity, with a review of sizes for nesting females from the Indian, Atlantic and Pacific Ocean basins. Herpetological Journal 17: 123–128.

Sullivan, S. K., and M. Field. 1991. Ion transport across mammalian small intestine. In: S. G. Schultz (ed.), Handbook of physiology—the gastrointestinal system, Vol 4. American Physiological Society, Bethesda, MD, USA, pp. 287–301.

Available in Comprehensive physiology, Wiley Online Library. doi: 10.1002/cphy.cp060410.
Wallace, B. P., A. D. DiMatteo, A. B. Bolten, M. Y. Chaloupka, B. J. Hutchinson, F. A. Abreu-Grobois, J. A. Mortimer, et al. 2011. Global conservation priorities for marine turtles. PloS ONE 6: e24510. doi: 10.1371.
Wallace, B. P., and T. T. Jones. 2008. What makes marine turtles go: a review of metabolic rates and their consequences. Journal of Experimental Marine Biology and Ecology 356: 8–24.
Wallace, B. P., S. S. Kilham, F. V. Paladino, and J. R. Spotila. 2006. Energy budget calculations indicate resource limitation in Eastern Pacific leatherback turtles. Marine Ecology Progress Series 318: 263–270.
Wallace, B. P., and V. S. Saba. 2009. Environmental and anthropogenic impacts on intra-specific variation in leatherback turtles: opportunities for targeted research and conservation. Endangered Species Research 7: 11–21.
Wallace, B. P., C. L. Williams, F. V. Paladino, S. J. Morreale, R. T. Lindstrom, and J. R. Spotila. 2005. Bioenergetics and diving activity of internesting leatherback turtles *Dermochelys coriacea* at Parque Nacional Marino Las Baulas, Costa Rica. Journal of Experimental Biology 208: 3873–3884.
Zug, G., and J. Parham. 1996. Age and growth in leatherback turtles, *Dermochelys coriacea* (Testudines: Dermochelyidae): a skeletochronological analysis. Chelonian Conservation and Biology 2: 244–249.

14

Movements and Behavior of Adult and Juvenile Leatherback Turtles

GEORGE L. SHILLINGER AND
HELEN BAILEY

Sea turtles were initially considered tropical or subtropical species, but increasing reports in northern waters began to raise suspicions that these were not merely accidental sightings (Bleakney 1965). Stranding records showed that leatherback turtles (*Dermochelys coriacea*) visited areas that were very distant from their known nesting colonies (Margaritoulis 1986). This included the stranding of the largest leatherback turtle ever recorded, a male weighing 916 kg with a curved carapace length of over 2.5 m, found dead on a beach in Wales (Eckert and Luginbuhl 1988). Live sightings and fisheries reports revealed that leatherbacks could survive in cool waters (Willgohs 1957; Lazell 1980). Large size and other physiological adaptations allow leatherbacks to maintain warm body temperatures even in northern waters (Paladino et al. 1990). Leatherbacks are dietary specialists feeding on gelatinous zooplankton (Bjorndal 1997), which are seasonally abundant in many temperate and coastal areas (Gibbons and Richardson 2009). The turtles regularly travel between tropical breeding grounds and temperate foraging habitats and are among the greatest animal migrants.

In this chapter we describe current knowledge about the remarkable migrations of leatherback turtles, their foraging strategies, and the role of dispersal in the movement of hatchlings. We also examine leatherbacks' finer-scale movements during the nesting season.

Migration

The large scale of leatherback turtle migrations was not understood until early capture-recapture tagging studies revealed that they were the longest journeys on record for any reptile. Pritchard (1976) described the recapture of six leatherbacks in the United States, Mexico, and Africa, five of which had traveled more than 5,000 km from tagging sites in Surinam and French Guiana. Although early tracking studies provided valuable insights into leatherback migration, these efforts were limited by very low return-recapture rates and the coarseness of the technology.

The advent of satellite telemetry enabled researchers to develop extensive tracking datasets documenting the movements and behaviors of leatherback turtles (for a review, see Godley et al. 2008). These technologies have the advantage of providing global coverage that allows turtles to be tracked in real time, or close to it, at any location. Incredibly, they show that leatherbacks travel across entire ocean basins in both the Atlantic (Hays, Houghton, and Myers 2004) and Pacific Oceans (Benson et al. 2011).

Leatherbacks exhibit homing to their natal nesting beaches (Dutton et al. 1999), and site fidelity to specific feeding grounds (James, Ottensmeyer, and Meyers 2005). These areas may be thousands of kilometers apart, suggesting that leatherbacks have a remarkable navigation sense. However, unlike on land, there are no visual landmarks in the open ocean. They must therefore be using other navigational cues. Sea turtles have a magnetic compass sense and leatherbacks detect the geomagnetic field even in the absence of visible light (Lohmann and Lohmann 1993). The magnetic field intensity and inclination angle vary predictably around the world and this provides positional information. Green turtles (*Chelonia mydas*) and loggerheads (*Caretta caretta*) employ bicoordinate geomagnetic maps (Lohmann et al. 2004, Lohmann et al. 2012) that provide both longitudinal and latitudinal information (Putnam et al. 2011).

Knowing their position and heading is important because ocean currents could drift leatherbacks off course (Gaspar et al. 2006; Sale and Luschi 2009). In the eastern Pacific, removing the drift effect of the ocean currents revealed that post-nesting female leatherbacks migrated in a relatively straight SSW heading (Shillinger et al. 2008). Turtles appeared to respond to current-induced drifting by increasing their southward velocity. Such a response provides support for the hypothesis that leatherbacks are not only capable of following a compass direction, but are also aware of their absolute position.

Leatherbacks may also use waves, wind, and olfactory or auditory cues as they approach their destination (Lohmann et al. 2008). Although experiments testing these mechanisms have not been performed, it is likely leatherbacks use multiple cues for navigation, and that different combinations may be used over long versus short distances (Lohmann et al. 2008). It is estimated that leatherback turtles are able to determine their position on average to within 96 km during the pelagic phase of their migration (Flemming et al. 2010).

Although we have a reasonable understanding of adult female leatherback migrations because they return to nest on land where they can be observed and tagged, very little is known about the movements of male and juvenile leatherback turtles. James, Eckert, and Myers (2005) deployed satellite tags on male leatherbacks on the foraging grounds off Nova Scotia and tracked them to waters near nesting colonies in the Caribbean Sea and Western Atlantic. This direct migration from temperate foraging grounds to tropical breeding areas suggests some mating may occur near the nesting colonies (Reina et al. 2005). Two males returned in consecutive years, indicating breeding site fidelity and potentially more frequent migrations to the breeding grounds than in female turtles (James, Eckert, and Myers 2005).

In contrast, juveniles and non-nesting adult females more commonly complete their migratory cycle in pelagic waters (James, Ottensmeyer, and Myers 2005). Similar loops to low latitude pelagic waters during the winter have been observed in the northeastern (Benson et al. 2011) and southeastern Pacific Ocean (Shillinger et al. 2011). These migrations are therefore more likely to be related to seasonal changes in temperature and/or prey abundance (James, Myers, and Ottensmeyer 2005).

We now describe in more detail what is known about leatherback migrations in the Atlantic, Indian, and Pacific Oceans (Bailey, Fossette, et al. 2012; fig. 14.1). These movements have mainly been derived from satellite tracking of adult females.

Atlantic Ocean

Pritchard (1976) noted that movement away from the North Atlantic nesting grounds occurred in all directions. Satellite tracking studies confirm these widely dispersed migratory movements throughout the Caribbean, Gulf of Mexico, and North Atlantic Ocean (Ferraroli et al. 2004; James, Ottensmeyer, and Meyers 2005; Eckert 2006; Eckert et al. 2006; Hays et al. 2006). Sighting and stranding records also show a year round presence of leatherbacks in the Mediterranean (Casale et al. 2003). Fossette, Hobson, et al. (2010) classified movements within the North Atlantic as involving round-trip, northern, and equatorial strategies. The round-trip strategy results in leatherbacks reaching high latitudes at the end of the summer/early autumn and then heading south again (Fossette, Hobson, et al. 2010). Leatherbacks using the northern strategy head northeast during the first post-nesting year, reaching the Irish Sea and Bay of Biscay the following spring. Finally, turtles using the equatorial strategy make eastward movements into the tropical Atlantic during the first post-nesting year and move north the next summer. Eckert (2006) reported similar movements across the North Atlantic for post-nesting leatherbacks tagged in Trinidad. However, turtles tagged at Florida nesting beaches tend to reside over the continental shelf for much of the first post-nesting year (Eckert et al. 2006).

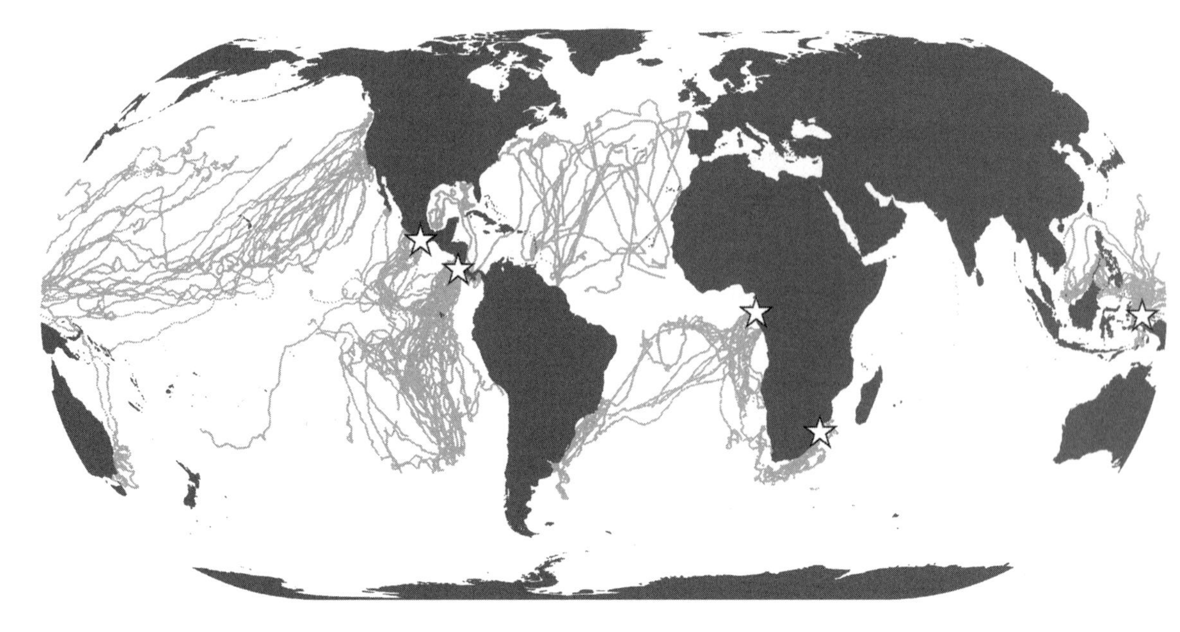

Fig. 14.1. Satellite tracks for 171 leatherback turtles within the Pacific, Atlantic, and Indian Oceans obtained from 1993 to 2010 demonstrating their large-scale migrations between nesting and foraging sites (Hays et al. 2006; Luschi et al. 2006; Fossette et al. 2010a; Witt et al. 2011; Bailey et al. 2012a). Stars indicate general nesting locations for each population featured within the image.

North and South Atlantic nesting populations appear to be spatially segregated (Fossette, Girard, et al. 2010). In the South Atlantic, tracks from 25 female leatherbacks tagged at their largest rookery in Gabon, Central Africa, exhibit three dispersal strategies with turtles moving to habitats in (1) the equatorial Atlantic, (2) temperate habitats off South America, and (3) temperate habitats off southern Africa (Witt et al. 2011). Leatherbacks captured off South America remain in the southwestern Atlantic (López-Mendilaharsu et al. 2009; Fossette, Girard, et al. 2010).

Indian Ocean

Limited data are available on leatherback movements in the Indian Ocean. Hughes et al. (1998), Luschi et al. (2003), Sale et al. (2006), and Lambardi et al. (2008) have tracked the circuitous offshore movements of leatherbacks around the tip of South Africa, and into the southeast Atlantic and Southern Ocean. Their findings suggest that leatherbacks utilize strong currents within the Agulhas Current system to passively transit large distances while searching for prey aggregated within mesoscale eddies and convergence zones (Luschi et al. 2003; Lambardi et al. 2008). They also share foraging areas within the productive Benguela Current with some leatherbacks from the South Atlantic (Witt et al. 2011).

Pacific Ocean

There are two regional populations that nest in the eastern and western Pacific (Dutton et al. 1999). Similar to the Atlantic, there appears to be spatial segregation between these two Pacific populations (Bailey, Benson, et al. 2012). In the eastern Pacific, Morreale et al. (1996) described a narrow migration corridor from nesting grounds in Playa Grande, Costa Rica, toward the Galápagos Islands. A later tracking study confirmed the presence of this migration corridor and the subsequent dispersal of leatherbacks into the Southeast Pacific Ocean (Shillinger et al. 2008; Shillinger et al. 2011). Turtles from nesting beaches in Mexico also migrate to these distant habitats in the Southeast Pacific (Eckert and Sarti 1997). Their migration pathways are similar, but also include coastal areas off South America (Eckert and Sarti 1997). Bycatch reports indicate frequent interactions with fisheries in these coastal areas (Donoso and Dutton 2010; Alfaro-Shigueto et al. 2011). Saba, Shillinger, et al. (2008) and Shillinger et al. (2008) hypothesized this coastal migration may represent a relic strategy as a result of intensive anthropogenic pressures (fisheries) within nearshore habitats off Central and South America that have eliminated many of the turtles that formerly used these areas as foraging grounds.

Satellite tracking of western Pacific leatherbacks revealed far more diverse migration strategies than

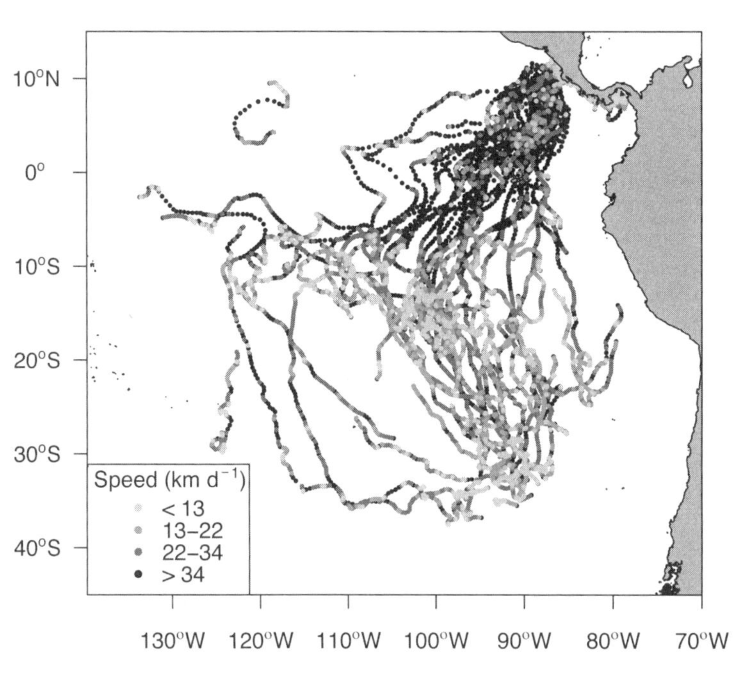

Fig. 14.2. Switching state-space model-derived daily positions for eastern Pacific leatherback turtles tagged at Playa Grande, Costa Rica (n = 46), shaded by speed (darker shading corresponds with higher speeds). Switching state models predict turtle positions on days when there is no signal fix. Slower speeds are indicative of foraging behavior. Redrawn from Shillinger et al. 2011.

in the eastern Pacific (Benson et al. 2011). Migratory pathways differed between females that nest during the boreal summer versus the winter. Winter nesters headed south into the tropical and temperate waters of the southwestern Pacific Ocean. Contrastingly, those nesting in the summer moved into the temperate North Pacific or the tropical waters of the South China Sea. Several of the turtles that migrated across the North Pacific headed to foraging grounds in the California Current ecosystem. Turtles tagged off California, United States, either returned to the western Pacific nesting beaches, or headed to the equatorial eastern Pacific for two to three months, returning to the California Current the following spring (Benson et al. 2011).

Foraging

Leatherback turtles are likely to rely on foraging areas that have dense concentrations of prey, because gelatinous zooplankton have a relatively low caloric content (Lutcavage and Lutz 1986; Fossette et al. 2012). Distribution of leatherback turtles shows a spatial association with estimates of gelatinous organisms in the northeastern Atlantic (Houghton et al. 2006; Witt et al. 2007). However, the patchy and ephemeral nature of gelatinous zooplankton distributions, together with the fragility of these organisms within plankton tow nets, have confounded efforts to establish a time series of abundance and distribution (Boero et al. 2008; Gibbons and Richardson 2009). It is therefore difficult to directly compare leatherback distributions with that of their prey. In many cases, chlorophyll a concentration is used as a proxy for productivity and hence prey availability.

Leatherback foraging behavior can be indirectly identified by changes in both horizontal and vertical movements (Jonsen et al. 2007). When prey are patchily distributed, a predator may be expected to remain within a patch once prey have been encountered by decreasing their speed and/or increasing their turning angle (Kareiva and Odell 1987; fig.14.2). Foraging behavior is also associated with shallower, shorter dives (Hays, Houghton, et al. 2004). These shallower dives (< 200 m) tend to occur at low and high latitudes around the edges of gyres (Fossette, Hobson, et al. 2010; Shillinger et al. 2011; Bailey, Fossette et al. 2012). Many foraging areas have high chlorophyll a concentrations (Saba, Spotila, et al. 2008; Bailey, Benson, et al. 2012) and high estimates of zooplankton biomass (Fossette, Hobson, et al. 2010), indicating high prey densities. They are also generally associated with current systems, and mesoscale features and convergences that act to aggregate prey (Eckert 2006; Doyle et al. 2008; Bailey, Benson, et al. 2012). An analysis of movement of Pacific leatherbacks in three dimensions, in relation to environmental covariates from a coupled biophysical oceanographic model, showed that movement patterns are related to patches of food at the surface and at depth (Schick et al. 2013).

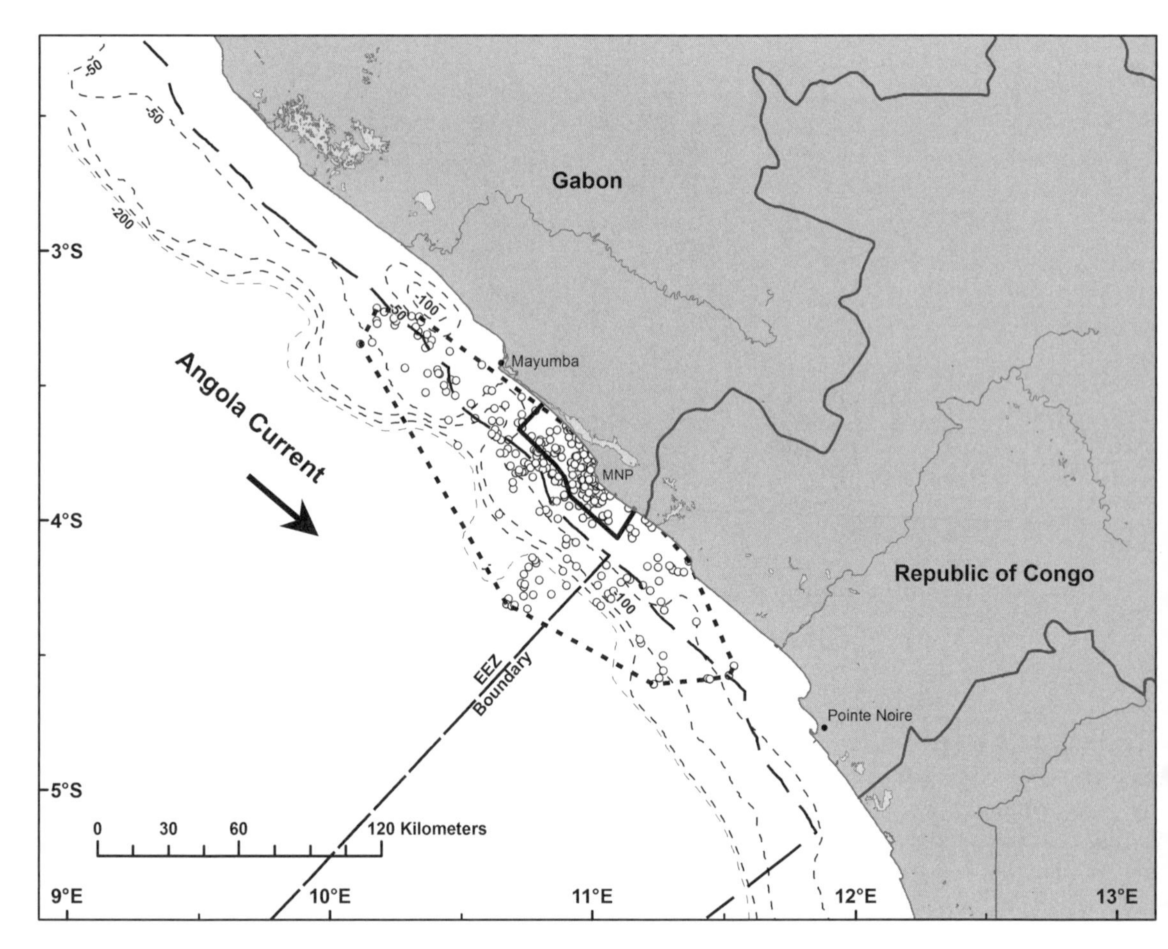

Fig. 14.3. Leatherback inter-nesting positions for turtles nesting in Mayumba National Park (black polygon labeled MNP) in Gabon on the equatorial West African coast. The 95% minimum convex polygon (MCP) is represented by the dashed black line with smaller dashes, and bathymetric contours by dashed gray lines. The dominant ocean current (Angola Current) is shown by the solid black arrow, and the Exclusive Economic Zone (EEZ) boundaries as long dashed black lines. Redrawn from Witt et al. (2008).

In the North Atlantic, foraging grounds exist off the northeastern United States, eastern Canada, Ireland, Spain, Portugal, and West Africa (James, Ottensmeyer, and Myers 2005; Eckert 2006; Eckert et al. 2006; Hays et al. 2006; Doyle et al. 2008). However, Hays et al. (2006) proposed that leatherbacks may modify their behavior during transit in response to local conditions, particularly during the first post-nesting year when there is a strong need to replenish energy reserves quickly. In the South Atlantic, high residence times indicating foraging grounds are identified off South Africa, South America, and in the South Equatorial Current (López-Mendilaharsu et al. 2009; Fossette, Girard, et al. 2010; Witt et al. 2011). On two separate occasions, males captured within pelagic longline fisheries spent long periods of time at the Rio de la Plata estuary off Uruguay, indicating that this may also be an important feeding area (López-Mendilaharsu et al. 2009; Fossette, Girard, et al. 2010).

Turtles nesting in the western Pacific forage within the South China Sea, Indonesian seas, the California Current, and the East Australian Current Extension (Benson et al. 2011). There are also pelagic foraging areas in the Kuroshio Extension and equatorial eastern Pacific (Benson et al. 2011). Eastern Pacific nesters have mainly pelagic foraging areas in the southeastern Pacific Ocean, although coastal foraging also occurs off Central and South America (Eckert and Sarti 1997; Shillinger et al. 2008; Shillinger et al. 2011).

Saba et al. (2007) reported a relationship between the El Niño Southern Oscillation and reproductive frequency of eastern Pacific leatherbacks, where El Niño events corresponded with a lower probability of leatherbacks remigrating to nest in Costa Rica. Warmer, less productive El Niño events when the thermocline is depressed result in less food available for leatherbacks. The ability of leatherbacks to build up sufficient en-

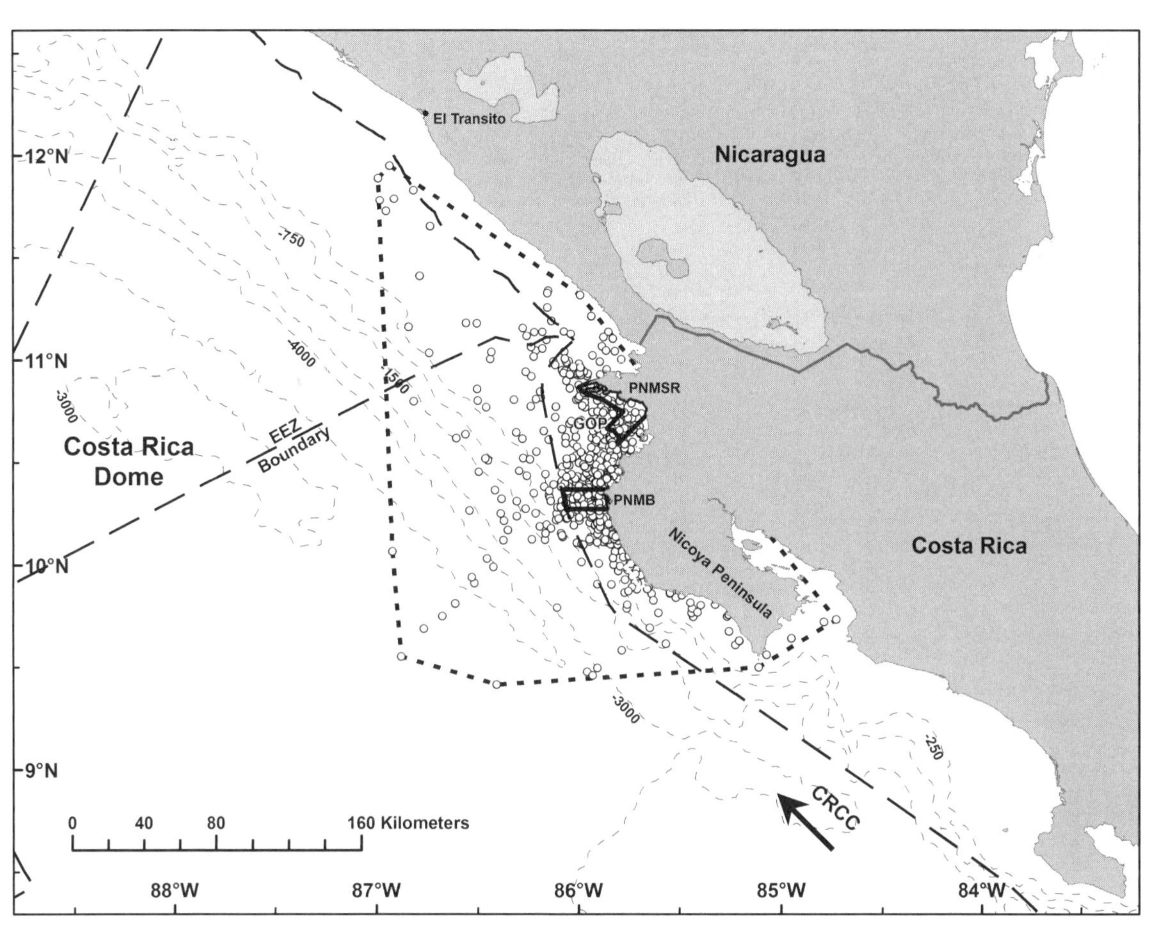

Fig. 14.4. Leatherback inter-nesting positions for turtles nesting on Playa Grande, Costa Rica, where Las Baulas National Marine Park (PNMB) and Santa Rosa National Marine Park (PNMSR) are shown as black polygons (Shillinger et al. 2010). The 95% minimum convex polygon (MCP) is represented by a dashed black line (short dashes) and bathymetric contours as dashed gray lines. The dominant ocean current (CRCC, Costa Rica Coastal Current) is shown by solid black arrow, and the Exclusive Economic Zone (EEZ) boundaries as long dashed black lines.

ergy reserves for reproduction therefore depends on their success at the foraging grounds (Bailey, Fossette, et al. 2012).

Inter-nesting

During the inter-nesting period, leatherbacks tend to occupy coastal habitats near their nesting beaches (Eckert et al. 2006; Benson et al. 2007; Georges et al. 2007; Byrne et al. 2009; Shillinger et al. 2010). Occasionally, more extended offshore movements occur (Georges et al. 2007; Hitipeuw et al. 2007), and this can vary interannually (Shillinger et al. 2010). In both Gabon and Pacific Costa Rica, females spend considerable time outside protected areas (figs. 14.3, 14.4). Wallace et al. (2005) proposed that moving between different water temperatures, combined with low activity levels, could allow leatherbacks to avoid overheating while in warm tropical waters.

Although sea turtles are commonly considered to be capital breeders, relying on body reserves stored prior to reproduction, leatherbacks may forage opportunistically during the inter-nesting period (Myers and Hays 2006; Fossette et al. 2008). Dive patterns suggest nocturnal foraging within the deep scattering layer of plankton (Eckert et al. 1989). Records of beak movement patterns showed rhythmic mouth opening that indicate turtles use gustatory cues to sense the surrounding water (Myers and Hays 2006). Dives associated with beak movement events were isolated in time, suggesting that leatherbacks forage opportunistically on dispersed prey (Fossette et al. 2008). It is most likely that this foraging allows leatherbacks to compensate for reproductive costs, but may vary between nesting

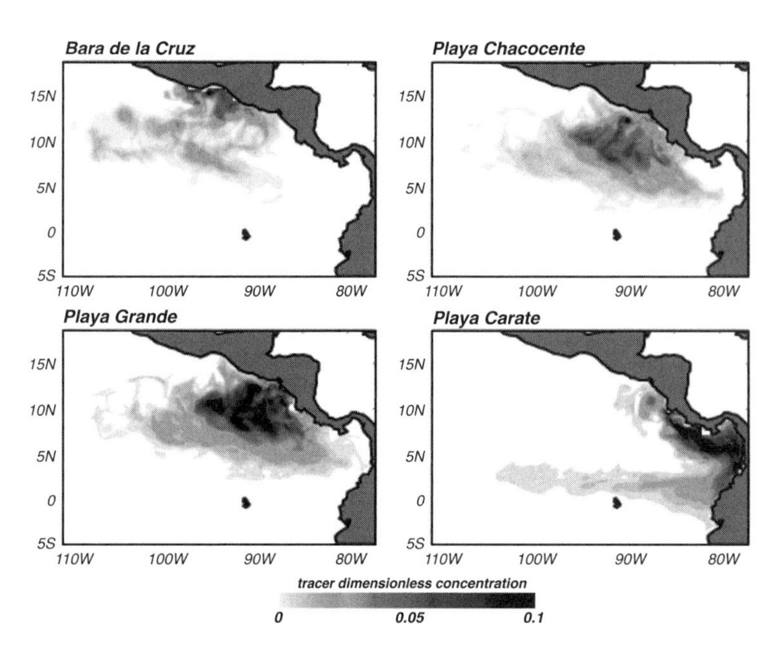

Fig. 14.5. Long-term (2000–2008) mean tracer concentrations on 1 June based on continuous tracer releases between 15 January and 15 April from nesting beaches at Barra de la Cruz, Playa Chacocente, Playa Grande, and Playa Carate. Black dots show tracer release locations. Redrawn from Shillinger et al. (2012).

locations depending on local food availability (Reina et al. 2005; Fossette et al. 2012).

The Lost Years

Migration pathways that adult leatherbacks use to return to the nesting beach are probably related to the dispersal pathways that they followed when they were hatchlings. Observations of sea turtles are generally lacking during life stages between hatchling and adult, the period coined the "lost years" by Carr (1980). Shillinger et al. (2012) used particle drift modeling to simulate leatherback hatchling dispersal patterns from four nesting beaches in the eastern Pacific, and found that hatchlings (e.g., inanimate particles) were more widely dispersed to productive oceanic habitats (e.g., the Costa Rica Dome) from beaches such as Playa Grande, Costa Rica, and Chacocente, Nicaragua, than from other beaches where nesting sometimes occurs north and south of Playa Grande (fig. 14.5). Modeled hatchlings from Playa Grande, Costa Rica, were most likely to be entrained and transported offshore by large-scale eddies coincident with the peak leatherback nesting and hatchling emergence period. These eddies potentially serve as "hatchling highways," providing a means of rapid offshore transport away from predation and toward a productive refuge within which newly hatched turtles can develop. The most important leatherback nesting beach remaining in the eastern Pacific (Playa Grande) was evolutionarily selected as an optimal nesting site owing to favorable ocean currents that enhance hatchling survival.

Ocean current variability determines the partition of leatherback hatchlings into different dispersal areas from western Pacific nesting beaches and largely influences juvenile survival rates at the population level (Gaspar et al. 2012). Simulations of drifting hatchlings combined with data from sightings, genetics, bycatch, and satellite tracking indicate that hatchlings emerging from the main New Guinea nesting beaches will be entrained into the North Pacific, South Pacific, or Indian Oceans. The animals that travel into the Indian Ocean are unlikely to survive.

Proximity to favorable ocean currents strongly influences sea turtle nesting distribution (Hamann et al. 2011). Loggerhead turtle nest density along the southeastern United States coast increases as distance to the Gulf Stream decreases (Putnam, Bane, and Lohmann 2010). Nesting distribution of Kemp's ridley turtles (*Lepidochelys kempii*) is associated with surface currents that facilitate successful migration to foraging grounds (Putnam, Shay, and Lohmann 2010). Variation in ocean circulation produces substantial annual variation in distribution and survival of simulated cohorts of Kemp's ridleys (Putnam et al. 2013). In the Mediterranean Sea, the pattern of adult loggerhead dispersion from nesting sites reflects the passive dispersion of hatchlings on ocean currents (Hays et al. 2010). Combining simulations from an ocean circulation model with data from genetic analysis, Putnam and Naro-Maciel (2013) predicted that transatlantic dispersal of green turtle hatchlings and juveniles is likely from Costa Rica, Guinea Bissau, and Ascension Island, and that many "lost year hotspots" are presently unstudied and unprotected.

Hatchlings don't just drift, but they swim as well, and that is important in their dispersal. Simulations suggest that offshore swimming aids newly hatched loggerheads to move past fronts and increases their probability of survival (Putnam, Scott, et al. 2012). Small amounts of oriented swimming in response to regional magnetic fields affects migratory routes of hatchling loggerheads in the Western Atlantic Ocean and helps them to remain in warm-water currents favorable for growth and survival (Putnam, Verley, et al. 2012).

Dispersal simulations of hawksbill (*Eretmochelys imbricata*) hatchlings from West Africa and northeastern Brazil indicated that passive drift alone would not explain the occurrence of juveniles at Ascension Island in the middle of the Atlantic. When oriented swimming was added to the model, simulated turtles reached the island in about five years, approximating the age of young turtles found there (Putnam et al. 2014). In the Pacific Ocean, juvenile leatherbacks have to swim toward lower latitudes before winter to stay warm and then swim back in spring to grow in higher latitudes, where food is abundant (Gaspar et al. 2012). Future efforts to simulate the movement of hatchling and juvenile sea turtles will be enhanced by including swimming behavior of animals and using ocean circulation models with resolution at the scale of days and tens of kilometers (Shillinger et al. 2012; Putnam and He 2013).

Conservation Implications

Movements of leatherback turtles throughout the stages of their life histories have important implications for their conservation (Shillinger et al. 2012). Their wide-ranging coastal and offshore movements expose them to considerable risk from anthropogenic pressures, such as being caught in fishing gear. There is an urgent need to develop integrated, enforceable regional conservation strategies that will protect these animals on land and at sea around the world (Dutton et al. 2005; Shillinger 2005; Georges et al. 2007; Shillinger et al. 2008). Understanding their migratory pathways and high-use areas provides essential information on where management efforts should be focused and where protection would be most effective.

LITERATURE CITED

Alfaro-Shigueto, J., J. C. Mangel, F. Bernedo, P. H. Dutton, J. A. Seminoff, and B. J. Godley. 2011. Small-scale fisheries of Peru: a major sink for marine turtles in the Pacific. Journal of Applied Ecology 48: 1432–1440.

Bailey, H., S. R. Benson, G. L. Shillinger, S. J. Bograd, P. H. Dutton, S. A. Eckert, S. J. Morreale, et al. 2012. Identification of distinct movement patterns in Pacific leatherback turtle populations influenced by ocean conditions. Ecological Applications 22: 735–747.

Bailey, H., S. Fossette, S. J. Bograd, G. L. Shillinger, A. M. Swithenbank, J. Y. Georges, P. Gaspar, et al. 2012. Movement patterns for a critically endangered species linked to foraging success and population status. PLoS ONE 7: e36401. doi: 36410.31371/journal.pone.0036401.

Benson, S. R., T. Eguchi, D. Foley, K. A. Forney, H. Bailey, C. Hitipeuw, B. P. Samber, et al. 2011. Large-scale movements and high use areas of western Pacific leatherback turtles, *Dermochelys coriacea*. Ecosphere 2(7): art84. doi: 10.1890/ES11-00053.1.

Benson, S. R., K. M. Kisokau, L. Ambio, V. Rei, P. H. Dutton, and D. Parker. 2007. Beach use, internesting movement, and migration of leatherback turtles, *Dermochelys coriacea*, nesting on the north coast of Papua New Guinea. Chelonian Conservation and Biology 6: 7–14.

Bjorndal, K. A. 1997. Foraging ecology and nutrition of sea turtles. In: P. L. Lutz and J. A. Musick (eds.), The biology of sea turtles. CRC Press, Boca Raton, FL, USA, pp. 199–231.

Bleakney, J. S. 1965. Reports of marine turtles from New England and eastern Canada. Canadian Field Naturalist 79: 120–128.

Boero, F., J. Bouillon, C. Gravili, M. P. Miglietta, T. Parsons, and S. Piraino. 2008. Gelatinous plankton: irregularities rule the world (sometimes). Marine Ecology Progress Series 356: 299–310.

Byrne, R., J. Fish, T. K. Doyle, and J.D.R. Houghton. 2009. Tracking leatherback turtles (*Dermochelys coriacea*) during consecutive inter-nesting intervals: further support for direct transmitter attachment. Journal of Experimental Marine Biology and Ecology 377: 68–75.

Carr, A. 1980. Some problems of sea turtle ecology. American Zoologist 20: 489–498.

Casale, P., P. Nicolosi, D. Freggi, M. Turchetto, and R. Argano. 2003. Leatherback turtles (*Dermochelys coriacea*) in Italy and in the Mediterranean Basin. Herpetological Journal 13: 135–139.

Donoso, M., and P. H. Dutton. 2010. Sea turtle bycatch in the Chilean pelagic longline fishery in the southeastern Pacific: opportunities for conservation. Biological Conservation 143: 2672–2684.

Doyle, T. K., J.D.R. Houghton, P. F. O'Súilleabháin, V. J. Hobson, F. Marnell, J. Davenport, and G. C. Hays. 2008. Leatherback turtles satellite-tagged in European waters. Endangered Species Research 4: 23–31.

Dutton, D. L., P. H. Dutton, M. Chaloupka, and R. H. Boulon. 2005. Increase of a Caribbean leatherback turtle *Dermochelys coriacea* nesting population linked to long-term nest protection. Biological Conservation 126: 186–194.

Dutton, P. H., B. W. Bowen, D. W. Owens, A. Barragán, and S. K. Davis. 1999. Global phylogeography of the leatherback turtle (*Dermochelys coriacea*). Journal of Zoology 248: 397–409.

Eckert, K. L., and C. Luginbuhl. 1988. Death of a giant. Marine Turtle Newsletter 43: 2–3.

Eckert, S. A. 2006. High-use oceanic areas for Atlantic leatherback sea turtles (*Dermochelys coriacea*) as identified using satellite telemetered location and dive information. Marine Biology 149: 1257–1267.

Eckert, S. A., D. Bagley, S. Kubis, L. Ehrhart, C. Johnson, K. Stewart, and D. Defreese. 2006. Internesting and postnesting movements and foraging habitats of leatherback sea turtles (*Dermochelys coriacea*) nesting in Florida. Chelonian Conservation and Biology 5: 239–248.

Eckert, S. A., K. L. Eckert, P. Ponganis, and G. L. Kooyman. 1989. Diving and foraging behavior by leatherback sea turtles (*Dermochelys coriacea*). Canadian Journal of Zoology 67: 2834–2840.

Eckert, S. A., and L. Sarti. 1997. Distant fisheries implicated in the loss of the world's largest leatherback nesting population. Marine Turtle Newsletter 78: 2–7.

Ferraroli, S., J. Y. Georges, P. Gaspar, and Y. Maho. 2004. Where leatherback turtles meet fisheries. Nature 429: 521–522.

Flemming, J. M., I. D. Jonsen, R. A. Myers, and C. A. Field. 2010. Hierarchical state-space estimation of leatherback turtle navigation ability. PLoS ONE 5: e14245. doi: 14210.11371/journal.pone.0014245.

Fossette, S., P. Gaspar, Y. Handrich, Y. Le Maho, and J. Y. Georges. 2008. Dive and beak movement patterns in leatherback turtles *Dermochelys coriacea* during internesting intervals in French Guiana. Journal of Animal Ecology 77: 236–246.

Fossette, S., C. Girard, M. López-Mendilaharsu, P. Miller, A. Domingo, D. Evans, L. Kelle, et al. 2010. Atlantic leatherback migratory paths and temporary residence areas. PLoS ONE 5: e13908. doi: 13910.11371/journal.pone.0013908.

Fossette, S., A. C. Gleiss, J. P. Casey, A. R. Lewis, and G. C. Hays. 2012. Does prey size matter? Novel observations of feeding in the leatherback turtle (*Dermochelys coriacea*) allow a test of predator-prey size relationships. Biology Letters 8: 351–354.

Fossette, S., V. J. Hobson, C. Girard, B. Calmettes, P. Gaspar, J. Y. Georges, and G. C. Hays. 2010. Spatio-temporal foraging patterns of a giant zooplanktivore, the leatherback turtle. Journal of Marine Systems 81: 225–234.

Gaspar, P., S. R. Benson, P. H. Dutton, A. Réveillère, G. Jacob, C. Meetoo, A. Dehecq, and S. Fossette. 2012. Oceanic dispersal of juvenile leatherback turtles: going beyond passive drift modeling. Marine Ecology Progress Series 457: 265–284.

Gaspar, P., J. Y. Georges, S. Fossette, A. Lenoble, S. Ferraroli, and Y. Maho. 2006. Marine animal behaviour: neglecting ocean currents can lead us up the wrong track. Proceedings of the Royal Society B 273: 2697–2702.

Georges, J. Y., S. Fossette, A. Billes, S. Ferraroli, J. Fretey, D. Grémillet, Y. Le Maho, et al. 2007. Meta-analysis of movements in Atlantic leatherback turtles during the nesting season: conservation implications. Marine Ecology Progress Series 338: 225–232.

Gibbons, M. J., and A. J. Richardson. 2009. Patterns of jellyfish abundance in the North Atlantic. Hydrobiologia 616: 51–65.

Godley, B. J., J. M. Blumenthal, A. C. Broderick, M. S. Coyne, M. H. Godfrey, L. A. Hawkes, and M. J. Witt. 2008. Satellite tracking of sea turtles: Where have we been and where do we go next? Endangered Species Research 4: 3–22.

Hamann, M., A. Grech, E. Wolanski, and J. Lambrechts. 2011. Modelling the fate of marine turtle hatchlings. Ecological Modeling 222: 1515–1521.

Hays, G. C., S. Fossette, K. A. Katselidis, P. Mariani, and G. Schofield. 2010. Ontogenetic development of migration: Lagrangian drift trajectories suggest a new paradigm for sea turtles. Journal of the Royal Society, Interface 7: 1319–1327.

Hays, G. C., V. J. Hobson, J. D. Metcalfe, D. Righton, and D. W. Sims. 2006. Flexible foraging movements of leatherback turtles across the North Atlantic ocean. Ecology 87: 2647–2656.

Hays, G. C., J.D.R. Houghton, C. Isaacs, R. S. King, C. Lloyd, and P. Lovell. 2004. First records of oceanic dive profiles for leatherback turtles, *Dermochelys coriacea*, indicate behavioural plasticity associated with long-distance migration. Animal Behaviour 67: 733–743.

Hays, G. C., J.D.R. Houghton, and A. E. Myers. 2004. Pan-Atlantic leatherback turtle movements. Nature 429: 522.

Hitipeuw, C., P. H. Dutton, S. Benson, J. Thebu, and J. Bakarbessy. 2007. Population status and internesting movement of leatherback turtles, *Dermochelys coriacea*, nesting on the northwest coast of Papua, Indonesia. Chelonian Conservation and Biology 6: 28–36.

Houghton, J.D.R., T. K. Doyle, M. W. Wilson, J. Davenport, and G. C. Hays. 2006. Jellyfish aggregations and leatherback turtle foraging patterns in a temperate coastal environment. Ecology 87: 1967–1972.

Hughes, G. R., P. Luschi, R. Mencacci, and F. Papi. 1998. The 7000-km oceanic journey of a leatherback turtle tracked by satellite. Journal of Experimental Marine Biology and Ecology 229: 209–217.

James, M. C., S. A. Eckert, and R. A. Myers. 2005. Migratory and reproductive movements of male leatherback turtles (*Dermochelys coriacea*). Marine Biology 147: 845–853.

James, M. C., R. A. Myers, and C. A. Ottensmeyer. 2005. Behaviour of leatherback sea turtles, *Dermochelys coriacea*, during the migratory cycle. Proceedings of the Royal Society B 272: 1547–1555.

James, M. C., C. A. Ottensmeyer, and R. A. Myers. 2005. Identification of high-use habitat and threats to leatherback sea turtles in northern waters: new directions for conservation. Ecology Letters 8: 195–201.

Jonsen, I. D., R. A. Myers, and M. C. James. 2007. Identifying leatherback turtle foraging behaviour from satellite-telemetry using a switching state-space model. Marine Ecology Progress Series 337: 255–264.

Kareiva, P., and G. Odell. 1987. Swarms of predators exhibit "preytaxis" if individual predators use area-restricted search. American Naturalist 130: 233–270.

Lambardi, P., J.R.E. Lutjeharms, R. Mencacci, G. C. Hays, and P. Luschi. 2008. Influence of ocean currents on long-distance movement of leatherback sea turtles in the Southwest Indian Ocean. Marine Ecology Progress Series 353: 289–301.

Lazell, J. D. 1980. New England waters: critical habitat for marine turtles. Copeia 1980: 290–295.

Lohmann, K. J., and C.M.F. Lohmann. 1993. A light-independent magnetic compass in the leatherback sea turtle. Biological Bulletin 185: 149–151.

Lohmann, K. J., C.M.F. Lohmann, L. M. Ehrhart, D. A. Bagley, and T. Swing. 2004. Geomagnetic map used in sea-turtle navigation. Nature 428: 909–910.

Lohmann, K. J., P. Luschi, and G. C. Hays. 2008. Goal navigation and island-finding in sea turtles. Journal of Experimental Marine Biology and Ecology 356: 83–95.

Lohmann, K. J., N. F. Putnam, and C.M.F. Lohmann. 2012. The magnetic map of hatchling loggerhead sea turtles. Current Opinion in Neurobiology 22: 336–342.

López-Mendilaharsu, M., C.F.D. Rocha, P. Miller, A. Domingo, and L. Prosdocimi. 2009. Insights on leatherback turtle movements and high use areas in the Southwest Atlantic Ocean. Journal of Experimental Marine Biology and Ecology 378 31–39.

Luschi, P., J.R.E. Lutjeharms, P. Lambardi, R. Mencacci, G. R. Hughes, and G. C. Hays. 2006. A review of migratory behaviour of sea turtles off southeastern Africa. South African Journal of Science 102: 51–58.

Luschi, P., A. Sale, R. Mencacci, G. R. Hughes, J.R.E. Lutjeharms, and F. Papi. 2003. Current transport of leatherback sea turtles (*Dermochelys coriacea*) in the ocean. Proceedings of the Royal Society B 270: S129–S132.

Lutcavage, M. E., and P. L. Lutz. 1986. Metabolic rate and food energy requirements of the leatherback sea turtle, *Dermochelys coriacea*. Copeia 1986: 796–798.

Margaritoulis, D. N. 1986. Captures and strandings of the leatherback sea turtles, *Dermochelys coriacea*, in Greece (1982–1984). Journal of Herpetology 20: 471–474.

Morreale, S. J., E. A. Standora, J. R. Spotila, and F. V. Paladino. 1996. Migration corridor for sea turtles. Nature 384: 319–320.

Myers, A. E., and G. C. Hays. 2006. Do leatherback turtles *Dermochelys coriacea* forage during the breeding season? A combination of data-logging devices provide new insights. Marine Ecology Progress Series 322: 259–267.

Paladino, F. V., M. P. O'Connor, and J. R. Spotila. 1990. Metabolism of leatherback turtles, gigantothermy, and thermoregulation of dinosaurs. Nature 344: 858–860.

Pritchard, P.C.H. 1976. Post-nesting movements of marine turtles (Cheloniidae and Dermochelyidae) tagged in the Guianas. Copeia 1976: 749–754.

Putnam, N. F., F. A. Abreu-Grobois, A. C. Broderick, C. Ciofi, A. Formia, B. J. Godley, S. Stroud, et al. 2014. Numerical dispersal simulations and genetics help explain the origin of hawksbill sea turtles on Ascension Island. Journal of Experimental Marine Biology and Ecology 450: 98–108.

Putnam, N. F., J. M. Bane, and K. J. Lohmann. 2010. Sea turtle nesting distributions and oceanographic constraints on hatchling migration. Proceedings of the Royal Society B 277: 3631–3637.

Putnam, N. F., C. S. Endres, C.M.F. Lohmann, and K. J. Lohmann. 2011. Longitude perception and bicoordinate magnetic maps in sea turtles. Current Biology 21: 463–466.

Putnam, N. F., and R. He. 2013. Tracking the long-distance dispersal of marine organisms: sensitivity to ocean model resolution. Journal of the Royal Society, Interface 10: 20120979. http://dx.doi.org/10.1098/rsif.2012.0979.

Putnam, N. F., K. L. Mansfield, R. He, D. J. Shaver, and P. Verley. 2013. Predicting the distribution of oceanic-stage Kemp's ridley sea turtles. Biology Letters 9: 20130345. http://dx.doi.org/10.1098/rsbl.2013.0345.

Putnam, N. F., and E. Naro-Maciel. 2013. Finding the "lost years" in green turtles: insights from ocean circulation models and genetic analysis. Proceedings of the Royal Society B 280: 20131468. http://dx.doi.org/10.1098/rspb.2013.1468.

Putnam, N. F., R. Scott, P. Verley, R. Marsh, and G. C. Hays. 2012. Natal site and offshore swimming influence fitness and long-distance ocean transport in young sea turtles. Marine Biology 159: 2117–2126.

Putnam, N. F., T. J. Shay, and K. J. Lohmann. 2010. Is the geographic distribution of nesting in the Kemp's ridley turtle shaped by the migratory needs of offspring? Integrative and Comparative Biology 50: 305–314.

Putman, N. F., P. Verley, T. J. Shay, and K. J. Lohmann. 2012. Simulating transoceanic migrations of young loggerhead sea turtles: merging magnetic navigation behavior with an ocean circulation model. Journal of Experimental Biology 215: 1863–1870.

Reina, R. D., K. J. Abernathy, G. J. Marshall, and J. R. Spotila. 2005. Respiratory frequency, dive behaviour and social interactions of leatherback turtles, *Dermochelys coriacea* during the inter-nesting interval. Journal of Experimental Marine Biology and Ecology 316: 1–16.

Saba, V. S., P. Santidrián-Tomillo, R. D. Reina, J. R. Spotila, J. A. Musick, D. A. Evans, and F. V. Paladino. 2007. The effect of the El Niño Southern Oscillation on the reproductive frequency of eastern Pacific leatherback turtles. Journal of Applied Ecology 44: 395–404.

Saba, V. S., G. L. Shillinger, A. M. Swithenbank, B. A. Block, J. R. Spotila, J. A. Musick, and F. V. Paladino. 2008. An oceanographic context for the foraging ecology of eastern Pacific leatherback turtles: consequences of ENSO. Deep-Sea Research I 55: 646–660.

Saba, V. S., J. R. Spotila, F. P. Chavez, and J. A. Musick. 2008. Bottom-up and climatic forcing on the worldwide population of leatherback turtles. Ecology 89: 1414–1427.

Sale, A., and P. Luschi. 2009. Navigational challenges in the oceanic migrations of leatherback turtles. Proceedings of the Royal Society B 276: 3737–3745.

Sale, A., P. Luschi, R. Mencacci, P. Lambardi, G. R. Hughes, G. C. Hays, S. Benvenuti, and F. Papi. 2006. Long-term monitoring of leatherback turtle diving behaviour during oceanic movements. Journal of Experimental Marine Biology and Ecology 328: 197–210.

Schick, R. S., J. J. Roberts, S. A. Eckert, P. N. Halpin, H. Bailey, F. Chai, L. Shi, and J. S. Clark. 2013. Pelagic movements of Pacific leatherback turtles (*Dermochelys coriacea*) highlight the role of prey and ocean currents. Movement Ecology 1: 11. http://www.movementecologyjournal.com/content/1/1/11.

Shillinger, G. L. 2005. The eastern tropical Pacific seascape: an innovative model for transboundary marine conservation. In: R. A. Mittermeier, C. F. Kormos, P.R.G. Mittermeier, T. Sandwith, and C. Besancon (eds.), Transboundary conservation: A new vision for protected areas. Conservation International, Washington, DC, USA, pp. 320–331.

Shillinger, G. L., H. Bailey, S. J. Bograd, E. L. Hazen, M. Hamann, P. Gaspar, B. J. Godley, et al. 2012. Tagging through the stages: technical and ecological challenges in observing life histories through biologging. Marine Ecology Progress Series 457: 165–170.

Shillinger, G. L., D. M. Palacios, H. Bailey, S. J. Bograd, A. M. Swithenbank, P. Gaspar, B. P. Wallace, et al. 2008. Persistent leatherback turtle migrations present opportunities for conservation. PLoS Biology 6: e171. doi: 110.1371/journal.pbio.0060171.

Shillinger, G. L., A. M. Swithenbank, H. Bailey, S. J. Bograd, M. R. Castelton, B. P. Wallace, J. R. Spotila, et al. 2011. Vertical and horizontal habitat preferences of post-nesting leatherback turtles in the South Pacific Ocean. Marine Ecology Progress Series 422: 275–289.

Shillinger, G. L., A. M. Swithenbank, S. J. Bograd, H. Bailey, M. R. Castelton, B. P. Wallace, J. R. Spotila, et al. 2010. Identification of high-use internesting habitats for eastern Pacific leatherback turtles: role of the environment and implications for conservation. Endangered Species Research 10: 215–232.

Wallace, B. P., C. L. Williams, F. V. Paladino, S. J. Morreale, R. T. Lindstrom, and J. R. Spotila. 2005. Bioenergetics and diving activity of internesting leatherback turtles *Dermochelys coriacea* at Parque Nacional Marino Las Baulas, Costa Rica. Journal of Experimental Biology 208: 3873–3884.

Willgohs, J. F. 1957. Occurrence of the leathery turtle in the northern North Sea and off western Norway. Nature 179: 163–164.

Witt, M. J., E. A. Bonguno, A. C. Broderick, M. S. Coyne, A. Formia, A. Gibudi, G. A. Mounguengui Mounguengui, et al. 2011. Tracking leatherback turtles from the world's largest rookery: assessing threats across the South Atlantic. Proceedings of the Royal Society B 278: 2338–2347.

Witt, M. J., A. C. Broderick, M. Coyne, A. Formia, S. Ngouessono, R. J. Parnell, G. P. Sounguet, et al. 2008. Satellite tracking highlights difficulties in the design of effective protected areas for leatherback turtles during the internesting period. Oryx 42: 296–300.

Witt, M. J., A. C. Broderick, D. J. Johns, C. Martin, R. Penrose, M. S. Hoogmoed, and B. J. Godley. 2007. Prey landscapes help identify potential foraging habitats for leatherback turtles in the NE Atlantic. Marine Ecology Progress Series 337: 231–244.

15

Relation of Marine Primary Productivity to Leatherback Turtle Biology and Behavior

VINCENT S. SABA,
CHARLES A. STOCK, AND
JOHN P. DUNNE

The Leatherback Diet: A Life Full of Jelly

Among sea turtles, leatherbacks (*Dermochelys coriacea*) have the fastest growth rates (chapter 13); achieve the largest body size (chapter 13); have the largest reproductive output (chapter 6); and are capable of the fastest travel speeds, deepest dives, and vast migrations that are sometimes across entire ocean basins (chapter 14). Leatherbacks have particularly large energy demands and substantial resource requirements (chapter 13) yet forage almost exclusively on gelatinous zooplankton (jellyfish), an energy-poor group of organisms consisting of 96% water (Doyle et al. 2007). Assuming leatherbacks mature at 15 to 20 years of age, an individual turtle would need to consume about 300 metric tons of gelatinous zooplankton from hatching to age of maturity (chapter 13). To achieve this foraging rate, leatherbacks must either feed on large organisms or on high-density patches of smaller organisms (Fossette et al. 2011). Mature females have the greatest resource requirements when nesting due to the vast amount of energy required for vitellogenesis (chapter 5) and the terrestrial nesting process (chapter 6).

The energy to support gelatinous zooplankton and leatherbacks must ultimately arise from primary producers at the base of the marine food web. In the following sections, we will discuss (1) marine primary productivity as the base of the marine food web; (2) the biophysical controls on gelatinous zooplankton; (3) leatherback foraging area associations with large phytoplankton productivity; (4) climate variability, large phytoplankton productivity, and the leatherback nesting response in the eastern tropical Pacific (ETP); and (5) bottom-up forcing on leatherback phenotype, fecundity, and nesting trends among populations worldwide.

Marine Primary Productivity and Bottom-up Forcing

Photosynthesis is the primary mechanism that fuels the base of food webs in terrestrial, aquatic, and marine ecosystems. The rate of photosynthesis (i.e., the rate at which organic carbon is synthesized from inorganic carbon using energy from light) accounts for the vast majority of net primary productivity (NPP) in both terrestrial and aquatic systems. Unlike phototrophs in the terrestrial environment that are predominantly macroscopic plants, those in the marine environment are typically microscopic and encompass a wider range of taxonomic groups that include plants, protists, and bacteria. Referred to as marine phytoplankton, recent estimates suggest that roughly one-third to one-half of Earth's photosynthetically fixed carbon derives from these organisms (Field et al. 1998; Carr et al. 2006; Welp et al. 2011).

Phytoplankton require sunlight and nutrients (nitrogen, phosphorous, iron and other trace metals) to fix carbon, and thus global patterns of NPP depend on the confluence of these two factors. The necessity of sunlight restricts productivity to a "euphotic zone," which generally extends over the upper 100 m or so of the water column. Nutrients in the euphotic zone are often cycled through plankton communities multiple times before sinking; therefore, the majority of production in most marine ecosystems is "recycled" (Eppley and Peterson 1979). However, sinking of a fraction of nutrient-laden organic material from the euphotic zone creates a persistent need for the resupply of nutrients from deeper waters to maintain NPP. In the open ocean, diverse physical processes, including wind-driven vertical mixing, upwelling, and circulation associated with mesoscale eddies, can accomplish this. In coastal waters, additional processes, such as riverine discharge, coastal runoff, mixing along shallow topography that allows nutrient resupply from sediments, and retention of nutrients near surface layers due to shallow water depths, can further augment NPP.

In marine food webs, the efficiency of carbon transfer from primary producers (phytoplankton) to upper trophic levels (zooplankton, fish, turtles, birds, whales) is a function of the magnitude of primary productivity, the community structure of the ecosystem, and the efficiency of trophic linkages (Ryther 1969). Ryther posited that the size of organisms plays a central role: "Since the size of an organism is an essential criterion of its potential usefulness to man, we have the following relationship: the larger the plant cells at the beginning of the food chain, the fewer the trophic levels that are required to convert the organic matter to a useful form." Therefore, ecosystems that have a high abundance of large phytoplankton (i.e., diatoms, dinoflagellates, cell sizes often >5 μm) relative to small phytoplankton (i.e., cryptophytes, cyanobacteria, cell sizes often < 5 μm) will have fewer trophic steps in the food chain up to the top predators and generally higher overall energy transfer efficiencies to higher trophic levels.

Large phytoplankton species are generally more prevalent in regions with large nutrient inputs (Falkowski et al. 2004). This pattern is thought to arise in part from the smaller surface area to volume ratio of large versus small cells that hinders the ability of large cells to obtain nutrients in nutrient-poor ecosystems; these ecosystems include subtropical gyres and the western Pacific equatorial warm pool that together account for about 50% of the global ocean. High proportions of large phytoplankton are thus common in eastern ocean boundary coastal upwelling systems, eastern equatorial systems (Pacific and Atlantic), western boundary eutrophic coastal systems, and temperate pelagic systems. Such ecosystems also produce some of the largest fishery yields (www.seaaroundus.org) and are also foraging hotspots for top predators (Block et al. 2011), including leatherback turtles (Saba, Spotila, et al. 2008).

Biophysical Controls on Gelatinous Zooplankton

Time-series studies of gelatinous zooplankton biomass and distribution in the world's oceans are extremely rare and typically restricted to coastal waters (Purcell 2012). A global review study of available data on epipelagic (0–200 m water depth) gelatinous zooplankton biomass reported that integrated biomass decreased with water column depth and that this relationship was driven by bottom-up forcing via NPP (Lilley et al. 2011). Essentially, shallow coastal areas are typically higher in NPP than offshore pelagic waters due to enhanced eutrophication from riverine input and coastal runoff, proximity to iron dust sources, coastal upwelling, mixing along shallow topography, and retention of nutrients near surface layers due to shallow water depths. Therefore, food resources for gelatinous zooplankton (phytoplankton, crustacean zooplankton, fish larvae) are likely in greater abundance in eutrophic coastal waters. In the northwestern Mediterranean Sea, a multidecadal study from 1974 to 2003 reported that bottom-up control was the primary mechanism controlling the annual standing stocks of gelatinous zooplankton over the entire time series (García-Comas et al. 2011). There is an association between the North Atlantic Oscillation (NAO) and winter vertical mixing, suggesting that the NAO climate index may be a proxy for gelatinous zooplankton within this ecosystem. Continuous plankton recorder data from the North Sea (Attrill et al. 2007) indicates that gelatinous zooplankton frequency is positively correlated to NAO and suggests a possible link to NPP. In the North Atlantic, positive NAO phases result in stronger westerly winds (and a deeper mixed layer) and thus delay the start of the subpolar spring phytoplankton bloom by two to three weeks (Henson et al.

2009). Large blooms and high biomass of gelatinous zooplankton, such as *Cnidaria* and *Ctenophora*, typically occur a few months after phytoplankton blooms (Saba, Spotila, et al. 2008 and references therein). This delay may reflect the time scales of trophic transfer in seasonal seas where leatherbacks are commonly observed in their foraging areas in the summer and fall months (chapter 14).

While bottom-up forcing via NPP is an essential control mechanism for gelatinous zooplankton biomass, the physical characteristics of the marine environment also exert direct influences on gelatinous zooplankton populations. Temperature can be a key constraint on gelatinous zooplankton growth at multiple life history stages (Purcell 2012). Physical retention areas, such as thermal fronts, haloclines, current boundaries, mesoscale eddies, and shelf breaks, are all associated with aggregations of gelatinous zooplankton. These retention areas have been shown to be targeted foraging areas for leatherback turtles (Benson et al. 2011). The accumulation of high gelatinous zooplankton biomass in such areas, however, is still ultimately subject to constraints imposed by the productivity of the ecosystem (Lilley et al. 2011).

Leatherback Foraging Hotspots: Swimming for Diatoms

Diatoms are commonly associated with the most productive marine ecosystems around the world. These often large, silica-bearing phytoplankton thrive in waters that receive large pulses of nutrients, such as coastal upwelling and high-latitude pelagic regions, where the diatoms can outcompete other phytoplankton groups due to their large vacuoles used for nutrient storage (Falkowski et al. 2004). The strong role of NPP as a determinant of gelatinous zooplankton production (Lilley et al. 2011), and phytoplankton size as a key determinant of the transfer efficiency of primary production to higher trophic levels (Ryther 1969), together suggest that estimates of diatom production are a useful metric for assessing the quality of leatherback foraging regions.

In situ measurements of marine NPP, let alone diatom productivity, are extremely sparse both spatially and temporally. Therefore, scientists must rely on models based on biogeochemical and physical principles to provide spatially and temporally continuous regional- and global-scale NPP estimates. The two common types of models used to estimate marine NPP are satellite ocean color models and biogeochemical ocean general circulation models (BOGCMs). The design of these two types of models is very different. Ocean color models use biological data measured by satellite sensors (sea surface chlorophyll a) combined with observed physical data (e.g., sea surface temperature) to derive NPP estimates. The BOGCMs, in contrast, are embedded in ocean circulation models with specified atmospheric forcing that provide estimates of physical ocean conditions as inputs to biogeochemical algorithms.

A satellite-derived ocean color model estimate (Uitz et al. 2010) of the mean proportion of large phytoplankton NPP for the global ocean from 1998 to 2007 (fig. 15.1) showed that diatoms are estimated to be relatively more productive (darker areas) in coastal upwelling waters (west coastlines of Africa and the Americas), coastal shelf waters (east coastlines of the Americas and Asia), pelagic subpolar and polar waters, and the Arabian Sea. Similarly, a BOGCM, the Tracers of Phytoplankton with Allometric Zooplankton (TOPAZ) model (Dunne et al. 2010 in Sarmiento et al. 2010), highlights similar regions that favor diatom growth (fig. 15.2). The main feature that all of these marine ecosystems have in common are substantial nutrient input into the euphotic zone.

Satellite tracking studies of post-nesting female leatherbacks from nesting beaches worldwide (chapter 14) show that most leatherbacks are foraging (based on high-use analysis) in waters where estimated diatom NPP is relatively high (fig. 15.1). Western Atlantic females that nest in St. Croix, Costa Rica, French Guiana, and Trinidad forage in the waters of the pelagic and coastal North Atlantic, the Mauritania coastal upwelling system, and the river deltas of the Gulf of Mexico (Fossette et al. 2010). Eastern Atlantic leatherbacks forage in the eastern equatorial Atlantic, the east coast of South America, and the Benguela upwelling system (Witt et al. 2011). Those nesting in South Africa forage in the Agulhas Current system and the Benguela upwelling system (Luschi et al. 2006). Most striking, western Pacific leatherbacks nesting in Papua Barat, Indonesia, migrate across the Pacific Ocean (for 10 to 12 months) to forage in the highly productive California Current upwelling system (Benson et al. 2011). Other populations from the western Pacific forage in the South China Sea, Tasman Front, East Australia Current Extension, and Kuroshio Extension (Benson et al. 2011). In the eastern Pacific, post-nesting females forage within and on the boundaries of the eastern portion of the South Pacific Gyre (Shillinger et al. 2011) and possibly, to a much lesser extent, the eastern equatorial Pacific upwelling system (Saba, Shillinger, et al. 2008) and the coastal upwelling systems of Peru (Alfaro-Shigueto et al. 2007; Alfaro-Shigueto et al. 2011) and Chile (Eckert and Sarti 1997).

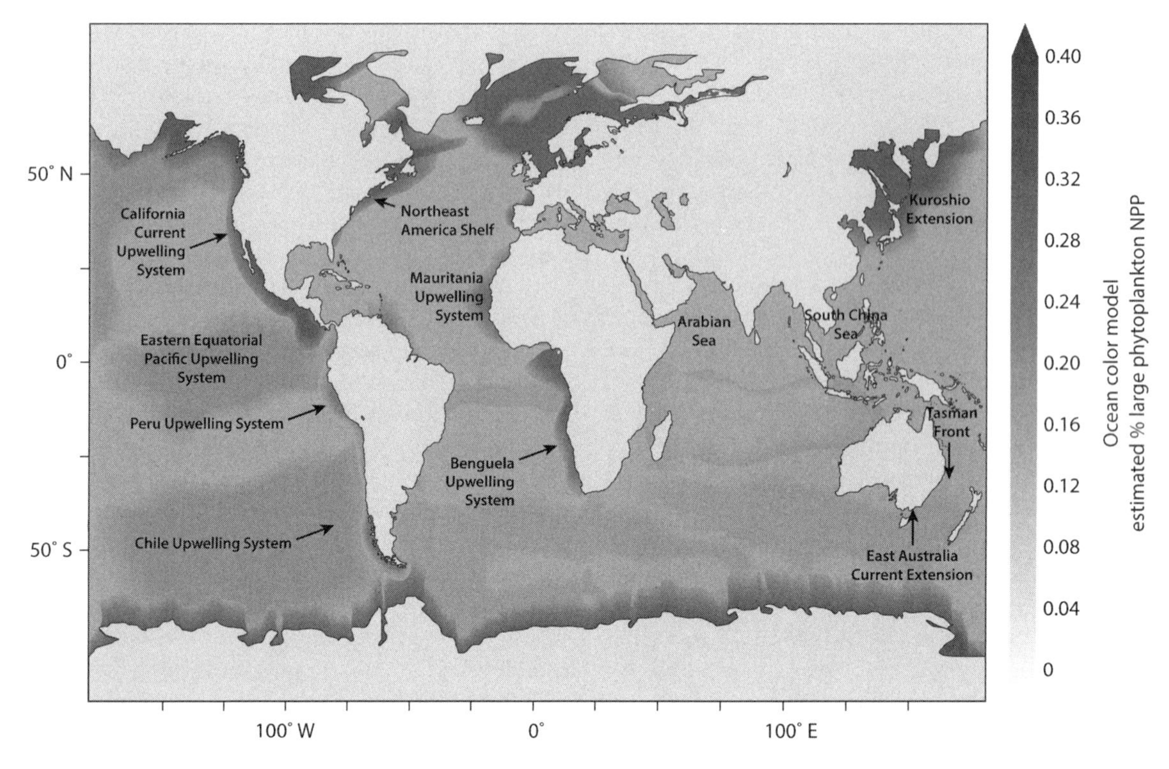

Fig. 15.1. Mean proportion of large phytoplankton (>20 μm cell size) NPP relative to small phytoplankton (<20 μm cell size) from 1997 to 2007 estimated by a phytoplankton size-class ocean color model (Uitz et al. 2010). The model estimates are based on level-3 SeaWiFS surface chlorophyll a data. Some of the waters discussed in this chapter are highlighted. Composite of NPP provided by Julia Uitz.

It is clear from these tracking studies (chapter 14) that most post-nesting females are typically foraging in waters that are rich in diatoms, consistent with the high potential trophic transfer efficiency of these ecosystems as described by Ryther (1969). Although data regarding the gelatinous zooplankton biomass in these waters are poor, the prevalence of foraging activity suggests that the leatherback prey items there are either very large in body size, smaller individuals in large aggregations, or both. This would explain the vast amount of energy that is expended by these turtles in order to reach these diatom rich waters—an expense best exemplified by the 10 to 12 month post-nesting migrations of western Pacific females across the entire Pacific basin to reach the diatom rich waters of the California Current upwelling system.

The Eastern Pacific Leatherback Nesting Model

A prime example of the connection between large phytoplankton NPP and leatherback ecology derives from a nesting model for the eastern Pacific leatherback population in Costa Rica (Saba et al. 2007). The model predicts the number of returning nesting females (remigrants) based on physical and biological oceanographic variables that are associated with the El Niño Southern Oscillation (ENSO). The ENSO is a coupled ocean-atmosphere natural climate oscillation in the tropical Pacific that is associated with global climate variability. The La Niña phase of ENSO is associated with cooler sea surface temperature (SST) and a shallower thermocline in the eastern equatorial Pacific. The shoaling of the thermocline is accompanied by a shoaling of the nutricline, increasing the amount of nutrients upwelled into the euphotic zone. This results in higher than average NPP due primarily to a substantial increase in the relative abundance of large phytoplankton (Chavez et al. 1999). The opposite occurs during the El Niño phase when SST is warmer than average and large phytoplankton NPP is below average. These phases of ENSO can last between nine months and two years (fig. 15.3, inset).

The Saba et al. (2007) nesting model estimates the annual number of remigrant nesting females at Playa Grande based on the Multivariate ENSO Index (MEI)

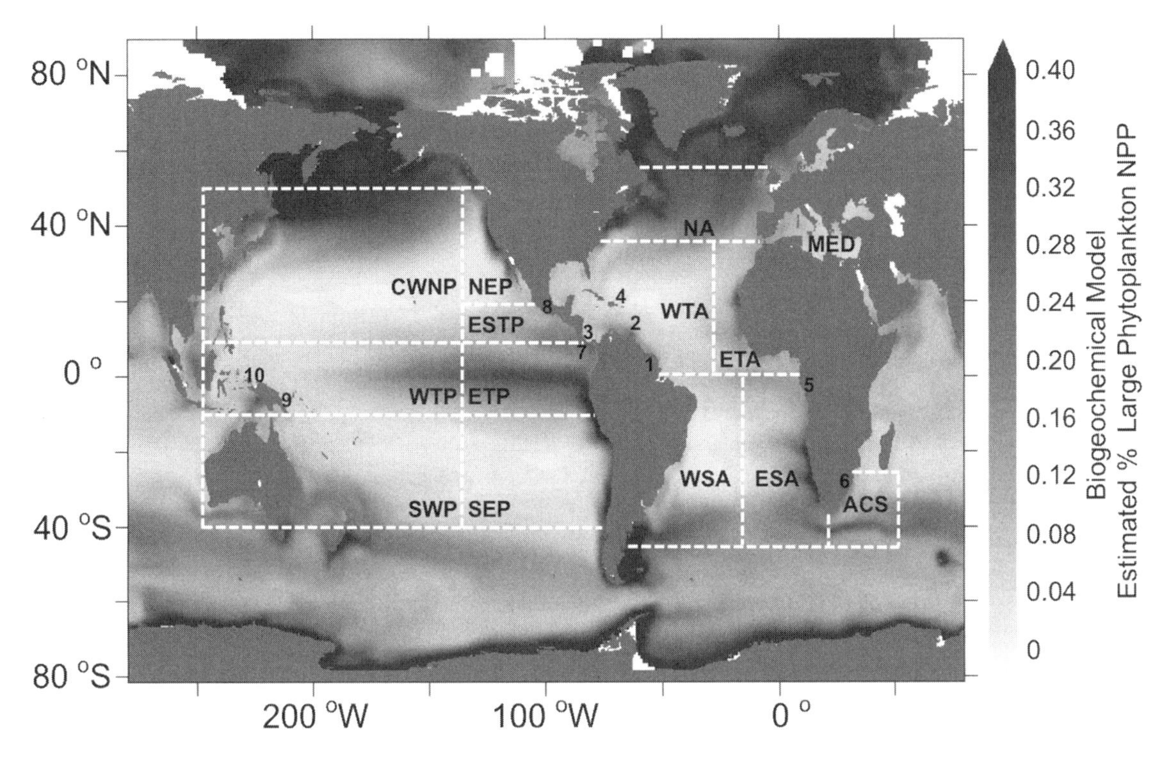

Fig. 15.2. Mean proportion of large phytoplankton (>5 μm cell size) NPP relative to small phytoplankton (<5 μm cell size) from 1948 to 2007 estimated by the TOPAZ BOGCM (Sarmiento et al. 2010). Mature female leatherback migration areas are outlined by white-dashed lines and the nesting beaches are numbered—see tables 15.1 and 15.2 (Saba, Spotila, et al. 2008; chapter 14). Nesting beaches in the Atlantic and western Indian are numbered as follows: for the western Atlantic Ocean population, (1) French Guiana and Suriname, (2) Trinidad, (3) Caribbean Costa Rica and Panama, (4) St. Croix; for the Atlantic Ocean population, (5) Gabon; and for the western Indian population, (6) South Africa. Nesting complexes in the Pacific are numbered as follows: for the eastern Pacific population, (7) Pacific Costa Rica, (8) Pacific Mexico; for the western Pacific population, (9) Papua New Guinea, and (10) Papua.

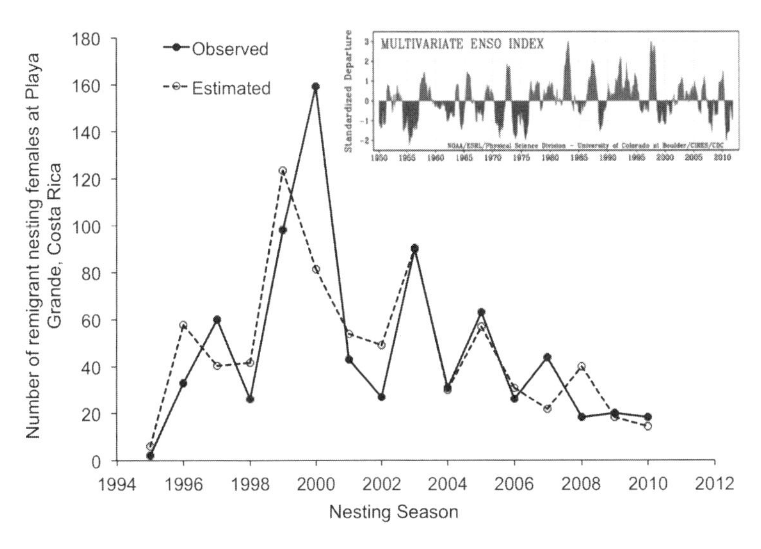

Fig. 15.3. Estimated and observed number of remigrant nesting female leatherbacks at Playa Grande, Costa Rica, from 1995 to 2010. Estimate is based on the Saba et al. (2007) model using the Multivariate ENSO Index (MEI) and SeaWiFS surface chlorophyll a in the eastern equatorial Pacific Ocean.

and satellite-derived sea surface chlorophyll a (ocean color) data averaged over the twelve months prior to each nesting season (fig. 15.3). The MEI is based on six ocean-atmosphere variables in the tropical Pacific (Wolter and Timlin 2011). Satellite-derived ocean color data can be used as a proxy for both phytoplankton biomass and size class (Uitz et al. 2010). The statistical significance of the leatherback nesting model's estimate is considerable over nearly fifteen years of remigrant data (correlation = 0.77; $P < 0.05$), and each yearly projection is thus used to prepare for beach monitoring efforts at Playa Grande. When La Niña conditions persist over the year before a nesting season, the probability that previously tagged females will return to nest increases and the reverse occurs for El Niño conditions. Therefore, it appears that when large phytoplankton are in higher abundance in the eastern equatorial Pacific, the foraging conditions for mature females are more suitable and these turtles have a greater chance of accumulating the threshold level of fat reserve required for migration, vitellogenesis, and the nesting process.

Large Phytoplankton Produce Large Turtles: Interpopulation Differences in Body Size, Fecundity, and Nesting Trends

A comprehensive review of leatherback nesting populations worldwide reveals substantial interpopulation differences in mature female body size, fecundity, and nesting trends (Saba, Spotila, et al. 2008). This nesting population review was combined with an analysis of ocean color model NPP in the migration and foraging areas of post-nesting females. The results suggested that differences in the estimated magnitude and variability of NPP among the migration and foraging areas of post-nesting females are driving the interpopulation trait differences. Essentially, smaller and less fecund mature females are associated with migration and foraging areas that are lower in NPP and/or those that have a higher proportion of interannual variability in NPP. The smallest and least fecund females are from the eastern Pacific population (nesting in Mexico and Costa Rica), where NPP variability in the equatorial waters is primarily interannual (associated with ENSO) and does not have the predictable bloom period of foraging areas of other populations. Saba, Spotila, et al. (2008) postulated that the reduced reproductive output of the eastern Pacific population increases their sensitivity to anthropogenic mortality (i.e., fisheries bycatch, egg poaching) via lower and more variable recruitment rates, likely explaining their precipitously declining population trends in Costa Rica and Mexico (chapter 10) despite continued beach protection.

The study by Saba, Spotila, et al. (2008) is revisited in this chapter using the TOPAZ model that, unlike the 2008 study, allows explicit consideration of phytoplankton size-class estimates from a BOGCM (Sarmiento et al. 2010). Figure 15.2 highlights the general migration areas of post-nesting females, as determined by satellite telemetry (chapter 14), overlaid on a TOPAZ composite of the estimated proportion of NPP from large phytoplankton between the years 1948 and 2007. The spatial extent of each migration area is the same as in the Saba, Spotila, et al. (2008) study. Within these migration areas, table 15.1 describes the estimated magnitude and temporal variability of large phytoplankton NPP based on TOPAZ (Sarmiento et al. 2010).

As with the ocean color model estimates of NPP from Saba, Spotila, et al. (2008), it is clear that there are substantial differences in both the estimated magnitude and temporal variability of large phytoplankton NPP among leatherback migration areas. In terms of the spatial extent of resources, the resource availability ratio (RAR is large phytoplankton NPP g C $\times 10^{12}$ day^{-1}: migration area size km^2 $\times 10^6$) in table 15.1 is a simple way of indicating the extent to which a turtle must migrate within a migration area to reach favorable foraging waters; essentially the RAR is an indicator of prey patchiness. Theoretically, leatherbacks within small migration areas that are rich in large phytoplankton (high RAR) expend less energy on migration to reach favorable foraging waters and can thus expend more energy on growth and reproduction. For example, the Agulhas Current System (ACS) is a very small migration area (fig. 15.2) that has a high rate of large phytoplankton NPP during the austral spring bloom (Oct.–Nov.) and thus has the highest RAR (table 15.1). Consequently, western Indian Ocean post-nesting female leatherbacks from nesting beaches in South Africa that utilize this migration area are the largest in body size and produce the largest clutch size and highest reproductive output relative to every other population worldwide (table 15.2). The favorability of this region is further facilitated by the warm Agulhas Current, which enables leatherback nesting beaches in South Africa to be at the highest latitude for this species. Thus, mature females do not have to travel as far between nesting beaches and temperate foraging areas, unlike those from other nesting beaches that, in some cases, travel across an entire ocean basin. Nesting females in the western Atlantic are also relatively large in body size and have a reproductive output that is twice that of females in the eastern Pacific (table 15.2). The most common forag-

Table 15.1. Magnitude and temporal variability of large phytoplankton NPP estimated by the Tracers of Phytoplankton with Allometric Zooplankton (TOPAZ) biogeochemical model (Sarmiento et al. 2010) among leatherback migration areas worldwide (Saba, Spotila, et al. 2008; chapter 14).

Migration area(s)	Large phytoplankton maximum bloom period[a]	Mean proportion of large phytoplankton integrated productivity at max bloom[a]	Mean integrated productivity of large phytoplankton at max bloom (10^{12} g C day^{-1})[a]	Resource Availability Ratio (RAR)[c]	Large phytoplankton bloom cycle[d]
Western Atlantic					
North Atlantic (NA)	Apr–May	36% (±2%)	6.2 (±0.8)	0.40	Seasonal (96%)
Western Tropical Atlantic (WTA)	Apr–May	9% (±0%)	1.0 (±0.1)	0.05	Seasonal (99%)
Eastern Tropical Atlantic (ETA)	Apr–May	13% (±1%)	1.9 (±0.3)	0.22	Seasonal (81%)
Mediterranean (MED)	May–Jun	9% (±0%)	0.9 (±0.1)	0.28	Seasonal (98%)
Eastern Atlantic					
Eastern South Atlantic (ESA)	Sep–Oct	18% (±2%)	2.2 (±0.4)	0.11	Seasonal (76%)
Western South Atlantic (WSA)	Sep–Oct	17% (±3%)	1.8 (±0.5)	0.12	Seasonal (91%)
Western Indian					
Agulhas Current System (ACS)	Oct–Nov	20% (±2%)	3.1 (±0.3)	0.51	Seasonal (71%)
Eastern South Atlantic (ESA)	Sep–Oct	18% (±2%)	2.2 (±0.4)	0.11	Seasonal (76%)
Eastern Pacific					
Eastern Subtropical Pacific (ESTP)	Dec–Jan[b]	13% (±2%)	0.6 (±0.2)	0.10	Interannual (33%)
Eastern Tropical Pacific (ETP)	Aug–Sep[b]	24% (±3%)	4.7 (±0.6)	0.31[e]	Interannual (19%)
Southeastern Pacific (SEP)	Sep–Oct	9% (±0%)	0.8 (±0.1)	0.03	Seasonal (68%)
Western Pacific					
Central/Western North Pacific (CWNP)	Apr–May	24% (±2%)	2.5 (±0.3)	0.05	Seasonal (99%)
Northeastern Pacific (NEP)	Apr–May	15% (±2%)	1.2 (±0.2)	0.15	Seasonal (70%)
Western Tropical Pacific (WTP)	Sep–Oct	11% (±2%)	1.5 (±0.4)	0.06	Interannual (7%)
Southwestern Pacific (SWP)	Oct–Nov	14% (±0%)	1.6 (±0.2)	0.05	Seasonal (83%)

[a] TOPAZ model estimate.

[b] Maximum bloom is not always during this period due to interannual variability.

[c] Ratio of large phytoplankton productivity at maximum bloom to migration area size.

[d] Based on Fourier regression of the large phytoplankton NPP time-series. Percent represents the proportion of seasonal variability.

[e] Resource availability ratio in the ETP during the 1997–98 El Niño = 0.18.

ing area for the western Atlantic nesting population is the North Atlantic (NA) (fig. 15.2), particularly the continental shelf of northeastern North America (fig. 15.1), where the spring and fall phytoplankton blooms fuel some of the most productive fisheries in the world (www.seaaroundus.org).

The temporal consistency of the maximum annual bloom of large phytoplankton NPP may also be an important factor linked to leatherback phenotype and fecundity. Table 15.1 describes the amount of seasonal variability of large phytoplankton NPP within each leatherback migration area. Based on a Fourier analysis of each migration area's time series of TOPAZ estimated large phytoplankton NPP from 1948–2007, the migration areas of eastern Pacific mature females have the lowest proportion of seasonal variability (table 15.1). The high proportion of interannual variability in estimated large phytoplankton NPP derives from ENSO, as discussed in the previous section. Although the RAR is quite high in the eastern tropical Pacific (ETP), the low proportion of seasonal variability of large phytoplankton NPP (19%) renders the maximum bloom periods temporally inconsistent (table 15.1). However, the RAR of the ETP during the strong 1997–1998 El Niño was 0.18, which is six times higher than the RAR for the southeastern Pacific (SEP) during the maximum bloom (table 15.1). Therefore, it would seem that the turtles would have a higher foraging success in the ETP than SEP, even during a strong El Niño.

Given the paucity of leatherback foraging in the ETP based on the first year of post-nesting tracking data, along with the infrequent observation of equatorial foraging among other populations worldwide (chapter 14), it is possible that temperature may also play a substantial role in the foraging success of leatherbacks. Foraging in cooler waters may benefit leather-

Table 15.2. Mean body size, reproductive output, and most common RAR (resource availability ratio) among the major leatherback nesting populations worldwide. From Saba, Spotila, et al. 2008 and references therein.

Population	Nesting complex	Population size (females/year)	Population trend	Remigration interval (years)	Clutch size (eggs/clutch)	Clutch frequency (clutches/season)	Reproductive output (eggs/5 years)	Body size CCL (cm)	Most common RAR[e]
Western Atlantic	St. Croix	150	Stable	2.5	80	5.3	848	152	0.40
	Trinidad	4,300	Stable	2.5[a]	84	6.4[b]	1075	156	0.40
	French Guiana & Suriname	5,000	Stable	2.5	85	7.5	1275	155	0.40
	Caribbean Costa Rica and Panama	1,900	Stable	2.5	82	6.4[b]	1050	154	0.40
Eastern Atlantic	Gabon	13,000	Stable	–	73	5.0	–	151	0.11
Western Indian	South Africa	125	Stable	2.5	104	7.3	1518	160	0.51
Eastern Pacific	Mexico	200	Decreasing	3.7[c]	62	5.5	461	144	0.03
	Costa Rica	200	Decreasing	3.7	64	6.1	528	145	0.03
Western Pacific	Papua[d]	1,250	Decreasing	–	78	–	–	–	–
	Papua New Guinea[d]	250	Decreasing	–	–	–	–	–	–

[a] Used mean remigration interval from Western Atlantic.

[b] Used mean clutch frequency from Western Atlantic.

[c] Used mean remigration interval from Costa Rica.

[d] Nesting complex has not been extensively studied.

[e] Based on most commonly observed migration area.

backs due to a variety of physiological factors such as a greater frequency of larger gelatinous organisms and more efficient food assimilation. Higher energy transfer efficiencies from phytoplankton to upper trophic levels in high latitude ecosystems may be due to (1) suppressed rates of microbial herbivory in colder water that increases the proportion of NPP available to zooplankton; and/or (2) the lower maturity of high latitude ecosystems that results in simpler, more efficient food webs that derive primarily from a high abundance of large phytoplankton (Friedland et al. 2012). One must note, however, that although the tracking data do not report high use of the ETP within the first year after nesting, the strong association between ENSO and the nesting variability in Costa Rica suggests that these females may be relying on the ETP as a critical foraging area just before the nesting season.

Another exclusive trait of the eastern Pacific nesting population is the rare occurrence of migration to coastal waters (chapter 14). Of the 46 post-nesting females tracked from nesting beaches in Costa Rica, only one individual remained in coastal waters within the first year after nesting (Shillinger et al. 2011). This finding is quite extraordinary considering that the highly productive Peru upwelling system (fig. 15.1) is relatively close to the nesting beaches in Costa Rica (Fig. 15.2). Moreover, coastal waters are common foraging areas for every other leatherback population worldwide. As previously discussed, coastal waters are typically high in large phytoplankton NPP, due to the high concentration of nutrients, and may also host a higher biomass of gelatinous zooplankton (Lilley et al. 2011). Leatherbacks that are genetically distinct from those nesting in Mexico and Costa Rica have been reported as fishery bycatch among small-scale Peruvian coastal drift net fisheries (Alfaro-Shigueto et al. 2011). However, a large portion of these incidentally caught leatherbacks were juveniles and subadults while only a small portion appeared to be mature turtles (Alfaro-Shigueto et al. 2007). The mortality rate of leatherbacks incidentally caught among the Peruvian drift net fisheries appears to be quite high (20%), mostly due to fishermen retaining live turtles for consumption (Alfaro-Shigueto et al. 2011). Moreover, Eckert and Sarti (1997) suggested that high leatherback mortality rates among leatherbacks caught as bycatch in the coastal drift net Chilean swordfish fishery were unsustainable. Therefore, historically high mortality rates of leatherbacks along the highly productive coastal upwelling waters of Peru and Chile (fig. 15.1) may have led to the present-day cohort that appears to be strictly pelagic (Saba, Shillinger, et al. 2008; chapter 14). This is based on the theory that leatherbacks return to the same foraging areas year after year (James et al. 2005). Natural selection has favored this pelagic foraging majority that completely bypasses the highly productive Peru and Chile upwelling systems en route to the SEP (Wallace and Saba 2009). These pelagic foragers are not actively avoiding the coastal waters of Peru and Chile due to high rates of fishery mortality, but are rather evolved to forage in alternative waters that happen to have lower mortality rates. The ontogenetic mechanisms responsible for determining sea turtle foraging site fidelity are still under debate among the research community.

Examining the trends of the major nesting populations worldwide (table 15.2; chapters 9, 10, 11), a major dichotomy becomes apparent. Populations in the Atlantic and Indian Oceans are either stable or increasing while those in the Pacific are declining. While nesting data in the western Pacific are limited, the eastern Pacific dataset is quite extensive and the decline rates in nesting populations in Mexico and Costa Rica are robust trends (chapter 10). Beach protection has been a proven success strategy in the recovery of leatherback nesting populations in both St. Croix (Dutton et al. 2005) and South Africa (Hughes 1996). The prevailing theory to explain the rates of decline in the eastern Pacific despite long-term beach protection is that this population has a substantially lower reproductive output (table 15.2) and thus a much lower recruitment rate than populations in the Atlantic and Indian Oceans (Saba, Spotila, et al. 2008). Therefore, population recovery time in the eastern Pacific may be prolonged, or recovery may not even possible given continued high mortality in coastal waters. Populations in the Atlantic and Indian Oceans have recovered or are stable, due to their recruitment rates that are likely double that of the eastern Pacific population. Ultimately, the higher rates of large phytoplankton NPP among Atlantic and Indian Ocean leatherback migration areas, more frequent coastal migrations, and lower propensities for ecosystem collapse associated with El Niño events, are the probable mechanisms responsible for their enhanced reproductive output, stable or recovering populations, and thus their higher tolerance of anthropogenic mortality relative to the eastern Pacific population. Populations in the western Pacific also appear to be declining (chapter 10), but data are limited regarding their long-term trends, body sizes, and reproductive outputs. One might surmise that mature females nesting in Papua Barat that migrate across the Pacific to the California Current upwelling system have a reduced reproductive output compared to those migrating to closer foraging areas in the South China Sea and other more local waters. Nonetheless, more data are required from the western Pacific nesting populations in order to make any inferences regarding bottom-up forcing and their reproductive output.

In summary, leatherback migration data combined with model estimates of phytoplankton size-class NPP support a compelling argument for strong connections between large phytoplankton and leatherback ecology. This chapter focused primarily on mature females because this demographic represents the largest dataset in terms of migration. It is likely that juveniles, subadults, and mature males are also influenced by large phytoplankton NPP in terms of their foraging success, bioenergetics, and growth.

LITERATURE CITED

Alfaro-Shigueto, J., P. H. Dutton, M. Van Bressem, and J. Mangel. 2007. Interactions between leatherback turtles and Peruvian artisanal fisheries. Chelonian Conservation and Biology 6: 129–134.

Alfaro-Shigueto, J., J. C. Mangel, F. Bernedo, P. H. Dutton, J. A. Seminoff, and B. J. Godley. 2011. Small-scale fisheries of Peru: a major sink for marine turtles in the Pacific. Journal of Applied Ecology 48: 1432–1440.

Attrill, M. J., J. Wright, and M. Edwards. 2007. Climate-related increases in jellyfish frequency suggest a more gelatinous future for the North Sea. Limnology and Oceanography 52: 480–485.

Benson, S. R., T. Eguchi, D. G. Foley, K. A. Forney, H. Bailey, C. Hitipeuw, B. P. Samber, et al. 2011. Large-scale movements and high-use areas of western Pacific leatherback turtles, *Dermochelys coriacea*. Ecosphere 2(7): art84. doi: 10.1890/ES11-00053.1.

Block, B. A., I. D. Jonsen, S. J. Jorgensen, A. J. Winship, S. A. Shaffer, S. J. Bograd, E. L. Hazen, et al. 2011. Tracking apex marine predator movements in a dynamic ocean. Nature 475: 86–90.

Carr, M. E., M.A.M. Friedrichs, M. Schmeltz, M. N. Aita, D. Antoine, K. R. Arrigo, I. Asanuma, et al. 2006. A comparison of global estimates of marine primary production from ocean color. Deep-Sea Research II 53: 741–770.

Chavez, F. P., P. G. Strutton, G. E. Friederich, R. A. Feely, G. C. Feldman, D. G. Foley, and M. J. McPhaden. 1999. Biological and chemical response of the Equatorial Pacific Ocean to the 1997–98 El Niño. Science 286: 2126–2131.

Doyle, T., J. Houghton, R. McDevitt, J. Davenport, and G. Hays. 2007. The energy density of jellyfish: estimates from bomb-calorimetry and proximate-composition. Journal of Experimental Marine Biology and Ecology 343: 239–252.

Dunne, J. P., A. Gnanadesikan, J. L. Sarmiento, and R. D. Slater. 2010. Technical description of the prototype version (v0) of Tracers of Phytoplankton with Allometric Zooplankton (TOPAZ) ocean biogeochemical model as used in the Princeton IFMIP_model. Electronic supplement. In: Sarmiento, J. L., R. D. Slater, J. P Dunne, A. Ganadesikan, and M. R. Hiscock. 2010. Efficiency of small scale carbon mitigation by patch iron fertilization. Biogeosciences 7: 3593–3624. http://www.biogeosciences.net/7/3593/2010/bg-7-3593-2010-supplement.pdf.

Dutton, D. L., P. H. Dutton, M. Chaloupka, and R. H. Boulon. 2005. Increase of a Caribbean leatherback turtle *Dermochelys coriacea* nesting population linked to long-term nest protection. Biological Conservation 126: 186–194.

Eckert, S. A., and L. Sarti. 1997. Distant fisheries implicated in the loss of the world's largest leatherback nesting population. Marine Turtle Newsletter 78: 2–7.

Eppley, R. W., and B. J. Peterson. 1979. Particulate organic matter flux and planktonic new production in the deep ocean. Nature 282: 677–680.

Falkowski, P. G., M. E. Katz, A. H. Knoll, A. Quigg, J. A. Raven, O. Schofield, and F.J.R. Taylor. 2004. The evolution of modern eukaryotic phytoplankton. Science 305: 354–360.

Field, C. B., M. J. Behrenfeld, J. T. Randerson, and P. Falkowski. 1998. Primary production of the biosphere: integrating terrestrial and oceanic components. Science 281: 237–240.

Fossette, S., C. Girard, M. López-Mendilaharsu, P. Miller, A. Domingo, D. Evans, L. Kelle, et al. 2010. Atlantic leatherback migratory paths and temporary residence areas. PLoS One 5: e13908. doi: 10.1371.

Fossette, S., A. C. Gleiss, J. P. Casey, A. R. Lewis, and G. C. Hays. 2011. Does prey size matter? Novel observations of feeding in the leatherback turtle (*Dermochelys coriacea*) allow a test of predator-prey size relationships. Biology Letters 8: 351–354. doi: 10.1098/rsbl.2011.0965.

Friedland, K. D., C. A. Stock, K. F. Drinkwater, J. S. Link, R. T. Leaf, B. V. Shank, J. M. Rose, et al. 2012. Pathways between primary production and fisheries yields of large marine ecosystems. PLoS One 7: e28945. doi: 10.1371.

García-Comas, C., L. Stemmann, F. Ibanez, L. Berline, M. G. Mazzocchi, S. Gasparini, M. Picheral, and G. Gorsky. 2011. Zooplankton long-term changes in the NW Mediterranean Sea: Decadal periodicity forced by winter hydrographic conditions related to large-scale atmospheric changes? Journal of Marine Systems 87: 216–226.

Henson, S. A., J. P. Dunne, and J. L. Sarmiento. 2009. Decadal variability in North Atlantic phytoplankton blooms. Journal of Geophysical Research 114: C04013. doi: 10.1029/2008JC005139.

Hughes, G. R. 1996. Nesting of the leatherback turtle (*Dermochelys coriacea*) in Tongaland, KwaZulu-Natal, South Africa, 1963–1995. Chelonian Conservation Biology 2: 153–158.

James, M. C., R. A. Myers, and C. A. Ottensmeyer. 2005. Behaviour of leatherback sea turtles, *Dermochelys coriacea*, during the migratory cycle. Proceedings of the Royal Society B 272: 1547–1555.

Lilley, M.K.S., S. E. Beggs, T. K. Doyle, V. J. Hobson, K.H.P. Stromberg, and G. C. Hays. 2011. Global patterns of epipelagic gelatinous zooplankton biomass. Marine Biology 158: 2429–2436.

Luschi, P., J. R. Lutjeharms, P. Lambardi, R. Mencacci, G. R. Hughes, and G. C. Hays. 2006. A review of migratory behaviour of sea turtles off southeastern Africa. South African Journal of Science 102: 51–58.

Purcell, J. 2012. Jellyfish and ctenophore blooms coincide with human proliferations and environmental perturbations. Annual Review of Marine Science 4: 209–235.

Ryther, J. H. 1969. Photosynthesis and fish production in the sea. Science 166: 72–76.

Saba, V. S., P. Santidrián-Tomillo, R. D. Reina, J. R. Spotila, J. A. Musick, D. A. Evans, and F. V. Paladino. 2007. The effect of the El Niño Southern Oscillation on the reproductive frequency of eastern Pacific leatherback turtles. Journal of Applied Ecology 44: 395–404.

Saba, V. S., G. L. Shillinger, A. M. Swithenbank, B. A. Block, J. R. Spotila, J. A. Musick, and F. V. Paladino. 2008. An oceanographic context for the foraging ecology of eastern Pacific leatherback turtles: consequences of ENSO. Deep Sea Research Part I: Oceanographic Research Papers 55: 646–660.

Saba, V. S., J. R. Spotila, F. P. Chavez, and J. A. Musick. 2008. Bottom-up and climatic forcing on the worldwide population of leatherback turtles. Ecology 89: 1414–1427.

Sarmiento, J. L., R. D. Slater, J. P Dunne, A. Ganadesikan, and M. R. Hiscock. 2010. Efficiency of small scale carbon mitigation by patch iron fertilization. Biogeosciences 7: 3593–3624. http://www.biogeosciences.net/7/3593/2010/bg-7-3593-2010.

Shillinger, G. L., A. M. Swithenbank, H. Bailey, S. J. Bograd, M. R. Castelton, B. P. Wallace, J. R. Spotila, et al. 2011. Vertical and horizontal habitat preferences of post-nesting leatherback turtles in the South Pacific Ocean. Marine Ecology Progress Series 422: 275–289.

Uitz, J., H. Claustre, B. Gentili, and D. Stramski. 2010. Phytoplankton class-specific primary production in the world's oceans: seasonal and interannual variability from satellite observations. Global Biogeochemical Cycles 24: GB3016.

Wallace, B. P., and V. S. Saba. 2009. Environmental and anthropogenic impacts on intra-specific variation in leatherback turtles: opportunities for targeted research and conservation. Endangered Species Research 7: 11–21.

Welp, L. R., R. F, Keeling, H.A.J. Meijer, A. F. Bollenbacher, S. C. Piper, K. Yoshimura, R. J. Francey, et al. 2011. Interannual variability in the oxygen isotopes of atmospheric CO_2 driven by El Niño. Nature 477: 579–582.

Witt, M. J., E. Augowet Bonguno, A. C. Broderick, M. S. Coyne, A. Formia, A. Gibudi, G. A. Mounguengui Mounguengui, et al. 2011. Tracking leatherback turtles from the world's largest rookery: assessing threats across the South Atlantic. Proceedings of the Royal Society B 278: 2338–2347.

Wolter, K., and M. S. Timlin. 2011. El Niño / Southern Oscillation behavior since 1871 as diagnosed in an extended multivariate ENSO index (MEI.ext). International Journal of Climatology 31: 1074–1087.

Part V THE FUTURE OF THE LEATHERBACK TURTLE

16

Warming Climate

A New Threat to the Leatherback Turtle

JAMES R. SPOTILA,
VINCENT S. SABA,
SAMIR H. PATEL, AND
PILAR SANTIDRIÁN TOMILLO

In 1956, when Archie Carr published his classic book, *The Windward Road* (Carr 1956), about sea turtles and the people who interact with them, the CO_2 concentration in Earth's atmosphere was about 314 ppm and Earth's mean temperature was about 0.8°C lower than today. In 1969, when Porter and Gates published their classic monograph on the biophysical ecology of animals (Porter and Gates 1969), the CO_2 concentration was 325 ppm and Earth's temperature was 0.1°C higher than in 1956. Today (2014) the CO_2 concentration of the atmosphere is 398 ppm and the planet is the warmest it has been since the mid-Pliocene (3 million years ago). Earth is warming up due to the anthropogenic addition of long-lived greenhouse gases (CO_2, CH_4, and others) that trap heat in the atmosphere (Gates 1962), and as a result the climate is changing (fig. 16.1). There is an increasing accumulation of data that indicate climate change is affecting plants (Miller-Rushing and Primack 2008) and animals (Parmesan and Yohe 2003). If the climate continues to warm, salamanders in the eastern United States may undergo a mass extinction during this century (Milanovich et al. 2010) and lizards may experience dramatic declines throughout the world (Sinervo et al. 2010). Will leatherbacks and other sea turtles suffer the same fate?

From egg to adult, on both land and in water, sea turtles face a wide array of threats due to a warming planet and changing climate (Hawkes et al. 2009). In this chapter, we assess those threats, consider the ability of leatherback turtles to adapt to them, and discuss the need for biologists and conservationists to take action to help leatherbacks survive into the twenty-second century and beyond.

Temperature Regulation

Major components of Earth's climate system include sea surface temperature (SST), air temperature (T_a), and precipitation; these are major drivers in sea turtle ecology. Spotila and Standora (1985) detailed the environmental constraints on the thermal energetics of sea turtles (fig. 16.2). These constraints are a matter of simple physics and physiology. Biophysical ecology, the field developed by David Gates (Gates 1980),

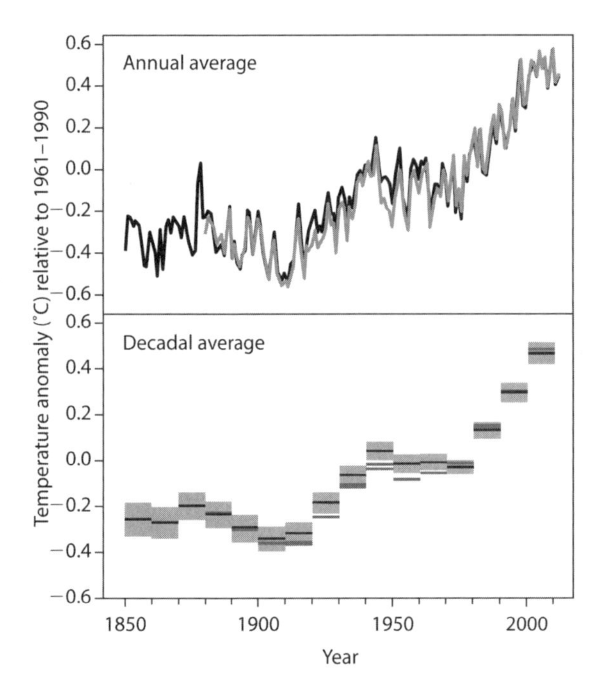

Fig. 16.1. Observed global mean combined land and ocean surface temperature anomalies, from 1850 to 2012, from three data sets. *Top panel:* annual mean values. *Bottom panel:* decadal mean values including the estimate of uncertainty for one dataset (black). Anomalies are relative to the mean of 1961–1990. From IPCC Working Group I 2013.

allows us to calculate all of the factors that add heat to an animal, including its metabolism, heat radiation from the sun, and heat radiation from its surroundings. We can also calculate factors that take heat away, including reradiation from the animal, wind blowing over it, evaporation from the skin, and conduction of heat to the ground or water. From these dynamics we can determine the animal's heat balance and predict where it can live.

Sea turtles spend most of their time in water and gain little heat from basking at the surface. Conduction of heat is about 100 times faster in water than in air so the body temperature of small sea turtles is at most about 1°C above water temperature. Typically sea turtles move between water of different temperatures to regulate their body temperature. Large green turtles are regionally endothermic; that is, they produce metabolic heat in their active pectoral muscles when swimming, and retain that heat through thermal inertia (Standora et al. 1982), like a Komodo dragon (McNab and Auffenberg 1976) or alligator (Spotila et al. 1973). Adult leatherback turtles are warmer than the cold water in which they swim, both at great depths and high latitudes, because they are gigantotherms, which produce heat in their swimming muscles, have a thick layer of fatty insulation, shunt blood between the body core and skin, and have considerable thermal inertia due to large size (Paladino et al. 1990). They also avoid overheating in tropical water and when on the beach by shunting blood to their surface. Leatherbacks dive such that they cool off during the inter-nesting period when surface waters are warm in the tropics (Shillinger et al. 2010). Dudley and Porter (2014) used a mechanistic biophysical/physiological model and found that adult leatherbacks will not be heat stressed during the inter-nesting period, though at some beaches might experience high body temperatures that are known to stress hatchlings. Leatherback turtles will not die of overheating in the water as the oceans warm up. Direct heat exchange will not be a problem. Instead, leatherbacks will probably increase their range into higher latitudes (McMahon and Hays 2006).

Large leatherbacks are common in cold North Atlantic waters in summer. They range as far north as Alaska in the Pacific Ocean and Labrador in the North Atlantic. They have elevated body temperatures in seas as cold as 5–15°C (Bleakney 1965; James and Mrosovsky 2004; James et al. 2006) and typically occur in waters with a temperature of 15°C and above. Juvenile leatherbacks are restricted to warmer waters (Eckert 2002). McMahon and Hays (2006) used long-term satellite telemetry to define habitat utilization of leatherbacks in the North Atlantic. They calculated weekly and monthly SSTs from data supplied by US and UK government agencies for the geographical area bounded by 45–60°N and 0–15°W and for 1970–2003. The latitudinal position of the 15°C isotherm defined the northerly distribution limit of telemetered leatherbacks. Because the summer

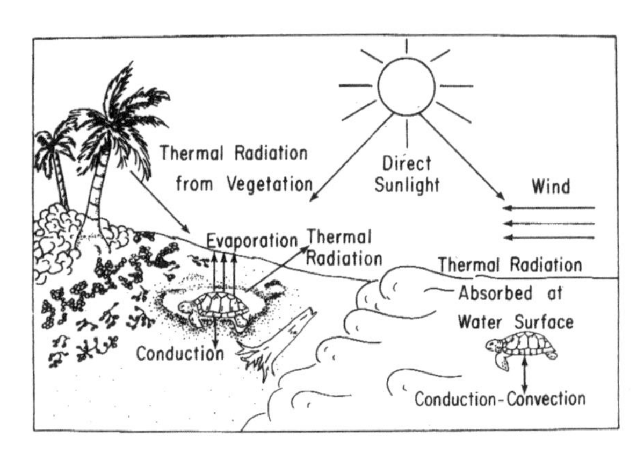

Fig. 16.2. Heat exchange between sea turtles and their environment. On land, heat transfer occurs via radiation, convection (wind), evaporation, and conduction. In water, heat transfer occurs by conduction-convection. Redrawn from Spotila and Standora 1985, with permission from *Copeia*.

position of this isotherm moved north by 330 km in the North Atlantic between 1986 and 2003 leatherbacks moved north as well.

Warming Oceans

Despite the apparent range extension of leatherbacks, these turtles could face critical problems as oceans warm up. First, their food supply could be threatened, and second, their reproductive success is in doubt. Although the oceans are warming three times slower than temperatures on land, marine species are shifting distribution and phenology at a greater rate than terrestrial species (Poloczanska et al. 2013). For example, the distribution of plankton (Beaugrand et al. 2002; Perry et al. 2005) is changing as the oceans warm up. Seagrass meadows are declining in many areas (Duarte 2002; Marbá and Duarte 2010). Increased acidification of the oceans will affect many species that form shells from $CaCO_3$ and coral reefs will be greatly affected (Hoegh-Guldberg et al. 2007).

Leatherback foraging hotspots are typically waters with high primary productivity of large phytoplankton (chapter 15). These areas have upwellings of deep, nutrient-rich waters (Hays et al. 2004; Shillinger et al. 2008; Shillinger et al. 2011); the coastal zones off Peru and Chile are an example (Alfaro-Shigueto et al. 2007; Saba, Shillinger, et al. 2008). Extensive satellite telemetry of leatherbacks has located foraging areas in the Pacific, Atlantic, and Indian Oceans (Bailey, Fossette, et al. 2012). In the Pacific (Bailey et al. 2007; Shillinger et al. 2008; Benson et al. 2011; Bailey, Benson, et al. 2012), leatherbacks forage off California, in the North Pacific, South Pacific, along the equator, and in the China Sea. In the Atlantic, hotspots include the area of the Grand Banks (James and Herman 2001; James et al. 2005; Eckert 2006), coastal regions along Europe (Houghton et al. 2006), and areas off Africa (Hays et al. 2004). In the Indian Ocean, they encompass areas off South Africa (Luschi et al. 2006), including the Agulhas Current (Luschi et al. 2003; Sale et al. 2006).

At this time, there is a great deal of uncertainty in the projected effects of rising ocean temperatures on upwellings, currents, and productivity. However, changes in the oceans are occurring. For example, climate fluctuations affect nesting frequency of sea turtles. The number of green turtles (*Chelonia mydas*) nesting in northern Australia varies substantially from year to year (Limpus and Nicholls 1988, 2000). Interannual fluctuations in numbers of nesting turtles are in phase at widely separated rookeries. These fluctuations are correlated with an index of the Southern Oscillation, a coherent pattern of atmospheric pressure, temperature, and rainfall fluctuations that dominates interannual variability of the climate of the tropical Pacific. Major fluctuations in numbers of turtles breeding occur two years after major fluctuations in the Southern Oscillation. Solow et al. (2002) reported that variation in SST is associated with variation in numbers of nesting green turtles at Tortuguero along the Caribbean coast of Costa Rica. In the case of leatherbacks, we know that El Niño Southern Oscillation (ENSO) has a great effect on the nesting ecology of turtles that forage in the eastern tropical Pacific. In addition, Willis-Norton et al. (2014) modeled the effect of climate change on pelagic foraging areas of leatherbacks in the South Pacific. They predicted that there would be a 15% loss of habitat in this century, further stressing the already highly threatened eastern Pacific population. This may be a harbinger of changes to come over the next century.

The best data set and analyses of the relationship between ocean conditions, foraging success, and reproductive success of leatherbacks come from the long-term study of leatherbacks at Parque Nacional Marino Las Baulas (PNMB), on the Pacific coast in northwest Costa Rica. Foraging success and reproductive frequency of mature females that nest at PNMB are enhanced after periods (about one year) of high primary productivity in the eastern equatorial Pacific associated with ENSO (Saba et al. 2007; Reina et al. 2009). The La Niña phase of ENSO is associated with higher primary productivity of large phytoplankton and cooler sea surface temperature (SST) in the eastern equatorial Pacific (Chavez et al. 1999). Nesting seasons that follow La Niña events thus result in peaks in the number of nesting females whereas the opposite holds true for El Niño events (Saba et al. 2007). This is seen in the periodic fluctuation in numbers of nesting females at PNMB that is superimposed on an exponential decline (fig. 16.3) due to egg harvesting and fisheries interactions (Spotila et al. 2000; Santidrián Tomillo et al. 2008).

Saba et al. (2007) investigated the effects of interannual climate variability on leatherback nesting ecology at PNMB. Nesting females at PNMB exhibit a strong sensitivity to ENSO, as reflected in nesting remigration probabilities. Cool La Niña events correspond to a higher remigration probability and warm El Niño events correspond to a lower remigration probability. Saba's model is very successful in predicting the number of leatherbacks that will return to nest at PNMB in a given year (chapter 15). This phenomenon may render eastern Pacific leatherbacks more vulnerable to anthropogenic mortality than other populations.

This type of modeling approach is extremely useful for understanding the effects of climatic variation on

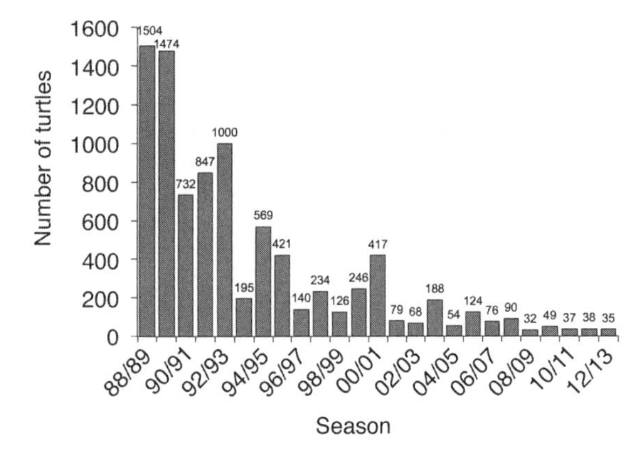

Fig. 16.3. The population of leatherback turtles nesting at Parque Nacional Marino Las Baulas on the Pacific coast of Costa Rica has declined exponentially since 1990 and fluctuates with El Niño. From P. Santidrián Tomillo, unpublished data.

the population dynamics of sea turtles. The remigration probability model can be applied to any monitored sea turtle nesting population where nesting site fidelity and beach monitoring coverage is high. It can help nesting beach monitoring programs forecast remigrant numbers based on prior climate data, and can further quantify anthropogenic mortality by validating survival estimates.

Saba, Shillinger, et al. (2008) analyzed some of the primary biological and physical dynamics within the eastern Pacific leatherback turtle migration area in relation to ENSO and leatherback nesting ecology at PMNB. They applied a net primary production (NPP) model to data from remote sensing to calculate resource availability and analyzed sea surface temperature fronts in the migration area. Nesting peaks of leatherbacks at PNMB are associated with cool, highly productive La Niña events (fig. 16.4). There are also some years when nesting peaks follow El Niño years. In those cases unusually large-scale blooms follow some El Niño events, triggered by the shoaling of the New Guinea Coastal Undercurrent (NGCUC). That results in a joining together of the New Guinea shelf and Equatorial Undercurrent in the western Pacific, which leads to enhanced iron concentrations in the euphotic zone of the central and eastern equatorial Pacific, producing large-scale blooms. The ENSO phenomenon significantly influences the nesting ecology of leatherbacks at PNMB because the majority of the population consists of pelagic foragers that strictly rely on the eastern equatorial Pacific for prey consumption prior to the nesting season. Coastal foragers that rely upon the upwelling off of South America may be a minority in the population because of high mortality rates associated with coastal gill net fisheries along Central and South America (Frazier and Montero 1990; Kaplan 2005; Alfaro-Shigueto et al. 2007).

Saba and colleagues (Saba, Spotila, et al. 2008) explored bottom-up forcing and the responding reproductive output of nesting leatherbacks worldwide. They did an extensive review of leatherback nesting and migration data, and analyzed the spatial, temporal, and quantitative nature of resources as indicated by NPP. Leatherbacks in the eastern Pacific are the smallest in body size and have the lowest reproductive output due to less productive and inconsistent resources within their migration and foraging areas. This is because of natural interannual and multidecadal climate variability as well as the influence of anthropogenic climate warming that may be affecting these natural cycles. The reproductive output of leatherbacks in the Atlantic and western Indian Oceans is nearly twice that of turtles in the eastern Pacific (fig. 16.5). The eastern Pacific leatherback population is more sensitive to anthropogenic mortality due to recruitment rates that are lower and more variable, thus accounting for much of the population differences compared to Atlantic and western Indian Ocean leatherback turtles.

Reina et al. (2009) used the long-term tagging dataset from PNMB to investigate alterations in leatherback reproductive ecology in response to changes in climatic conditions. They calculated the remigration interval (RI) and reproductive output of individual turtles to determine whether these variables changed after a transition from El Niño to La Niña conditions. The reproductive schedule of turtles changed after the end of El Niño conditions. Numbers of turtles nesting

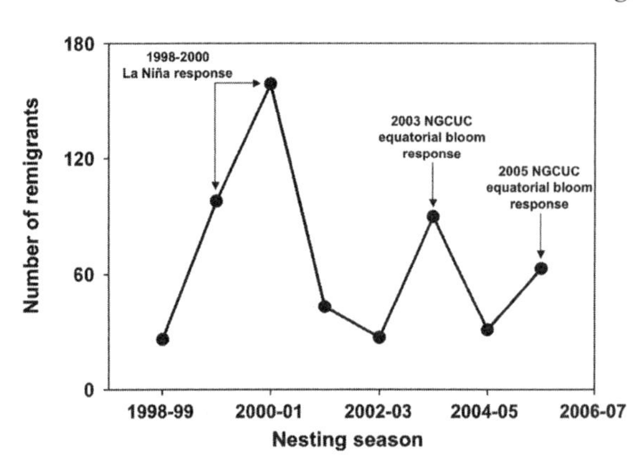

Fig. 16.4. Number of remigrant leatherback turtles at Playa Grande, PNMB, from 1998–1999 to 2005–2006, and their response to ENSO and NGCUC (New Guinea Coastal Undercurrent) induced primary production transitions in the eastern equatorial Pacific. Redrawn from Saba, Shillinger, et al. 2008, with permission from Elsevier.

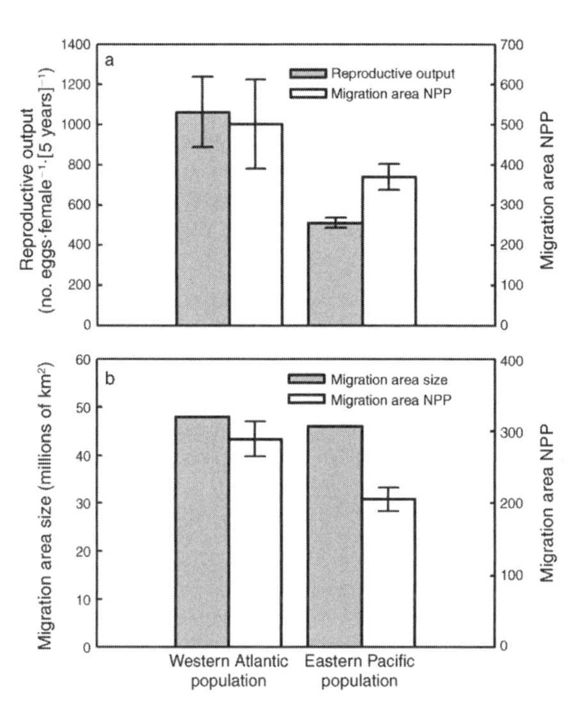

Fig. 16.5. (a) Reproductive output of western Atlantic (WA) and eastern Pacific (EP) females and mean monthly NPP (mg C m^{-2} day^{-1}) at their respective total migration areas. (b) Size and NPP (Tg C 12-$days^{-1}$) of the total migration area of WA and EP females. There are no error bars for migration area sizes because they are constant. In both panels, data are shown as mean ± standard deviation.

with short RIs increased, but their reproductive output did not change. In addition, RIs of individuals that nested before, during, and after the transition changed coincident with the oceanic productivity changes caused by the climatic events. Reina and colleagues then used the Euler equation to model those data and found that a dominance of "bad" El Niño conditions reduced population feasibility (the likelihood that the population would be viable) through increases in age at maturity and RI, while a dominance of "good" La Niña conditions caused the reverse. Their results suggest that future warming of the Pacific climate will result in reductions in the reproduction and migration of individual animals, thus threatening the survival of the eastern Pacific leatherback population.

Loggerhead turtles, *Caretta caretta*, are also threatened by increasing temperatures in their foraging areas in the Pacific Ocean (Chaloupka et al. 2008). There is an inverse correlation between nesting abundance, and the mean annual sea surface temperature in the core foraging region during the year prior to the nesting season. Cooler ocean temperatures in foraging habitat are presumably associated with increased ocean productivity and prey abundance, and consequently increased loggerhead breeding capacity. Warming regional ocean temperatures could lead to long-term decreased food supply and reduced nesting and recruitment. However, ocean productivity is also linked to temperature variability. If variability increases with warming, then productivity could increase. If variability decreases, then productivity could decrease. At this time we cannot predict which of these scenarios will take place.

Eggs and Hatchlings in a Warming World

Leatherback turtles exhibit temperature-dependent sex determination (TSD) (Binckley et al. 1998; chapter 8), so we would expect that an increase in the temperature of their nest environment would affect that phenomenon. One of the causes of the extinction of dinosaurs may have been TSD (Paladino et al. 1991), and Janzen (1994) first hypothesized that TSD may be unable to evolve rapidly enough to counteract the negative fitness consequences of rapid global temperature change. Modest increases in mean global temperature would drastically skew sex ratios and an increase of 4°C would effectively eliminate production of male offspring. Schwanz and Janzen (2008) argued that it is unlikely that individual plasticity in timing of nesting will be sufficient to offset the effects of global warming on sex ratios of turtles. Recruitment of female hatchlings in a freshwater turtle population was determined by stochastic variation in nest depredation, annual climate, and total nest production (Schwanz et al. 2010). Population demography depended on annual variation in temperature and predation of nests and hatchlings rather than nest-specific behavior of adult females. That is, female nest-site choice could not compensate for those factors.

Primary sex ratios in sea turtles are often female biased (Standora and Spotila 1985; Mrosovsky and Provancha 1992; Marcovaldi et al. 1997). Green turtles and leatherback turtles in Suriname had more female-biased sex ratios in 1993 than in 1982. Estimates of sex ratios for green turtles there, based on monthly rainfall, ranged from 20 to 90% female annually over 12 years (Godfrey et al. 1996). Predicted sex ratios of hawksbill turtles from Brazil were more than 90% female from 1991–1997 (Godfrey et al. 1999). Loggerhead turtle hatchlings are highly female biased in the eastern Mediterranean (Broderick et al. 2000; Zbinden et al. 2007). However, there is large spatial variation, depending on the beach, in sex ratios of hatchlings in the nesting colony on Zakynthos Island, Greece (Zbinden et al. 2007). In the United States, loggerhead turtle hatchlings in Florida are highly female biased while populations that nest in North Carolina and other more northern beaches produce more even sex ratios (Hawkes et al. 2007).

Only one leatherback turtle nesting beach produces hatchlings with a male-biased sex ratio. Nesting beaches on the Huon coast of Papua New Guinea area experience large amounts of rainfall due to the intertropical convergence. As a result, sand temperatures are low and more male hatchlings are produced (Steckenreuter et al. 2010). More typical is the case at Playa Grande in PNMB in Costa Rica, where primary sex ratios are highly female biased (Binckley et al. 1998; Sieg et al. 2011). However, sand temperatures at Playa Grande are so high in El Niño years that there is a high mortality of eggs and hatchlings (Santidrián Tomillo et al. 2009; Santidrián Tomillo et al. 2012). High temperatures are partly due to lower rainfall. A nine-year study indicated that production of female hatchlings increased with incubation temperature as it reached the upper limit of the transitional temperature range that produced both male and female hatchings (~30°C) and decreased afterward because high temperatures increased mortality of "female clutches" (Santidrián Tomillo et al. 2014).

Saba et al. (2012) projected the response of the leatherback turtle population that nests at PNMB to climate change by combining an Earth system model, climate model projections assessed by the Intergovernmental Panel on Climate Change, and a population dynamics model. The authors estimated a 7% per decade decline in the Costa Rica nesting population over the twenty-first century; whereas changes in ocean conditions will have a small effect on the population, the approximately 2.5°C warming of the nesting beach will be the primary driver of the decline through reduced hatching success and hatchling emergence rate. Saba et al. (2012) predicted hatchling sex ratios will not substantially change because they are already highly female biased. The major effect would be that eggs and hatchlings will die from overheating in the hot and dry sand, as is already happening in El Niño years (Santidrián Tomillo et al. 2014). Dudley and Porter (2014) used a Bayesian regression and mechanistic microclimate model to predict that leatherback clutches will have greatly reduced hatching success in West Papua and Gabon by the end of the twenty-first century.

Results such as these for the leatherback turtles at Playa Grande in Costa Rica, and other nesting beaches create a conundrum for conservation biologists and managers. As Archie Carr used to say, the development of flipper tags for sea turtles produced a "tag response" in people who wanted to do something for turtles on their local nesting beach. Not knowing what else to do, they often put flipper tags on sea turtles without much thought about what the biological question was that they wanted to answer. More recently we have seen the development of a hatchery response in people on many sea turtle nesting beaches. Not knowing exactly what to do to protect the turtles and their eggs, people automatically move clutches of eggs into hatcheries. Sometimes this is done to attract tourists for "turtle releases" and other times it becomes a competition between local conservation groups to justify their existence. Taking action, such as moving eggs to a hatchery, can have unintended consequences, including the production of all male hatchlings in cool incubators (Morreale et al. 1982). Therefore, we caution that moving eggs to a hatchery should not be an automatic response to female-biased primary sex ratios on a given beach (Santidrián Tomillo et al. 2014). Moving eggs to a hatchery can not only increase the production of male hatchlings, but can also reduce overall hatching success (Sieg et al. 2011). In addition, different beaches in the same area or region can produce different sex ratios among hatchlings (Hawkes et al. 2007; Zbinden et al. 2007).

It is true that current female-biased sex ratios in sea turtle populations are "normal" for current climate conditions (Santidrián Tomillo et al. 2014). However, current climate cannot be considered "normal" as compared to the historic period before 1878, after which a rise in Earth's temperature due to anthropogenic warming (global warming) was first noted. It is possible that selection for an adjustment in nesting phenology or change in nest-site selection may offset the decline rate at locations such as Playa Grande (Saba et al. 2012). However, that is unlikely given the long generation time of leatherback turtles and the lack of selection observed in phenology of nesting and nest-site selection in freshwater turtles (Janzen 1994; Schwanz and Janzen 2008). Sinervo et al. (2010) calculated that estimates of evolutionary rates required to keep pace with global change compromises population growth rates, thereby precipitating extinctions. They reported that species extinction probabilities for lizard species of 6% by 2050 and 20% by 2080.

Therefore, it is important to document temperatures in clutches of eggs in nesting beaches throughout the nesting season and for several nesting seasons to determine the effect of climate variation on the long-term sex ratios of sea turtle populations. If data show a continued trend in highly female-biased sex ratios in a regional population or a long-term decline in hatching success and emergence rate on a given beach, then anthropogenic climate mitigation of nests (for example, shading, irrigation) may preserve the nesting population (Patiño-Martínez et al. 2012; Saba et al. 2012; Santidrián Tomillo et al. 2014).

To put it simply, if temperatures in the nesting beach are so high that few if any hatchlings survive, then one should go ahead and shade or water the nests to prevent an extreme mortality event. Otherwise, one should be cautious and not manipulate the clutches in natural nests without adequate data to understand the scope of the problem and a well thought-out plan to solve that problem. This approach is supported by Mitchell and Janzen (2010), who recommended a precautionary approach to management of reptiles identified as being at high risk of extinction due to climate change. They suggested that management should aim to neutralize directional sex ratio bias by manipulating incubation temperature and by promoting adaptive processes through genetic supplementation of populations. They hope that such measures will buy time for research directed at more accurate predictions of species' vulnerability to climate change. We anticipate that improved science on the nesting beaches of the world will more clearly define the problems facing eggs and hatchlings due to rising temperatures. The goal should be to employ cost-effective management strategies to allow sea turtles, and especially leatherbacks, to survive long enough so that their natural variability in nesting beach selection finds future populations on beaches in cooler regions of the world. Perhaps they will move north to Maryland, Delaware, and New Jersey in the United States when beaches in Florida, Georgia, and South Carolina become too hot for successful reproduction. But if we do not produce viable male and female hatchlings from current nesting beaches over the next 100 years, there will be no adults out in the ocean to find new beaches.

To further complicate the situation, a warming ocean has the potential to change the distribution and strength of local and regional currents. The location of successful nesting beaches for leatherbacks is tied to the distribution of hatchlings in offshore currents (Shillinger et al. 2012). There are only a few beaches from Mexico to Costa Rica where offshore currents take leatherback hatchlings to locations such as the Costa Rica Dome, where upwellings and convergence zones provide adequate food for their growth and development. If a warming climate changes those currents, then hatchling leatherbacks may have difficulty in reaching habitats well suited to juvenile development.

Rising Seas

In addition to all of these threats to leatherback turtles, there is another that is quite serious: seas are rising. Ocean level has increased about 20 cm between 1960 and 2014. Levels will continue to increase as much as 1 to 2 m by 2100 (IPCC Working Group I 2013). As a result, there will be shoreline erosion from daily increased sea levels, higher waves and tides, greater storm surges, and saline intrusion into the water table. Sea turtle nesting beaches are very vulnerable to these physical impacts because beaches can be lost and salt water can penetrate nests and kill developing eggs. Projected sea level rise along the northern Great Barrier Reef in Australia will inundate up to 38% of available nesting areas for green turtles (Fuentes et al. 2010). A study by the World Wildlife Fund at Playa Grande, Costa Rica, projected that a one meter rise in sea level will cause significant flooding of the beach and cause it to move many meters inland (Drews 2008).

Katselidis et al. (2014) modeled sea-level rise scenarios of 0.2 to 1.3 m for Zakynthos Island in Greece. They projected that even under the most conservative scenario (0.2 m rise) about 38% of the total nesting beach area will be lost; under the 0.9 m scenario about 4 km of nesting area encompassing 85% of nesting activity will be lost. Cliffs prevent landward beach migration on some beaches and the 2 km of beach that can migrate backward because of sand dunes are already threatened by urban development.

Fish et al. (2008) proposed that management of sea turtle beaches should include setback regulations to account for sea level rise. Where coastal structures such as walls and buildings prevent landward migration of beaches, sea turtles will lose their nesting sites. Typical examples include some urbanized beaches in Florida, Georgia, and the Carolinas in the United States, and beaches such as Rethymno on Crete in Greece. One of the possibilities discussed above that would alleviate the problem of female-biased sex ratios is that turtle nesting locations could shift to higher latitudes through natural selection. In the United States, sea turtles may encounter sea walls as they migrate northward to beaches in Maryland, Delaware, and New Jersey where residents are currently considering building structures to protect homes and businesses from rising sea levels. Setback regulations, which prohibit construction within a set distance from the sea, need to be established for all current and potential nesting beaches for leatherbacks and other sea turtles in order to allow the natural movement of beaches in response to rising seas.

Conclusions

For many years, biologists, conservationists, and managers have focused on the current threats to leatherbacks and other sea turtles. These threats have involved taking of eggs, destruction of nesting habitat, and loss of turtles at sea due to fishing activities. It is only in the last few years that threats due to human-caused climate change have become apparent. We will have to solve problems in the twenty-first century involving loss of habitats and food supplies in the oceans, threats of high temperatures on nesting beaches to the process of TSD and the production of hatchlings in general, and threats of rising seas to the existence of nesting beaches themselves. Certainly the challenges faced by all of those concerned with the survival of leatherback turtles are great indeed. Only increased management of all aspects of the turtles' life history will ensure that they survive into the twenty-second century. Sometimes young biologists fear that all of the interesting questions have been answered and all of the interesting challenges in conservation have been addressed. That is certainly not the case when the looming crises of climate change are considered. The question is, will future generations of sea turtle biologists and conservationists succeed in asking the right questions and in taking the correct actions to address the great challenges that our actions have caused and will create? The appropriate questions and actions will be essential as Earth's climate changes more rapidly and more extensively than it has in several million years.

LITERATURE CITED

Alfaro-Shigueto, J., P. H. Dutton, M. Van Bressem, and J. Mangel. 2007. Interactions between leatherback turtles and Peruvian artisanal fisheries. Chelonian Conservation and Biology 6: 129–134.

Bailey, H., S. R. Benson, G. L. Shillinger, S. J. Bograd, P. H. Dutton, S. A. Eckert, S. J. Morreale, et al. 2012. Identification of distinct movement patterns in Pacific leatherback turtle populations influenced by ocean conditions. Ecological Applications 22: 735–747.

Bailey, H., S. Fossette, S. J. Bograd, G. L. Shillinger, A. M. Swithenbank, J.-Y. Georges, P. Gaspar, et al. 2012. Movement patterns for a critically endangered species, the leatherback turtle (*Dermochelys coriacea*), linked to foraging success and population status. PLoS One 2012:7(5): e36401. doi: 10.1371/journal.pone.0036401.

Bailey, H., G. Shillinger, D. Palacios, S. Bograd, J. Spotila, F. Paladino, and B. Block. 2007. Identifying and comparing phases of movement by leatherback turtles using state-space models. Journal of Experimental Marine Biology and Ecology 356: 128–135.

Beaugrand G., P. C. Reid, F. Ibañez, J. A. Lindley, and M. Edwards. 2002. Reorganization of North Atlantic marine copepod biodiversity and climate. Science 296: 1692–1694.

Benson, S. R., T. Eguchi, D. G. Foley, K. A. Forney, H. Bailey, C. Hitipeuw, B. P. Samber, et al. 2011. Large-scale movements and high-use areas of western Pacific leatherback turtles, *Dermochelys coriacea*. Ecosphere 2(7): art 84. doi: 10.1890/ES11-00053.1.

Binckley, C. A., J. R. Spotila, K. S. Wilson, and F. V. Paladino. 1998. Sex determination and sex ratios of Pacific leatherback turtles, *Dermochelys coriacea*. Copeia 1998: 291–300.

Bleakney, J. S. 1965. Reports of marine turtles from New England and Eastern Canada. Canadian Field Naturalist 79: 120–128.

Broderick, A. C., B. J. Godley, S. Reece, and J. R. Downie. 2000. Incubation periods and sex ratios of green turtles: highly female biased hatchling production in the eastern Mediterranean. Marine Ecology Progress Series 202: 273–281.

Carr, A. 1956. The windward road, adventures of a naturalist on remote Caribbean shores. A. A. Knopf, New York, NY, USA.

Chaloupka, M., N. Kamezaki, and C. Limpus. 2008. Is climate change affecting the population dynamics of the endangered Pacific loggerhead sea turtle? Journal of Experimental Marine Biology and Ecology 356: 136–142.

Chavez, F. P., P. G. Strutton, G. E. Friederich, R. A. Feely, G. C. Feldman, D. G. Foley, and M. J. McPhaden. 1999. Biological and chemical response of the equatorial Pacific Ocean to the 1997–98 El Niño. Science 286: 2126–2131.

Drews, C. 2008. Letter from World Wildlife Fund Costa Rica to H. M. Durán, Comisión Permanente Especial de Ambiente, Asamblea Legislativa, San José, Costa Rica, 20 July 2008, 6 pages.

Duarte, C. M. 2002. The future of seagrass meadows. Environmental Conservation 29: 192–206.

Dudley, P. N., and W. P. Porter. 2014. Using empirical and mechanistic models to assess global warming threats to leatherback sea turtles. Marine Ecology Progress Series 501: 265–278.

Eckert, S. A. 2002. Distribution of juvenile leatherback sea turtle (*Dermochelys coriacea*) sightings. Marine Ecology Progress Series 230: 289–293.

Eckert, S. A. 2006. High-use oceanic areas for Atlantic leatherback sea turtles (*Dermochelys coriacea*) as identified using satellite telemetered location and dive information. Marine Biology 149: 1257–1267.

Fish, M. R., I. M. Cote, J. A. Horrocks, B. Mulligan, A. R. Watkinson, and A. P. Jones. 2008. Construction setback regulations and sea-level rise: mitigating sea turtle nesting beach loss. Ocean & Coastal Management 51: 330–341.

Frazier, J., and J.L.B. Montero. 1990. Incidental capture of marine turtles by the swordfish fishery at San Antonio, Chile. Marine Turtle Newsletter 49: 8–13.

Fuentes, M.M.P.B., C. J. Limpus, M. Hamann, and J. Dawson. 2010. Potential impacts of projected sea-level rise on sea turtle rookeries. Aquatic Conservation: Marine and Freshwater Ecosystems 20: 132–139.

Gates, D. M. 1962. Energy exchange in the biosphere. Harper and Row, New York, NY, USA.

Gates, D. M. 1980. Biophysical ecology. Springer-Verlag, New York, NY, USA.

Godfrey, M. H., A. F. D'Amato, M. Ã. Marcovaldi, and N. Mrosovsky. 1999. Pivotal temperature and predicted sex ratios for hatchling hawksbill turtles from Brazil. Canadian Journal of Zoology 77: 1465–1473.

Godfrey, M. H., N. Mrosovsky, and R. Barreto. 1996. Estimating past and present sex ratios of sea turtles in Suriname. Canadian Journal of Zoology 74: 367–277.

Hawkes, L. A., A. C. Broderick, M. H. Godfrey, and B. J. Godley. 2007. Investigating the potential impacts of climate change on a marine turtle population. Global Change Biology 13: 923–932.

Hawkes, L. A., A. C. Broderick, M. H. Godfrey, and B. J. Godley. 2009. Climate change and marine turtles. Endangered Species Research 7: 137–154.

Hays, G. C., J.D.R. Houghton, and A. E. Myers. 2004. Pan-Atlantic leatherback turtle movements. Nature 429: 522.

Hoegh-Guldberg, O., P. J. Mumby, A. J. Hooten, R. S. Steneck, P. Greenfield, E. Gomez, C. D. Harvell, et al. 2007. Coral reefs under rapid climate change and ocean acidification. Science 318: 1737–1742.

Houghton, J.D.R., T. K. Doyle, M. W. Wilson, J. Davenport, and G. C. Hays. 2006. Jellyfish aggregations and leatherback turtle foraging patterns in a temperate coastal environment. Ecology 87: 1967–1972.

IPCC Working Group I. 2013. Climate change 2013, the physical science basis. Working Group 1 Contribution to the Fifth Assessment Report of the Intergovernmental Panel on Climate Change. Cambridge University Press, Cambridge, UK, and New York, NY, USA.

James, M. C., J. Davenport, and G. C. Hays. 2006. Expanded thermal niche for a diving vertebrate: a leatherback turtle diving into near-freezing water. Journal of Experimental Marine Biology and Ecology 335: 221–226.

James, M. C., and T. B. Herman. 2001. Feeding of *Dermochelys coriacea* on Medusae in the Northwest Atlantic. Chelonian Conservation and Biology 4: 202–205.

James, M. C., and N. Mrosovsky. 2004. Body temperatures of leatherback turtles (*Dermochelys coriacea*) in temperate waters off Nova Scotia, Canada. Canadian Journal of Zoology 82: 1302–1306.

James, M. C., C. A. Ottensmeyer, and R. A. Myers. 2005. Identification of high-use habitat and threats to leatherback sea turtles in northern waters: new directions for conservation. Ecology Letters 8: 195–201.

Janzen, F. J. 1994. Climate change and temperature-dependent sex determination in reptiles. Proceedings of the National Academy of Science USA 91: 7487–7490.

Kaplan, I. C. 2005. A risk assessment for Pacific leatherback turtles (*Dermochelys coriacea*). Canadian Journal of Fisheries and Aquatic Sciences 62: 1710–1719.

Katselidis, K. A., G. Schofield, G. Stamou, P. Dimopoulos, and J. Pantis. 2014. Employing sea-level rise scenarios to strategically select sea turtle nesting habitat important for long-term management at a temperate breeding area. Journal of Experimental Marine Biology and Ecology 450: 47–54.

Limpus, C. J., and N. Nicholls. 1988. The southern oscillation regulates the annual numbers of green turtles (*Chelonia mydas*) breeding around northern Australia. Australian Wildlife Research 15: 157–161.

Limpus, C. J., and N. Nicholls. 2000. ENSO regulation of Indo-Pacific green turtle populations. In: G. L. Hammer, N. Nicholls, and C. Mitchell (eds.), Applications of seasonal climate forecasting in agricultural and natural ecosystems. The Australian experience. Kluwer Academic Publishers, Dordrecht, Netherlands.

Luschi, P., J.R.E. Lutjeharms, P. Lambardi, R. Mencacci, G. R. Hughes, and G. C. Hays. 2006. A review of migratory behavior of sea turtles off southeastern Africa. South African Journal of Science 102: 51–58.

Luschi, P., A. Sale, R. Mencacci, G. R. Hughes, J. R. Lutjeharms, and F. Papi. 2003. Current transport of leatherback sea turtles (*Dermochelys coriacea*) in the Indian Ocean. Proceedings of the Royal Society B 270: 129–132.

Marbá, N., and C. M. Duarte. 2010. Mediterranean warming triggers seagrass (*Posidonia oceanica*) shoot mortality. Global Change Biology 16: 2366–2375.

Marcovaldi, M. Ã., M. H. Godfrey, and N. Mrosovsky. 1997. Estimating sex ratios of loggerhead turtles in Brazil from pivotal incubation durations. Canadian Journal of Zoology 75: 755–770.

McMahon, C. R., and G. C. Hays. 2006. Thermal niche, large-scale movements and implications of climate change for a critically endangered marine vertebrate. Global Change Biology 12: 1–9.

McNab, B. K., and W. Auffenburg. 1976. The effect of large body size on the temperature regulation of the Komodo dragon, *Varanus komodoensis*. Comparative Biochemistry and Physiology 55A: 345–350.

Milanovich, J. R., W. E. Peterman, N. P. Nibbelink, and J. C. Maerz. 2010. Projected loss of a salamander diversity hotspot as a consequence of projected global climate change. PLOS One 5: e12189. doi: 10.1371/journal.pone.0012189.

Miller-Rushing, A. J., and R. B. Primack. 2008. Global warming and flowering times in Thoreau's Concord: a community perspective. Ecology 89: 332–341.

Mitchell, N. J., and F. J. Janzen. 2010. Temperature-dependent sex determination and contemporary climate change. Sexual Development 4: 129–140.

Morreale, S. J., G. J. Ruiz, J. R. Spotila, and E. A. Standora 1982. Temperature-dependent sex determination: current practices threaten conservation of sea turtles. Science 216: 1245–1247.

Mrosovsky, N., and J. Provancha. 1992. Sex ratio of hatchling loggerhead sea turtles: data and estimates from a 5-year study. Canadian Journal of Zoology 70: 530–538.

Paladino, F. V., P. Dodson, J. K. Hammond, and J. R. Spotila. 1991. Temperature dependent sex determination in dinosaurs: implications for population dynamics and

extinction. Geological Society of America Special Paper 238: 63–70.
Paladino, F. V., M. P. O'Connor, and J. R. Spotila. 1990. Metabolism of leatherback turtles, gigantothermy, and thermoregulation of dinosaurs. Nature 344: 858–860.
Parmesan, C., and G. Yohe. 2003. A globally coherent fingerprint of climate change impacts across natural systems. Nature 421: 37–42.
Patiño-Martínez, J., A. Marco, L. Quiñones, and L. Hawkes. 2012. A potential tool to mitigate the impacts of climate change to the Caribbean leatherback sea turtle. Global Change Biology 18: 401–411.
Perry, A. L., P. J. Low, J. R. Ellis, and J. D. Reynolds. 2005. Climate change and distribution shifts in marine fishes. Science 308: 1912–1915.
Poloczanska, E. S., C. J. Brown, W. J. Sydeman, W. Kiessling, D. S. Schoeman, P. J. Moore, K. Brander, et al. 2013. Global imprint of climate change on marine life. Nature Climate Change 3: 912–925.
Porter, W. P., and D. M. Gates. 1969. Thermodynamic equilibria of animals with environment. Ecological Monographs 39: 227–244.
Reina, R. D., J. R. Spotila, F. V. Paladino, and A. E. Dunham. 2009. Changed reproductive schedule of eastern Pacific leatherback turtles *Dermochelys coriacea* following the 1997–98 El Niño to La Niña transition. Endangered Species Research 7: 155–161.
Saba, V. S., P. Santidrián-Tomillo, R. D. Reina, J. R. Spotila, J. A. Musick, D. A. Evans, and F. V. Paladino. 2007. The effect of the El Niño Southern Oscillation on the reproductive frequency of eastern Pacific leatherback turtles. Journal of Applied Ecology 44: 395–404.
Saba, V. S., G. L. Shillinger, A. M. Swithenbank, B. A. Block, J. R. Spotila, J. A. Musick, and F. V. Paladino. 2008. An oceanographic context for the foraging ecology of eastern Pacific leatherback turtles: consequences of ENSO and coastal gillnet fisheries. Deep Sea Research. 55: 646–660.
Saba, V. S., J. R. Spotila, F. P. Chavez, and J. A. Musick. 2008. Bottom-up and climatic forcing on the worldwide population of leatherback turtles. Ecology 89: 1414–1427.
Saba, V. S., C. A. Stock, J. R. Spotila, F. V. Paladino, and P. Santidrián Tomillo. 2012. Projected response of an endangered marine turtle to climate change. Nature Climate Change 2: 814–820. doi: 10.1038/nclimate1582.
Sale, A., P. Luschi, R. Mencacci, P. Lambardi, G. Hughes, G. Hays, S. Benvenuti, and F. Papi. 2006. Long-term monitoring of leatherback turtle diving behavior during oceanic movements. Journal of Experimental Marine Biology and Ecology 328: 197–210.
Santidrián Tomillo, P., D. Oro, F. V. Paladino, R. Piedra, A. E. Sieg, and J. R. Spotila. 2014. High beach temperatures increased female-biased primary sex ratios but reduced output of female hatchlings in the leatherback turtle. Biological Conservation 176: 71–79.
Santidrián Tomillo, P., V. S. Saba, G. S. Blanco, C. A. Stock, F. V. Paladino, and J. R. Spotila. 2012. Climate driven egg and hatchling mortality threatens survival of eastern Pacific leatherback turtles. PloS One 7: e37602. doi: 10.1371/journal.pone.0037602.
Santidrián Tomillo, P., V. S. Saba, R. Piedra, F. V. Paladino, and J. R. Spotila. 2008. Egg poaching: a major factor in the population decline of leatherback turtles, *Dermochelys coriacea*, at Parque Nacional Marino Las Baulas, Costa Rica. Conservation Biology 22: 1216–1224.
Santidrián Tomillo, P., J. S. Suss, B. P. Wallace, K. D. Magrini, G. Blanco, F. V. Paladino, and J. R. Spotila. 2009. Influence of emergence success on the annual reproductive output of leatherback turtles. Marine Biology 156: 2021–2031.
Schwanz, L. E., and F. J. Janzen. 2008. Climate change and temperature-dependent sex determination: Can individual plasticity in nesting phenology prevent extreme sex ratios? Physiological and Biochemical Zoology 81: 826–834.
Schwanz, L. E., R.-J. Spencer, R. M. Bowden, and F. J. Janzen. 2010. Climate and predation dominate juvenile and adult recruitment in a turtle with temperature-dependent sex determination. Ecology 91: 3016–3026.
Shillinger, G. L., E. Di Lorenzo, H. Luo, S. J. Bograd, E. L. Hazen, H. Bailey, and J. R. Spotila. 2012. On the dispersal of leatherback turtle hatchlings from Mesoamerican nesting beaches. Proceedings of the Royal Society B. 279: 2391–2395. doi: 10.1098/rspb.2011.2348.
Shillinger, G. L., D. M. Palacios, H. Bailey, S. J. Bograd, A. M. Swithenbank, P. Gaspar, B. P. Wallace, et al. 2008. Persistent leatherback turtle migrations present opportunities for conservation. PLOS Biology 6(7): e171. doi: 10.1371/journal.pbio.0060171.
Shillinger, G. L., A. M. Swithenbank, H. Bailey, S. J. Bograd, M. R. Castelton, B. P. Wallace, J. R. Spotila, et al. 2011. Vertical and horizontal habitat preferences of post-nesting leatherback turtles in the South Pacific Ocean. Marine Ecology Progress Series 422: 275–289.
Shillinger, G. L., A. M. Swithenbank, S. J. Bograd, H. Bailey, M. R. Castelton, B. P. Wallace, J. R. Spotila, et al. 2010. Identification of high-use internesting habitats for eastern Pacific leatherback turtles: role of the environment and implications for conservation. Endangered Species Research 10: 215–243.
Sieg, A. E., C. A. Binckley, B. P. Wallace, P. Santidrián Tomillo, R. D. Reina, F. V. Paladino, and J. R. Spotila. 2011. Sex ratios of leatherback turtles: hatchery translocations decreases metabolic heating and female-bias. Endangered Species Research 15: 195–204.
Sinervo, B., F. Méndez-de-La-Cruz, D. B. Miles, B. Heulin, E. Bastiaans, M. Villagrán-Santa Cruz, R. Lara-Resendiz, et al. 2010. Erosion of lizard diversity by climate change and altered thermal niches. Science 328: 894–899.
Solow, A. R., K. A. Bjorndal, and A. B. Bolten. 2002. Annual variation in nesting numbers of marine turtles: the effect of sea surface temperature on re-migration intervals. Ecology Letters 5: 742–746.
Spotila, J. R., P. W. Lommen, G. S. Bakken, and D. M. Gates. 1973. A mathematical model for body temperatures of

large reptiles: implications for dinosaur ecology. American Naturalist 197: 391–404.

Spotila, J. R., R. R. Reina, A. C. Steyermark, P. T. Plotkin, and F. V. Paladino. 2000. Pacific leatherback turtles face extinction. Nature 405: 529–530.

Spotila, J. R., and E. A. Standora. 1985. Environmental constraints on the thermal energetics of sea turtles. Copeia 1985: 694–702.

Standora, E. A., and J. R. Spotila. 1985. Temperature dependent sex determination in sea turtles. Copeia 1985: 711–722.

Standora, E. A., J. R. Spotila, and R. E. Foley. 1982. Regional endothermy in the sea turtle, *Chelonia mydas*. Journal of Thermal Biology 7: 159–165.

Steckenreuter, A., N. Pilcher, B. Krüger, and J. Ben. 2010. Male-biased primary sex ratio of leatherback turtles (*Dermochelys coriacea*) at the Huon Coast, Papua New Guinea. Chelonian Conservation and Biology 9: 123–128.

Willis-Norton, E., E. L. Hazen, S. Fossette, G. Shillinger, R. R. Rykaczewski, D. G. Foley, J. P. Dunne, and S. J. Bograd. 2014. Climate change impacts on leatherback turtle pelagic habitat in the southeast Pacific. Deep-Sea Research II. http://dx.doi.org/10.1016/j.dsr2.2013.12.019.

Zbinden, J. A., C. Davy, D. Margaritoulis, and R. Arlettaz. 2007. Large spatial variation and female bias in the estimated sex ratio of loggerhead sea turtle hatchlings of a Mediterranean rookery. Endangered Species Research 3: 305–312.

17

Impacts of Fisheries on the Leatherback Turtle

REBECCA L. LEWISON,
BRYAN P. WALLACE, AND
SARA M. MAXWELL

Fisheries bycatch, the unintended capture of nontarget species, is a documented and substantial source of mortality for sea turtle populations worldwide, and has been a major focus among sea turtle ecologists and fisheries managers since the 1980s (National Research Council 1990). Recently, there have been major advances in our ability to assess and mitigate sea turtle bycatch (Gilman 2011). Although data on leatherback turtle (*Dermochelys coriacea*) bycatch are relatively scarce compared to other sea turtle species, available data indicate that fisheries bycatch poses a substantial threat to leatherbacks in many ocean regions. In this chapter we review the current state of knowledge on leatherback bycatch, advances in leatherback bycatch mitigation, and innovative approaches that have been and need to be implemented to reduce the impacts fisheries bycatch are having on leatherback populations worldwide.

Evaluating Impact of Fisheries on Leatherbacks: Key Types of Information

To evaluate effects of fisheries on leatherback populations, three types of information are required: bycatch data, fishing effort, and turtle distribution. Bycatch data are most commonly collected in two ways: (1) data recorded by trained observers on fishing vessels (termed observer data), and (2) data collected during dockside interviews and surveys. Although less common, bycatch data can also be recorded in vessel logbooks. Observer data provide high resolution information by providing a more accurate and precise estimate of number of turtles caught as well as locations where bycatch occurs. Although observer data are an essential ingredient to characterizing and quantifying bycatch, precision of the data is influenced by the amount of fishing effort upon which data are based (Tuck 2011). Extreme turtle bycatch rates (either high or low) in gill nets in the northwest Atlantic Ocean tend to occur where relatively little fishing effort is observed (Sims et al. 2008), a finding confirmed by a similar assessment using global-scale bycatch data across geographic regions and different gear categories (Wallace, Lewison, et al. 2010).

Table 17.1. Data on nonuniformity of leatherback bycatch data within and across fishing gear types (1990–2011). Database from Project GloBAL (The Global Bycatch Assessment of Long-lived Species, http://bycatch.env.duke.edu). Note that records can contain more than one leatherback bycatch event.

Gear	Total number of records	Total number of records with observed effort	Number of different effort metrics reported	Total number of records with bycatch rate per unit effort	Number of different bycatch metrics reported
Gill net	103	45	14	75	22
Longline	157	130	6	142	10
Trawl	8	4	4	6	5

Observers normally monitor extremely small proportions of a fishing fleet's total effort (typically < 5% with some exceptions; Finkbeiner et al. 2011). Even when observer data are available for a fishing region, there is a lack of uniformity in how bycatch rates are reported (table 17.1). In this database example, each report of bycatch is listed as a separate record. Of the existing 268 records reporting leatherback bycatch from fishing grounds around the world, there are 37 different metrics used, with 21 different metrics used within a single gear category. For example, gill net bycatch is reported as individuals / set, individuals / soak time hour, individuals / 10 m of net, or individuals / day. This lack of uniformity can be overcome (Wallace, Lewison, et al. 2010; Wallace et al. 2011), but it presents a substantial challenge to comparing or assessing bycatch effects among fisheries, gear types, or ocean regions.

The second kind of information that is needed to evaluate fisheries impacts is the amount of fishing gear deployed in the water, also called fishing effort. The most commonly reported measure of fisheries production is the amount of catch. While catch data provide important information on quantity of target species harvested, they do not provide information on expended effort, which is likely to be a better indicator of bycatch of nontarget species like leatherbacks. Data on intensity, spatial locations, and timing of fishing effort are also needed to quantify and monitor fishery-related effects on leatherbacks and other nontarget species. Fishing effort measures that reflect detailed gear characteristics, such as soak time (length of time fished), depth of gear, or number of hooks, are even less common. These gear characteristics are extremely valuable in identifying the relationship between fishing practices, gear deployment patterns, and bycatch.

The final piece to the bycatch puzzle is sea turtle distribution. For leatherbacks, this means understanding their at-sea location as well as their diving behavior, because leatherbacks make deep and extended dives (chapters 3, 13). Telemetry is the most accurate approach used to describe leatherback horizontal and vertical distribution. For leatherbacks, as with other sea turtles, a combination of location transmitters, either Platform Transmitting Terminals (PTT) or Global Positioning System (GPS) transmitters, plus time-depth recorders (TDRs) have identified leatherback seasonal movement patterns and dive behavior (James, Meyers, and Ottensmeyer 2005; James, Ottensmeyer, and Meyers 2005; Shillinger et al. 2011). These data can be used to either identify the potential spatial and temporal overlap of turtles with fishing fleets or protected areas (James, Ottensmeyer, and Meyers 2005; Witt et al. 2011; Fossette et al. 2014; Roe et al. 2014) or to estimate the likelihood that turtles will come into contact with fishing gear based on the overlap of gear depth in the water column and turtle dive behavior (Polovina et al. 2003). Additionally, tracking can quantify the level of fisheries-related mortality of sea turtles, using movement data and changes in satellite transmissions as an indication of a mortality event (Hays et al. 2003). Tracking data also can be used to compare management options; for example, Hays et al. (2003) suggested that gear changes, rather than spatial management strategies such as closures, will be most effective in reducing leatherback bycatch.

Tracking serves as a powerful tool for conservation and management of turtles (Shillinger et al. 2008; Maxwell et al. 2011), but, despite increased use of tracking for leatherbacks, we still lack knowledge of at-sea distribution for most of their life stages (Hazen et al. 2012). Since much of the leatherback life cycle occurs in the open ocean, it is not surprising that most of our knowledge of leatherback movements comes largely from research on nesting females (Georges et al. 2007; Shillinger et al. 2010; Witt et al. 2011; Bailey et al. 2012), although in the Atlantic Ocean, research has explored movements of males and subadults (James, Myers, and Ottensmeyer 2005; James, Ottensmeyer, and Myers 2005; López-Mendilaharsu et al. 2009). Satellite tracking and diving research that focuses on underrepresented life stages (e.g., small and middle-sized juveniles) is critical to improving population-level conservation, management, and bycatch reduction (Godley et al. 2008).

Current Understanding of Fisheries Effects on Leatherbacks

Leatherback bycatch is documented primarily in pelagic longlines (Beverly and Chapman 2007; Donoso and Dutton 2010) and net fisheries in coastal waters (Alfaro-Shigueto et al. 2007, 2011). With both gear types, encounters are likely the result of entanglement. In longlines, leatherbacks typically become entangled in hooked branch lines that hang off the mainline, in contrast to other sea turtle species that tend to become hooked in the mouth after trying to consume bait (Watson et al. 2005; Gilman et al. 2006). Leatherbacks can become entangled in the mesh of stationary nets held in place in the water column by floats or anchors. While some fleets have adopted measures that may reduce leatherback entanglements (Kobayashi and Polovina 2005; Watson et al. 2005), leatherback bycatch remains a problem that fisheries management agencies or ministries need to address, especially in regions where declining leatherback populations occur (Wallace et al. 2011). In addition, leatherbacks, like other sea turtles, are targeted directly by sea turtle fisheries in some waters and opportunistic take of captured turtles is still prevalent in some regions (Alfaro-Shigueto et al. 2011).

Assessing impacts of fisheries on leatherbacks is hindered by a lack of comprehensive observer data and reporting on leatherback interactions with fisheries throughout their range. Lewison et al. (2004) used existing bycatch data (published and grey literature) to assess the magnitude and extent of sea turtle bycatch from pelagic longlines worldwide. Integrating catch data from over 40 nations and bycatch data from 13 international observer programs, they described the global nature of leatherback bycatch and concluded that while longline bycatch was not the only source of fisheries mortality, it was an important mortality source that required attention. Leatherbacks in the Pacific are particularly vulnerable to fisheries mortality given their dramatic nesting population declines (Spotila et al. 2000).

Kaplan (2005) used a risk assessment to determine the relative importance of longline fisheries and other sources of coastal mortality as drivers of the decline in Pacific leatherbacks. Using a Bayesian framework to account for uncertainty about parameter estimates and future outcomes, he evaluated the potential effect of policy alternatives on adult leatherback abundances (e.g., elimination of longline bycatch versus elimination of other mortality sources). His results highlighted the need for international efforts to protect leatherbacks from both longline and gill net fisheries bycatch. Kaplan's analysis also emphasized the importance of evaluating fisheries-related mortality in the context of nonfisheries-mediated mortality (i.e., harvest of adult females and eggs at sea turtle nesting colonies).

Comprehensive assessments of leatherback bycatch can take several different forms, ranging from expert opinion-based assessments, to regional or global scale assessments of bycatch patterns based on existing data. Regardless, any conservation status assessment, including those assessing impact of different threats on leatherback populations, must be conducted at spatial scales that permit population-based evaluations and subsequent management responses. In the case of sea turtles, establishment of regional management units (RMUs) for sea turtles worldwide (Wallace, DiMatteo, et al. 2010) provides an appropriate biogeographic and population framework for such assessments. Wallace et al. (2011) used expert evaluation of available data to assess the conservation status of all sea turtle RMUs according to population viability (e.g., abundance and trends) and relative impacts of various threats. Results for the seven different leatherback RMUs demonstrated the variation in conservation status among populations worldwide, ranging from high risk (to relative population viability) and high threat (relative impacts of anthropogenic threats) for East Pacific, West Pacific, and Southwest Atlantic RMUs to low risk and low threat for Northwest and Southeast Atlantic RMUs (fig. 17.1). This analysis revealed that fisheries bycatch presents the largest threat to leatherbacks globally.

To assess bycatch impacts in a more quantitative way and to facilitate comparisons within and among RMUs, Wallace et al. (2013) combined information on bycatch rates, fishing effort, mortality rates, and the reported sizes of leatherbacks caught (larger, older turtles have higher reproductive value; Wallace et al. 2008) to calculate a "bycatch impact score" for each RMU. This data integration identifies gear of particular conservation concern for each RMU as well as data gaps—areas where data are unavailable or insufficient to allow quantitative assessments of bycatch impacts (fig. 17.2). A review of leatherback data presented by Wallace et al. (2013) yielded several important findings (table 17.2). First, there were relatively few records of leatherback bycatch in trawls compared with longlines and nets, suggesting that leatherbacks have little overlap with trawl gear. Second, although there were wide variations in observed leatherback bycatch rates, rates were relatively low overall across gears. Third, mortality

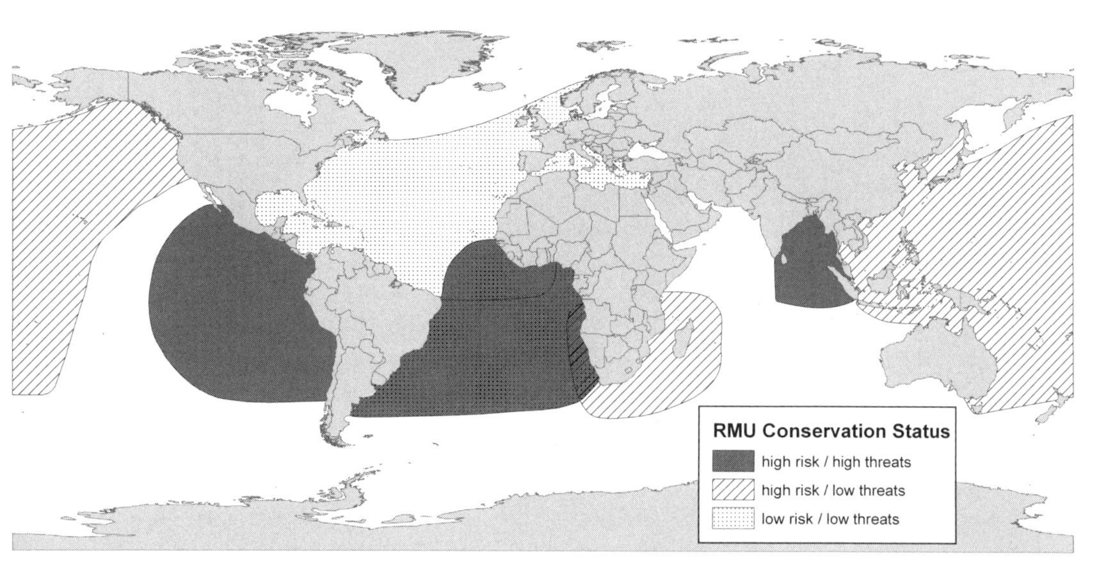

Fig. 17.1. Leatherback regional management units (RMUs), shown according to conservation status assessments based on relative population viability scores (i.e., "Risk") and relative impacts of anthropogenic threats (i.e., "Threat"). See Wallace et al. (2011) for details.

rates were generally higher in nets than in longlines. Although data limitations only allowed for a complete RMU comparison for longline fisheries, net gear had a higher impact for two of three leatherback RMUs (fig. 17.2). Taken together, these results highlight the importance of nets as a gear of concern for leatherbacks globally, at least as much and perhaps more so than longlines.

Advances in Bycatch Reduction

Given the impact fisheries are having on leatherbacks worldwide, bycatch reduction is essential to improve fisheries sustainability and to save leatherbacks from extinction. Bycatch reduction can come from different sources: governance structure and regulatory frameworks, spatial fisheries management (i.e., reduction of fishing effort in high bycatch areas), switching from high bycatch gear to lower bycatch gear, technological changes in gear or other equipment to make fishing more selective, changes in gear deployment or retrieval, and deterrents designed to avoid interactions with non-target species (Chuenpagdee et al. 2003; Gilman et al. 2006; Gilman 2011). Here, we focus on technological changes in gear and methods of gear deployment/retrieval that have been developed to reduce leatherback bycatch; many of these rely on partnerships with fishing organizations to increase both the success and compliance of management strategies (Gilman et al. 2006; Gilman et al. 2010).

Longline Fisheries

Longline fisheries, both pelagic and demersal, have been the focus of many of the efforts to reduce sea turtle bycatch. Experimental trials have tested changes in gear deployment, gear types or shapes, and distribution. Settling longline gear deeper than 100 m may reduce bycatch of sea turtles (Beverly et al. 2009). With deeper lines, leatherbacks may experience fewer encounters with gear, even though they often dive deeper than 100 m (Houghton et al. 2008; Fossette et al. 2010). However, entanglement may be more lethal because turtles will not be able to reach the surface to breath. Other tested mitigation approaches for longline fisheries include gear deployments only during the day when leatherback interactions are reduced, alterations in target attractants, and reducing soak time (Gilman et al. 2006). Restricting fishing to certain areas, such as those below a certain sea surface temperature where turtles are unlikely to occur, is a promising technique with hard-shell turtles in the Central Pacific (Howell et al. 2008).

Changes in shape and size of longline hooks have been one of the primary leatherback and other sea turtle bycatch reduction measures developed to date (Gilman et al. 2006). Circle hooks were developed to replace traditional J hooks, which have the barb of the hook angled outward to snag tissue more easily. In circle hooks, the eye of the hook is wider and the barb of hook is pointed back to the shaft of the hook. Designed

Table 17.2. Summary of bycatch data for all leatherback regional management units (RMUs) in longlines, nets, and trawls, from 1990–2011. No. recs: number of records. No. turtles taken: number of turtles reported as bycatch for that RMU-gear stratum. Total effort: sum of all gear reported/observed for that RMU-gear stratum, converted to "sets" (see Wallace, DiMatteo, et al. 2010 for details on methods). Weighted median BPUE: bycatch per unit effort calculated for each RMU-gear stratum accounting for amount of effort (see Wallace et al. 2013 for details on methods).

	Longlines					Nets					Trawls				
RMU	No. recs	No. turtles taken	Total effort (×1,000 hooks)	Weighted median BPUE (no. recs)	Median mortality rate (range)	No. recs	No. turtles taken	Total effort (# sets)	Weighted median BPUE (no. recs)	Median mortality rate (range)	No. recs	No. turtles taken	Total effort (# sets)	Weighted median BPUE (no. recs)	Median mortality rate (range)
NW ATL	77	4,362	50,214	0.0527 (69)	0 (0–0.72)	27	7,218	111,413	0.0129 (23)	0.21 (0–1)	5	18	31,149	0.0042 (5)	0
SW ATL	31	871	20,089	0.0951 (30)	0 (0–0.33)	2	2	1	1 (1)	1	2	13	17	1.3000 (2)	ND
SE ATL	31	871	20,089	0.0951 (30)	0 (0–0.33)	2	2	1	1 (1)	1	2	13	17	1.3000 (2)	ND
SW IND	8	213	35,586	0.0467 (8)	0 (0–0.20)	ND	ND	ND	ND	ND	ND	ND	ND	ND	ND
E PAC	9	643	23,893	0.0293 (8)	0 (0–0.05)	20	1949	304,318	0.0062 (18)	0.33 (0–0.64)	ND	ND	ND	ND	ND
W PAC	22	678	31,124	0.0171 (9)	0 (0–0.16)	38	93	22,015	0.0022 (27)	0 (0–1)	ND	ND	ND	ND	ND
NE IND	ND	ND	ND	ND	ND	2	438	120,000	0.0004 (2)	0.86	ND	ND	ND	ND	ND

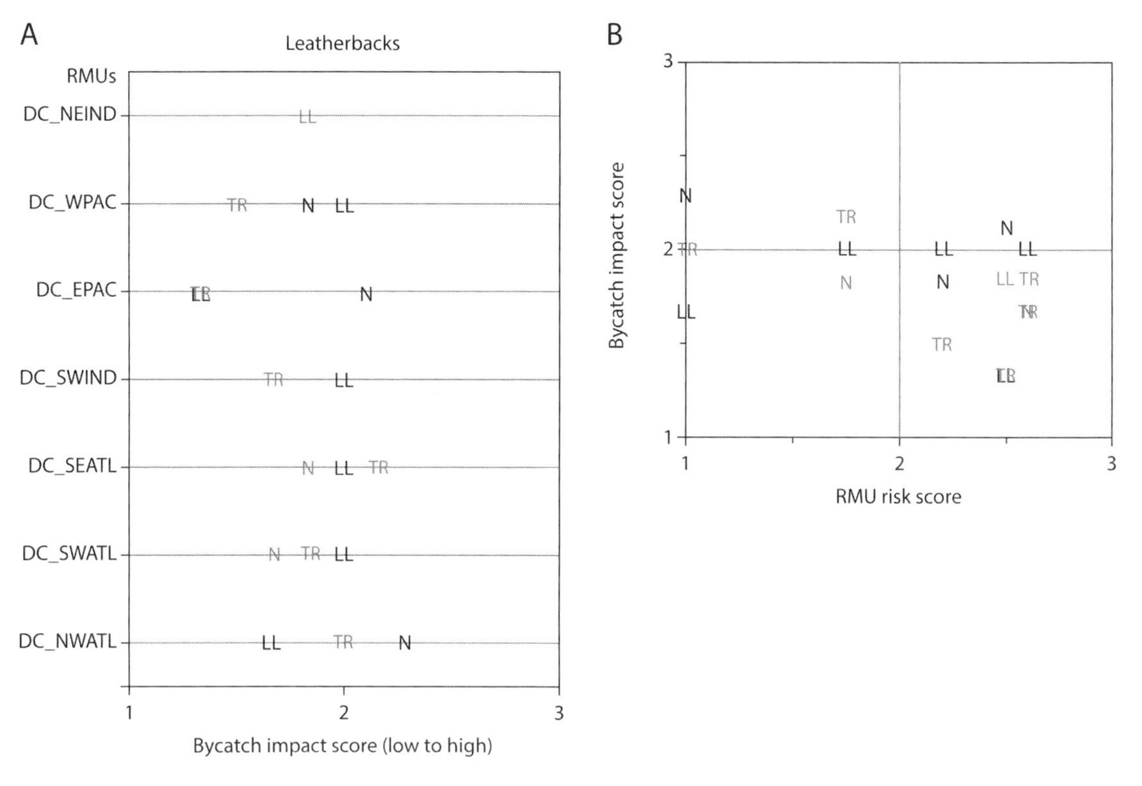

Fig. 17.2. Bycatch impact scores (average score of weighted median BPUEs [bycatch per unit effort], median mortality rates, and body sizes of turtles caught), increasing from low to high impact (see Wallace et al. 2013). Bycatch impact scores for longlines (LL), nets (N), and trawls (TR) are plotted for each (A) leatherback RMU (RMU geographic region code appears in y-axis labels) and against (B) RMU risk scores (i.e., indicators of population viability, increasing from low to high risk, and including population size and trends, rookery vulnerability, and genetic diversity; see Wallace et al. 2011). Larger, darker symbols indicate "higher confidence," because a higher number of records (≥ 3 records) are associated with bycatch parameters, while smaller, gray symbols indicate "lower confidence" due to a low numbers of records (≤ 2 records) associated with bycatch parameters.

to reduce both overall capture and ingestion of hooks by sea turtles, circle hooks could reduce postcapture mortality because they result in less severe internal injuries when ingested (Read 2007; Curran and Bigelow 2011). Increased hook size also decreases capture and mortality in experimental trials, because larger hooks are more difficult for turtles to ingest (Watson et al. 2005, Curran and Bigelow 2011). Using hooks without offsets, so that the barb is in line with the shaft of the hook, is also proposed as a bycatch reduction method (Swimmer et al. 2010).

Experimental trials demonstrate that circle hooks can significantly reduce bycatch of leatherbacks, particularly when used in conjunction with alterations to traditionally used bait, such as squid (Watson et al. 2005; Read 2007). Bait can be artificially color dyed or changed to a species that turtles are less likely to ingest (Echwikhi et al. 2011; Gilman 2011). A number of circle hook experiments have been conducted in several fisheries, including the North Atlantic pelagic longline swordfish fleet (Watson et al. 2005), the Hawaii-based longline swordfish fishery (Gilman et al. 2007), the western equatorial Atlantic pelagic longline fishery (Pacheco et al. 2011), and the Brazilian pelagic longline fishery (Sales et al. 2010). In all these fisheries, circle hooks and bait shifts significantly reduce bycatch more than 50% and with no or minimal effect on target catch (but see Gilman et al. 2007; Read et al. 2007).

Circle hooks and bait changes have already been mandated or promoted in many parts of the world in conjunction with other types of bycatch mitigation measures (Watson et al. 2005; Gilman et al. 2007; Curran and Bigelow 2011; Gilman 2011). The Hawaii-based pelagic longline fishery was closed in 2001 due to high sea turtle bycatch, but reopened in 2004, requiring fishermen to switch from J hooks and squid bait to large circle hooks and whole fish bait (Gilman et al. 2007). The Inter-American Tropical Tuna Commission (IATTC) encourages use of circle hooks and works with the World Wildlife Fund (WWF) on a large-scale project to

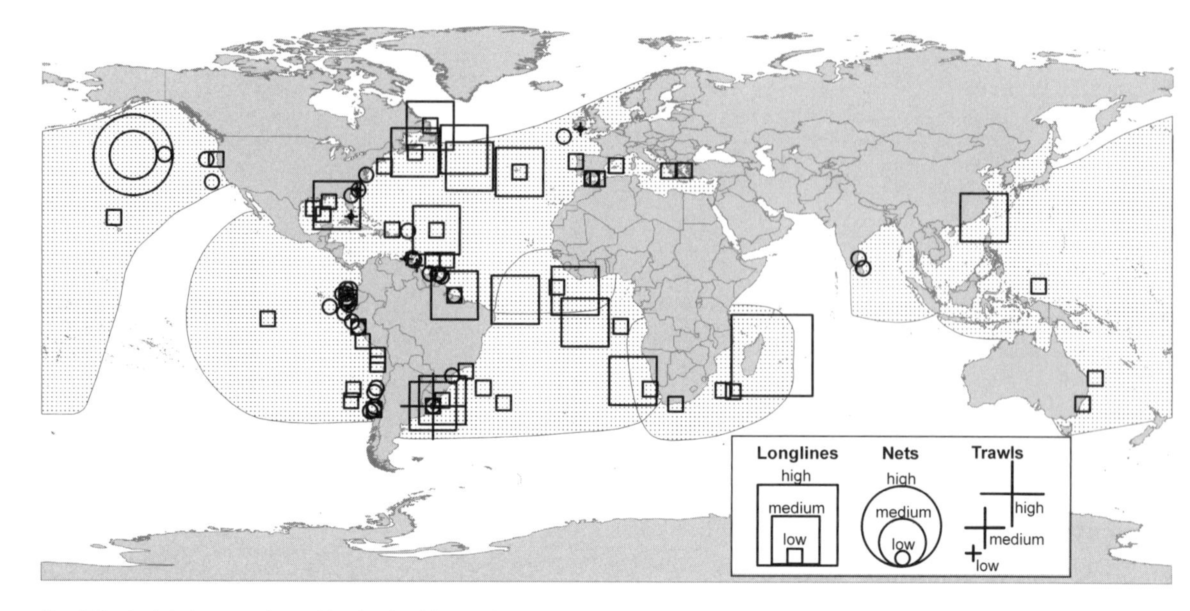

Fig. 17.3. A global comparison of leatherback bycatch rates (number of turtles per generalized set) for longlines (squares), nets (circles), and trawls (crosses). Symbols are scaled by size according to low (lowest 5%), medium (5–95%), or high (highest 5%) values within each gear category. See Wallace, Lewison, et al. (2010) for details.

introduce circle hooks to smaller longline vessels across many countries, including Mexico, Guatemala, El Salvador, Costa Rica, Panama, Colombia, Ecuador, and Peru (Gjertsen et al. 2010). The Western and Central Pacific Fisheries Commission, the regional fishery management organization that oversees longline fisheries in much of the Pacific high seas, requires longline fishermen to use one of three mitigation measures: large circle hooks, whole finfish bait, and/or development of a sea turtle bycatch mitigation plan (Curran and Bigelow 2011). Despite these important advances, many longline fleets do not require mitigation measures, existing measures fall short of "best practice," or compliance is typically low (Gilman 2011). The number of vessels that do not employ bycatch reduction measures likely dwarfs the number of longline vessels that do, and, as a result, leatherback bycatch in longline fisheries is still a globally significant problem.

Passive Nets and Trawls

Although longline fisheries have been the focus of most bycatch reduction efforts, leatherback bycatch in other gear types, such as trawls and gill nets, may pose an even greater threat to leatherbacks (fig. 17.3) (Lewison and Crowder 2007; Moore et al. 2009; Mancini et al. 2012; Wallace et al. 2013). Shrimp trawling causes high leatherback bycatch in some areas (Crouse et al. 1987), although use of turtle excluder devices (TEDs) significantly reduces bycatch of sea turtles (Crowder et al. 1994). The TEDs fit into the neck of a trawl, allowing shrimp to pass through to the hold of the trawl but deflecting turtles out (the device shoots the turtle "out the escape hatch"), reducing sea turtle bycatch by up to 97% (Henwood et al. 1992; Lewison et al. 2003). Use of TEDs has spread to many parts of the world (Chaboud and Vendeville 2011; Sala et al. 2011). High numbers of leatherbacks stranding along the coasts of North and South Carolina, Georgia, and Florida during winter to spring in the early 1990s suggested that leatherbacks were encountering shrimp trawls. Therefore, NMFS required TED openings enlarged to accommodate the larger size of leatherbacks (Federal Register 1994, 1995).

There are little data on incidence and impact of bycatch in passive nets, such as gill nets, yet this gear type is one of the most commonly used in small-scale, nearshore fisheries around the world (Gilman et al. 2010; Stewart et al. 2010). Studies on gill net bycatch suggest that it may pose the greatest impact to leatherbacks worldwide (Lee Lum 2006; Alfaro-Shigueto et al. 2007; Alfaro-Shigueto et al. 2008; Alfaro-Shigueto et al. 2011; Mancini et al. 2012; Wallace et al. 2013). To date, gill net bycatch reduction has been less effective than reduction measures in other gear types. Of the four modifications to gill net gear tested, only reduction of net profile (by reducing height, number of meshes and reducing length of or eliminating tie down lines) statistically reduces leatherback bycatch (Eckert et al. 2008;

Gilman et al. 2010; Wang et al. 2010). Gill net bycatch requires considerably more attention, especially where high leatherback densities (e.g., feeding or breeding areas) overlap areas of intense fishing pressure.

Ways Forward: The Future of Leatherback Bycatch Reduction

Quantifying impacts of fisheries bycatch on leatherback populations requires new approaches. Many fleets and countries are focused specifically on development of new fishing gear or practices that can reduce leatherback bycatch. However, development of management strategies to reduce bycatch is largely dependent upon a clearer understanding of the spatiotemporal overlap between fisheries and leatherbacks, as well as gear characteristics and fishing methods that are most likely to catch and potentially kill the turtles. To develop effective bycatch mitigation policy, new approaches and analyses are needed that make the most of and integrate existing data on bycatch, fishing effort, telemetry, population designations, and oceanography. The roles of community involvement, partnerships with fisheries, development of governance structures, and tackling bycatch in small-scale fisheries are also critical components to effective leatherback bycatch reduction.

Telemetry Data Informing Bycatch

One newer research direction that can directly inform leatherback bycatch management and mitigation is to use telemetry data to help characterize bycatch risk. Shillinger et al. (2011) described three-dimensional leatherback habitat use (i.e., horizontal movements as well as diving behavior) in an oceanographic context, which can provide explicit guidance for fisheries managers as to how gear configurations and spatiotemporal distribution of fishing effort could be modified to reduce interactions with leatherbacks.

There have also been other innovative approaches to using leatherback telemetry data to inform bycatch questions. Bailey et al. (2012) synthesized the largest leatherback telemetry data set ever assembled (135 Argos satellite tracks) from the two RMUs in the Pacific Ocean, and compared turtle movement patterns using a Bayesian state-space model. These models can differentiate between migratory and foraging behavioral states where foraging is characterized by area-restricted searching behaviors and migrations are inferred from transitory behaviors. These analyses highlight potential foraging "hotspots," where area-restricted search behavior is most likely to occur in the Pacific and where highly productive environmental conditions might also support greater fishing effort, thus putting leatherbacks and fishing gear in the same places at the same times. Roe at al. (2014) combined data on leatherback movements from telemetry and on pelagic longline fishing effort to define areas of potentially high interactions between leatherbacks and longline fisheries in the Pacific Ocean. Fossette et al. (2014) did a similar analysis for the Atlantic Ocean.

The Role of Governance

National and international policy can also play roles in achieving bycatch reduction of sea turtles, including leatherbacks. At the national level, a number of countries have implemented policies and regulations to protect leatherbacks. At the international level, conventions exist to protect critical sea turtle habitat (e.g., the Convention on the Conservation of Migratory Species of Wild Animals, CMS). At the regional level, several conventions and initiatives address sea turtle bycatch (e.g., The Inter-American Convention on the Protection and Conservation of Sea Turtles, IAC). A number of regional fishery management bodies (RFMOs) have turned their attention to sea turtle bycatch or have adopted measures to address bycatch. The Food and Agriculture Organization (FAO) International Plans of Action (IPOA) for sharks and seabirds and the recently published FAO Best Practice Technical Guidelines for National Plans of Action serve as templates for how policy instruments can support bycatch reduction for sea turtles.

Small-Scale Fisheries

One of the most critical knowledge gaps in leatherback bycatch is the magnitude of bycatch that occurs in small-scale fisheries and its consequences for leatherback populations. Within the coastal zone, a substantial proportion of fishing activity is described as small-scale or artisanal fisheries. The FAO estimates that more than 95% of fishermen worldwide operate in small-scale fisheries, and yet, the actual magnitude of fishing effort is difficult to assess because national, regional, and global fisheries statistics are imprecise.

Historically, bycatch assessments for leatherbacks were primarily focused on industrial, large-scale fisheries (Lewison et al. 2004). This was in part due to the long-held view that leatherbacks spend the vast majority of their lives in epipelagic areas. However, the focus of bycatch research has shifted toward important nearshore areas. Recent research finds relatively high leatherback bycatch and mortality in small-scale fisheries (Lee Lum 2006; Alfaro-Shigueto et al. 2007; Alfaro-Shigueto 2008; Alfaro-Shigueto 2011; Wallace et al. 2013), particularly in passive net fisheries, and several tracking studies highlight leatherback high-use areas in nearshore zones around the world (James, Ottensmeyer, and Myers 2005; Fossette et al. 2010; Benson et al. 2011). Efforts to assess and reduce leatherback bycatch should be enhanced worldwide, but particularly in small-scale fisheries where impacts are likely to be acute, severe, and more challenging to manage. The need for enhanced bycatch monitoring and community-based mitigation programs in small-scale gill net fisheries, in particular, is critical (Peckham et al. 2007; Moore et al. 2009; Alfaro-Shigueto et al. 2011; Mancini et al. 2012; Wallace et al. 2013).

The Power of Partnerships

Partnerships with industry, fishermen, and their communities are essential to reducing bycatch in all fisheries. For small-scale fisheries, in particular, command-and-control approaches, such as fisheries closures and mandated technological fixes, are often impractical (Berkes et al. 2001; Hilborn et al. 2005). In the context of small-scale fisheries, where management and enforcement are often limited, engaging fishermen and their communities is particularly important. Numerous studies show that engaging fishermen from the outset in bycatch research and reduction initiatives can augment development and adoption of long-term solutions (Hall et al. 2007), in part because investment in the conservation process may increase fishermen's adoption of conservation strategies. One example of this comes from the work of the Inter-American Tropical Tuna Commission (IATTC) and World Wildlife Fund (WWF) that have partnered with fishermen, exporters, fisheries agencies, and other conservation organizations to build a strong, regional network across eight countries, with the goal to reduce bycatch (Hall et al. 2007). In the industrial fisheries sector, industry involvement has been central to innovation and advances in bycatch mitigation innovation. The WWF's International Smart Gear Competition demonstrates the power of partnerships among fishing industries, universities, and governments to innovate fishing gear designs that reduce bycatch.

The Bycatch Seascape

Although bycatch of leatherbacks has been a concern for several decades, it is only in the last ten years that an accurate picture of the bycatch seascape for them and other sea turtle species has emerged. Characterizing the ubiquity and severity of bycatch globally (fig. 17.1) is an important step in galvanizing cooperation and coordination from fisheries agencies and management organization around the world. Likewise, synthesizing existing bycatch information (table 17.2) is essential to identifying the relative importance of gear types and other threats to leatherback populations. Global patterns of sea turtle bycatch indicate that gill nets have the highest bycatch intensity, followed by longlines and then trawls. Bycatch remains a significant threat to leatherbacks and other sea turtles because many nations and regional management agencies do not require or enforce use of proven bycatch reduction measures (Lewison et al. 2014).

While leatherback populations experience threats from many sectors, fisheries bycatch remains one of the most substantial challenges to leatherback survival. Quantifying bycatch, framing it within appropriate scales via RMUs, and determining appropriate mitigation measures are critical steps in effective management. Spatial integration of species distribution and spatial allocation of fishing effort within the oceanographic landscape, however, may be one of the most powerful tools for reducing leatherback bycatch via dynamic ocean management—that is, management that incorporates the shifting nature of the marine environment. A number of fisheries in the United States and Australia have begun to adopt this type of approach with success. TurtleWatch is a voluntary program in the central North Pacific that uses sea surface temperature to determine areas longliners should avoid to minimize loggerhead (*Caretta caretta*) sea turtle bycatch (Howell et al. 2008; Hobday et al. 2011). Management actions that integrate dynamic ocean features hold great promise for leatherback bycatch management.

While the future of leatherback populations worldwide is by no means secure, the challenges posed by incidental fisheries capture to leatherback populations are addressable in both small-scale and industrial-scale fisheries sectors. By combining decades of research and new approaches to understanding leatherback-fisheries interactions, ongoing data collection, potential for partnerships with communities and industry, and continuing development of governance structures and regulatory frameworks, the necessary tools are available to tackle the challenging conservation issue of fisheries bycatch.

LITERATURE CITED

Alfaro-Shigueto, J., P. H. Dutton, M. F. Van Bressem, and J. Mangel. 2007. Interactions between leatherback turtles and

Peruvian artisanal fisheries. Chelonian Conservation and Biology 6: 129–134.

Alfaro-Shigueto, J., J. C. Mangel, F. Bernedo, P. H. Dutton, J. A. Seminoff, and B. J. Godley. 2011. Small-scale fisheries of Peru: a major sink for marine turtles in the Pacific. Journal of Applied Ecology 48: 1432–1440.

Alfaro-Shigueto, J., J. C. Mangel, J. A. Seminoff, and P. A. Dutton. 2008. Demography of loggerhead turtles *Caretta caretta* in the southeastern Pacific Ocean: fisheries-based observations and implications for management. Endangered Species Research. 5: 129–135.

Bailey, H., S. R. Benson, G. L. Shillinger, S. J. Bograd, P. H. Dutton, S. A. Eckert, S. J. Morreale, et al. 2012. Identification of distinct movement patterns in Pacific leatherback turtle populations influenced by ocean conditions. Ecological Applications 22: 735–747.

Benson, S. R., T. Eguchi, D. G. Foley, K. A. Forney, H. Bailey, C. Hitipeuw, B. P. Samber, et al. 2011. Large-scale movements and high-use areas of western Pacific leatherback turtles, *Dermochelys coriacea*. Ecosphere 2(7): art84. doi: 10.1890/ES11-00053.1.

Berkes F., R. Mahon, P. McConney, R. Pollnac, and R. Pomeroy. 2001. Managing small-scale fisheries: alternative directions and methods. International Development Research Centre, Ottawa, Canada.

Beverly, S., and L. Chapman. 2007. Interactions between sea turtles and longline fisheries. Information Paper WCPFC-SC3-EB SWG/IP-01, Scientific Committee Third Regular Session. Western and Central Pacific Fisheries Commission, Kolonia, Pohnpei State, Federated States of Micronesia. http://www.wcpfc.int/system/files/SC3_EB_IP1.

Beverly, S., D. Curran, M. Musyl, and B. Molony. 2009. Effects of eliminating shallow hooks from tuna longline sets on target and non-target species in the Hawaii-based pelagic tuna fishery. Fisheries Research 96: 281–288.

Chaboud, C., and P. Vendeville. 2011. Evaluation of selectivity and bycatch mitigation measures using bioeconomic modelling. The cases of Madagascar and French Guiana shrimp fisheries. Aquatic Living Resources 24: 137–148.

Chuenpagdee, R., L. E. Morgan, S. M. Maxwell, E. A. Norse, and D. Pauly. 2003. Shifting gears: assessing collateral impacts of fishing methods in US waters. Frontiers in Ecology and the Environment 1: 517–524.

Crouse, D., L. Crowder, and H. Caswell. 1987. A stage-based population model for loggerhead sea turtles and implications for conservation. Ecology 68: 1412–1423.

Crowder, L. B., D. T. Crouse, S. S. Heppell, and T. Martin. 1994. Predicting the impact of turtle excluder devices on loggerhead sea turtle populations. Ecological Applications 4: 437–445.

Curran, D., and K. Bigelow. 2011. Effects of circle hooks on pelagic catches in the Hawaii-based tuna longline fishery. Fisheries Research 109: 265–275.

Donoso, M., and P. H. Dutton. 2010. Sea turtle bycatch in the Chilean pelagic longline fishery in the southeastern Pacific: opportunities for conservation. Biological Conservation 143: 2672–2684.

Echwikhi, K., I. Jribi, M. N. Bradai, and A. Bouain. 2011. Effect of bait on sea turtle bycatch rates in pelagic longlines: an overview. Amphibia-Reptilia 32: 493–502.

Eckert, S. A., J. Gearhart, C. Bergmann, and K. L. Eckert. 2008. Reducing leatherback sea turtle bycatch in the surface drift-gillnet fishery in Trinidad. Bycatch Communication Newsletter 8: 2–6.

Federal Register. 1994. Sea turtle conservation; approved turtle excluder devices. 18 May 1994. Pages 25827–25831 in NOAA Administration ed., Washington, DC, USA.

Federal Register. 1995. Sea turtle conservation; restrictions applicable to shrimp trawl activities; leatherback conservation zone. 14 September 1995. Pages 47713–47715 in NOAA Administration ed., Washington, DC, USA.

Finkbeiner, E. M., B. P. Wallace, J. E. Moore, R. L. Lewison, L. B. Crowder, and A. J. Read. 2011. Cumulative estimates of sea turtle bycatch and mortality in USA fisheries between 1990 and 2007. Biological Conservation 144: 2719–2727.

Fossette, S., A. C. Gleiss, A. E. Myers, S. Garner, N. Liebsch, N. M. Whitney, G. C. Hays, et al. 2010. Behaviour and buoyancy regulation in the deepest-diving reptile: the leatherback turtle. Journal of Experimental Biology 213: 4074–4083.

Fossette, S., M. J. Witt, P. Miller, M. A. Nalovic, D. Albareda, A. P. Almeida, A. C. Broderick, et al. 2014. Pan-Atlantic analysis of the overlap of a highly migratory species, the leatherback turtle, with pelagic longline fisheries. Proceedings of the Royal Society B 281: 20133065. doi: 10.1098/rspb.2013.3065.

Georges, J.-Y., S. Fossette, A. Billes, S. Ferraroli, J. Fretey, D. Gremillet, Y. Le Maho, et al. 2007. Meta-analysis of movements in Atlantic leatherback turtles during the nesting season: conservation implications. Marine Ecology Progress Series 338: 225–232.

Gilman, E. L. 2011. Bycatch governance and best practice mitigation technology in global tuna fisheries. Marine Policy 35: 590–609.

Gilman, E., J. Gearhart, B. Price, S. Eckert, H. Milliken, J. Wang, Y. Swimmer, et al. 2010. Mitigating sea turtle bycatch in coastal passive net fisheries. Fish and Fisheries 11: 57–88.

Gilman, E., D. Kobayashi, T. Swenarton, N. Brothers, P. Dalzell, and I. Kinan-Kelly. 2007. Reducing sea turtle interactions in the Hawaii-based longline swordfish fishery. Biological Conservation 139: 19–28.

Gilman, E., E. Zollett, S. Beverly, H. Nakano, K. Davis, D. Shiode, P. Dalzell, and I. Kinan. 2006. Reducing sea turtle by-catch in pelagic longline fisheries. Fish and Fisheries 7: 2–23.

Gjertsen, H., M. Hall, and D. Squires. 2010. Incentives to address bycatch issues. In: R. Allen, J. A. Joseph, and D. Squires (eds.), Conservation and management of transnational tuna fisheries. Wiley, Hoboken, NJ, USA, pp. 225–248.

Godley, B., J. Blumenthal, A. Broderick, M. Coyne, M. Godfrey, L. Hawkes, and M. Witt. 2008. Satellite tracking of sea turtles: Where have we been and where do we go next? Endangered Species Research 4: 3–22.

Hall, M. A., H. Nakano, S. Clarke, S. Thomas, J. Molloy, S. H. Peckham, J. Laudino-Santillán, et al. 2007. Working with fishers to reduce bycatches. In: S. J. Kennelly (ed.), Bycatch reduction in the world's fisheries. Springer-Verlag, New York, NY, USA, pp. 235–288.

Hays, G. C., A. C. Broderick, B. J. Godley, P. Luschi, and W. J. Nichols. 2003. Satellite telemetry suggests high levels of fishing-induced mortality in marine turtles. Marine Ecology Progress Series 262: 305–309.

Hazen, E. L., S. M. Maxwell, H. Bailey, S. J. Bograd, M. Hamann, P. Gaspar, B. J. Godley, and G. L. Shillinger. 2012. Ontogeny in marine tagging and tracking science: technologies and data gaps. Marine Ecology Progress Series 457: 221–240.

Henwood, T., S. Warren, and N. Thompson.1992. Evaluation of US turtle protective measures under existing TED regulations, including estimates of shrimp trawler related mortality in the wider Caribbean. NOAA Technical Memorandum NMFS-SEFSC-303. NOAA, Miami, FL, USA.

Hilborn, R., J. M. Orensanz, and A. M. Parma. 2005. Institutions, incentives and the future of fisheries. Philosophical Transactions of the Royal Society B 360: 47–57.

Hobday, A. J., J. R. Hartog, C. M. Spillman, and O. Alves. 2011. Seasonal forecasting of tuna habitat for dynamic spatial management. Canadian Journal of Fisheries and Aquatic Sciences 68: 898–911.

Houghton, J.D.R., T. K. Doyle, J. Davenport, R. P. Wilson, and G. C. Hays. 2008. The role of infrequent and extraordinary deep dives in leatherback turtles (*Dermochelys coriacea*). Journal of Experimental Biology 211: 2566–2575.

Howell, E., D. Kobayashi, D. Parker, and G. Balazs. 2008. TurtleWatch: a tool to aid in the bycatch reduction of loggerhead turtles *Caretta caretta* in the Hawaii-based pelagic longline fishery. Endangered Species Research 5: 267–278.

James, M. C., R. A. Myers, and C. A. Ottensmeyer. 2005. Behaviour of leatherback sea turtles, *Dermochelys coriacea*, during the migratory cycle. Proceedings of the Royal Society B 272: 1547–1555.

James, M. C., C. A. Ottensmeyer, and R. A. Myers. 2005. Identification of high-use habitat and threats to leatherback sea turtles in northern waters: new directions for conservation. Ecology Letters 8: 195–201.

Kaplan, I. C. 2005. A risk assessment for Pacific leatherback turtles (*Dermochelys coriacea*). Canadian Journal of Fisheries and Aquatic Sciences 62: 1710–1719.

Kobayashi, D. R., and J. J. Polovina. 2005. Evaluation of time-area closures to reduce incidental sea turtle take in the Hawaii-based longline fishery: generalized additive model (GAM) development and retrospective examination. Technical Memorandum NOAA-TM-NMFS-PIFSC-4. NOAA, Honolulu, Hawaii, USA.

Lee Lum, L. 2006. Assessment of incidental sea turtle catch in the artisanal gillnet fishery in Trinidad and Tobago, West Indies. Applied Herpetology 3: 357–368.

Lewison, R. L., and L. B. Crowder. 2007. Putting longline bycatch of sea turtles into perspective. Conservation Biology 21: 79–86.

Lewison, R. L., L. B. Crowder, and D. J. Shaver. 2003. The impact of turtle excluder devices and fisheries closures on loggerhead and Kemp's ridley strandings in the western Gulf of Mexico. Conservation Biology 17: 1089–1097.

Lewison, R. L., L. B. Crowder, B. P. Wallace, J. E. Moore, T. Cox, R. Zydelis, S. McDonald, et al. 2014. Global patterns of marine mammal, seabird, and sea turtle bycatch reveal taxa-specific and cumulative megafauna hotspots. Proceedings of the National Academy of Sciences, USA 111: 5271–5276. doi: 10.1073/pnas.1318960111.

Lewison, R. L., S. A. Freeman, and L. B. Crowder. 2004. Quantifying the effects of fisheries on threatened species: the impact of pelagic longlines on loggerhead and leatherback sea turtles. Ecology Letters 7: 221–231.

López-Mendilaharsu, M., C.F.D. Rocha, P. Miller, A. Domingo, and L. Prosdocimi. 2009. Insights on leatherback turtle movements and high use areas in the Southwest Atlantic Ocean. Journal of Experimental Marine Biology and Ecology 378: 31–39.

Mancini, A., V. Koch, J. A. Seminoff, and B. Madon. 2012. Small-scale gill-net fisheries cause massive green turtle *Chelonia mydas* mortality in Baja California Sur, Mexico. Oryx 46: 69–77.

Maxwell, S. M., G. A. Breed, B. A. Nickel, J. Makang-Bahouna, E. Pemo-Makaya, R. J. Parnell, A. Formia, et al. 2011. Using satellite tracking to optimize protection of long-lived marine species: olive ridley sea turtle conservation in Central Africa. PLoS ONE 6(5): e19905. doi: 10.1371/journal.pone.0019905.

Moore, J. E., B. R. Wallace, R. L. Lewison, R. Zydelis, T. M. Cox, and L. B. Crowder. 2009. A review of marine mammal, sea turtle and seabird bycatch in USA fisheries and the role of policy in shaping management. Marine Policy 33: 435–451.

National Research Council. 1990. Decline of the sea turtles: causes and prevention. National Academy Press, Washington, DC, USA.

Pacheco, J. C., D. W. Kerstetter, F. H. Hazin, R.S.S.L. Segundo, J. E. Graves, F. Carvalho, and P. E. Travassos. 2011. A comparison of circle hook and J hook performance in a western equatorial Atlantic Ocean pelagic longline fishery. Fisheries Research 107: 39–45.

Peckham, S. H., D. M. Diaz, A. Walli, G. Ruiz, L. B. Crowder, and W. J. Nichols. 2007. Small-scale fisheries bycatch jeopardizes endangered Pacific loggerhead turtles. PLoS ONE 2 (10): e1041. doi: 10.1371/journal.pone.0001041.

Polovina, J. J., E. Howell, D. M. Parker, and G. H. Balazs. 2003. Dive-depth distribution of loggerhead (*Caretta caretta*) and olive ridley (*Lepidochelys olivacea*) sea turtles in the central North Pacific: Might deep longline sets catch fewer turtles? Fishery Bulletin 101: 189–193.

Read, A. J. 2007. Do circle hooks reduce the mortality of sea turtles in pelagic longlines? A review of recent experiments. Biological Conservation 135: 155–169.

Roe, J. H., S. J. Morreale, F. V. Paladino, G. L. Shillinger, S. R. Benson, S. A. Eckert, H. Bailey, et al. 2014. Predicting bycatch hotspots for endangered leatherback turtles on longlines in the Pacific Ocean. Proceedings of the Royal Society B 281: 20132559. doi: 10.1098/rspb.2013.2559.

Sala, A., A. Lucchetti, and M. Affronte. 2011. Effects of turtle excluder devices on bycatch and discard reduction in the demersal fisheries of Mediterranean Sea. Aquatic Living Resources 24: 183–192.

Sales, G., B. B. Giffoni, F. N. Fiedler, V. G. Azevedo, J. E. Kotas, Y. Swimmer, and L. Bugoni. 2010. Circle hook effectiveness for the mitigation of sea turtle bycatch and capture of target species in a Brazilian pelagic longline fishery. Aquatic Conservation-Marine and Freshwater Ecosystems 20: 428–436.

Shillinger, G. L., D. M. Palacios, H. Bailey, S. J. Bograd, A. M. Swithenbank, P. Gaspar, B. P. Wallace, et al. 2008. Persistent leatherback turtle migrations present opportunities for conservation. PLoS Biol. 6: 1408–1416. doi:10.1371.

Shillinger, G. L., A. M. Swithenbank, H. Bailey, S. J. Bograd, M. R. Castelton, B. P. Wallace, J. R. Spotila, et al. 2011. Characterization of leatherback turtle post-nesting habitats and its application to marine spatial planning. Marine Ecology Progress Series 422: 275–289.

Shillinger, G., A. Swithenbank, S. Bograd, H. Bailey, M. R. Castelton, B. P. Wallace, J. R. Spotila, et al. 2010. Identification of high-use internesting habitats for eastern Pacific leatherback turtles: role of the environment and implications for conservation. Endangered Species Research 10: 215–232.

Sims, M., T. C. Cox, and R. L. Lewison. 2008. Modeling spatial patterns in fisheries bycatch: improving bycatch maps to aid fisheries management. Ecological Applications 18: 649–661.

Spotila, J. R., R. D. Reina, A. C. Steyermark, P. T. Plotkin, and F. V. Paladino. 2000. Pacific leatherback turtles face extinction. Nature 405: 529–530.

Stewart, K. R., R. L. Lewison, D. C. Dunn, R. H. Bjorkland, S. Kelez, P. N. Halpin, and L. B. Crowder. 2010. Characterizing fishing effort and spatial extent of coastal fisheries. PLoS ONE 5: e14451. doi: 10.1371/journal.pone.0014451.

Swimmer, Y., R. Arauz, J. Wang, J. Suter, M. Musyl, A. Bolaños, and A. López. 2010. Comparing the effects of offset and non-offset circle hooks on catch rates of fish and sea turtles in a shallow longline fishery. Aquatic Conservation-Marine and Freshwater Ecosystems 20: 445–451.

Tuck, G. N. 2011. Are bycatch rates sufficient as the principal fishery performance measure and method of assessment for seabirds? Aquatic Conservation: Marine and Freshwater Ecosystems 21: 412–422.

Wallace, B. P., A. D. DiMatteo, A. B. Bolten, M. Y. Chaloupka, B. J. Hutchinson, et al. 2011. Global conservation priorities for marine turtles. PLoS ONE 6: e24510. doi: 10.1371/journal.pone.0024510.

Wallace, B. P., A. D. DiMatteo, B. J. Hurley, E. M. Finkbeiner, A. B. Bolten, et al. 2010. Regional Management Units for marine turtles: a novel framework for prioritizing conservation and research across multiple scales. PLoS ONE 5(12): e15465. doi: 10.1371/journal.pone.0015465.

Wallace, B. P., S. S. Heppell, R. L. Lewison, S. Kelez, and L. B. Crowder. 2008. Impacts of fisheries bycatch on loggerhead turtles worldwide inferred from reproductive value analyses. Journal of Applied Ecology 45: 1076–1085.

Wallace, B. P., C. Y. Kot, A. D. DiMatteo, T. Lee, L. B. Crowder, and R. L. Lewison. 2013. Impacts of fisheries bycatch on marine turtle populations worldwide: toward conservation and research priorities. Ecosphere 4(3): 40. http://dx.doi.org/10.1890/ES12-00388.1.

Wallace, B. P., R. L. Lewison, S. McDonald, T. McDonald, C. Y. Kot, S. Kelez, R. K. Bjorkland, et al. 2010. Global patterns of marine turtle bycatch in fisheries. Conservation Letters 3: 131–142.

Wang, J. H., S. Fisler, and Y. Swimmer. 2010. Developing visual deterrents to reduce sea turtle bycatch in gill net fisheries. Marine Ecology-Progress Series 408: 241–250.

Watson, J. W., S. P. Epperly, A. K. Shah, and D. G. Foster. 2005. Fishing methods to reduce sea turtle mortality associated with pelagic longlines. Canadian Journal of Fisheries and Aquatic Sciences 62: 965–981.

Witt, M. J., E. A. Bonguno, A. C. Broderick, M. S. Coyne, A. Formia, A. Gibudi, G. A. Mounguengui Mounguengui, et al. 2011. Tracking leatherback turtles from the world's largest rookery: assessing threats across the South Atlantic. Proceedings of the Royal Society B: 278: 2338–2347.

18

Conclusion

Problems and Solutions

JAMES R. SPOTILA AND
PILAR SANTIDRIÁN TOMILLO

In 1989 the leatherback turtle (*Dermochelys coriacea*) was a mystery both in terms of its biology and its conservation. Not all of its major nesting beaches were known. Little was known about its metabolism and thermoregulation. Its migratory pathways were not yet discovered. There were only a few scientific articles published about the species. Most of those who studied sea turtles thought that the leatherback was one of the most secure sea turtle species in terms of extinction or extirpation risks. Estimates of its numbers were in the range of 90,000 females or more (Pritchard 1982). Major new populations had been discovered in Mexico (chapter 10) and rumor had it that there was a major nesting population in Costa Rica as well. In the Atlantic Ocean and Caribbean Sea we knew that there were large numbers of leatherbacks nesting in French Guiana and Suriname, but only a few female leatherbacks nested in Costa Rica (Hirth and Ogren 1987) with only an occasional turtle observed in Florida (chapter 9). The Malaysian population was robust and there was little information about leatherbacks on the island of New Guinea and surrounding areas. Little was known about interactions of leatherback turtles and fisheries. How time flies!

What We Know

In 2014 we ask the question: "What have we learned since 1989 and what else is left to discover?" We know where most leatherbacks nest; we know much more about their migration, diving, and physiology; there are extensive conservation programs on most nesting beaches; and technology is helping us to learn more details about the life of these magnificent animals every day. As is clear from the previous 17 chapters in this book, we have acquired a tremendous amount of information about the biology and conservation of this species. In 25 years, the veil has been pulled back from many mysteries about this animal, the greatest of all sea turtles. Yet the scientific discoveries reported here have not stopped the exponential decline of many populations, especially in the Pacific Ocean (chapter 10). We feel a bittersweet taste in our mouths because our success as scientists has taken place at the same time as the demise of many individuals of the species that we care so much about, both as a unique example of biodiversity on planet Earth and as an animal that provides us with a special connection to nature as it should be, without human interference.

Molecular genetics (chapter 2) has provided a clear picture of the relationships between leatherback populations in the different ocean basins and within those basins, as well as within populations. There is considerable genetic diversity among the populations, but leatherbacks went through a population bottleneck about three million years ago and leatherbacks in the eastern Pacific Ocean are now the most genetically isolated. Leatherbacks are among the deepest diving vertebrates (chapter 3) and have a unique anatomy that allows this (chapter 4). Their reproductive biology is like that of other sea turtles (chapter 5), as is their nesting ecology (chapter 6). Of course, being the largest sea turtles, they dig the deepest nests (chapter 6) and although they do not lay the most eggs in a clutch, they do have the largest biomass of eggs in a clutch and in a nesting season (5 kg). Yet leatherbacks have the lowest hatching success of all the sea turtles (chapter 7). Since they undergo temperature-dependent sex determination, the sex ratios of hatchlings produced on many beaches are skewed toward females (chapter 8). The physiology of both the egg and the adult are in many ways unique and in some ways similar to that of other sea turtles (chapters 12 and 13). The migration and feeding of leatherbacks takes them across the oceans (chapter 14) and is driven by ocean productivity (chapter 15).

The population biology of leatherbacks differs in the various ocean basins. In the Atlantic Ocean (chapter 9), leatherback populations are growing and spreading to new nesting beaches. The populations in French Guiana and Suriname are steady or perhaps declining, but the nesting populations on Trinidad and in Guyana are very large despite considerable bycatch in local fisheries. In 1989 there were only a few leatherback turtles nesting on the Caribbean coast of Costa Rica and today (2014) leatherbacks are common. They are also common in Florida, where they were rare in the late 1980s. The population on St. Croix, US Virgin Islands, underwent a considerable increase starting in the 1990s. The West Africa nesting population may be the largest in the world (Witt et al. 2009).

In contrast, the leatherback nesting population in Malaysia has been effectively extirpated and the nesting populations in Mexico and Costa Rica have undergone parallel exponential declines (chapter 10). The populations in Indonesia and Papua New Guinea were thought to be robust a few years ago, but have been undergoing a steady decline of nearly 6% a year for the last decade or more. The leatherback population nesting along the Indian Ocean coast of South Africa increased in the 1980s and 1990s after many years of beach protection, but seems to have plateaued in recent years (chapter 11). Why do these differences exist? What are the driving forces?

Problems

One important reason for the differences between the Atlantic and Pacific populations is that the former nest about every two to three years and the East Pacific leatherbacks nest about every four years. The latter are also the smallest leatherbacks. These differences are due to variations in ocean productivity and the predictive nature of that productivity (Saba et al. 2008; chapter 15). More food in more predictable locations allows Atlantic leatherbacks to grow larger, produce more eggs, and nest more frequently. Those populations thus have a considerable demographic edge over the eastern Pacific populations. The western Pacific leatherbacks grow larger than eastern Pacific leatherbacks and may have more productive areas in which to feed, but all leatherbacks in the Pacific must undertake enormous migrations (e.g., New Guinea to Monterrey Bay, California) to reach their feeding grounds. South African leatherbacks nesting along the Indian Ocean coast are the largest within the species and they have the shortest migrations. They only have to travel a short distance to the Agulhas Current to find reliable food.

The other problem facing leatherbacks is the impact of people at all phases of their life cycle. Poaching (illegal collection) of eggs was the major reason for the decline in the leatherback population in Malaysia (Liew 2011), Mexico (Sarti Martínez et al. 2007), and Costa Rica (Santidrián Tomillo et al. 2008; chapter 10). In addition, capture in fisheries, both intentional and indirect (so-called bycatch), has had a tremendous effect on populations (Eckert and Sarti 1997; Kaplan 2005; Alfaro-Shigueto et al. 2011; chapter 17). The combination of egg removal and capture in fisheries had a much greater effect on Pacific populations than on Atlantic populations because the latter probably grew faster, matured earlier, produced more eggs, and nested more frequently than their Pacific relatives.

If these problems are not enough, we only have to consider the effect of global warming on eggs, hatchlings, juveniles, and adults (Saba et al. 2012; chapter 16). Many beaches are heating up and now produce primarily female hatchlings. In the future they may be too hot to produce any hatchlings. Not only temperature, but rainfall is also changing. Eastern Pacific beaches are getting hotter and drier. At the same time, some western Pacific beaches may be growing wetter. Ocean currents may shift, taking hatchlings to areas that are not conducive for their growth and survival. Food supplies might become less abundant and less predictable. Sea levels will rise, changing beaches, with others disappearing. Where there are now houses and hotels, beaches may not be able to naturally shift inland and turtles will lose these beaches as nesting sites. Although leatherbacks as a species survived the last great warming period in the Pliocene, they did not then also face threats from people; in addition, changes in the Pliocene may have occurred over a much longer period of time, allowing turtles to adapt to the changing climate through natural selection. Our climate is now warming so rapidly that our grandchildren will face its consequences, and leatherbacks have only two or three generations to adjust to those changes.

Solutions

The necessity of more research on the biology and ecology of leatherback turtles is obvious. In particular, research into the best conservation strategies is sorely needed. How should we respond to changing beach temperatures? Should we manage the beaches, manage the eggs, or leave them so that natural selection can take its course? Unfortunately, the intensity of selective forces related to climate is too severe (Sinervo et al. 2010) to allow many, if any, eggs to survive on numerous beaches that are now major nesting sites for leatherbacks. So some action is needed. However, to avoid more harm than good, caution must be exercised. The overuse of hatcheries, what we have termed the "hatchery reflex," (chapter 16) has caused harm in the past; for example, movement to hatcheries has sometimes resulted in male-biased hatchlings and a reduction in hatching success. Therefore, quantitative studies of the situation on particular beaches are needed, as well as clear plans to improve the problem before people start moving eggs from natural nests to hatcheries. There is still time to find the best solution.

More research is also needed to discover where hatchlings go after they leave the beaches. Although we have placed satellite transmitters on many leatherback turtles, more data are needed to define migratory pathways and behaviors at sea. Additional data are needed on the locations where juveniles live, how they behave, and where they go throughout the year. Further work in genetics will better define local populations; improved understanding of the genetic diversity within those populations is badly needed.

Although there are good conservation and protection programs on many nesting beaches, further efforts are paramount. Local people must be invested in the conservation ethic, while learning to support protection of turtles, eggs, and the beaches that turtles use to reproduce. Protection will not be assured until we can be sure that when professional and volunteer outsiders leave an area or project, local residents will consider safeguarding beaches as part of their heritage and will be empowered to make this a central part of their lives.

The ocean, the ocean, what is happening in the ocean? After all these years of research and conservation, can we say that we have a complete picture of what is happening to leatherbacks in the oceans of the world? We know that there has been a heavy toll on leatherbacks from gill nets and longlines. Yet the picture is incomplete. We still need more information. What is the Spanish longline fleet doing? How many turtles are the Chinese fishing boats catching? How are artisanal fishermen now dealing with leatherbacks? We have some data and some ideas in regard to fishing, but we need more, including additional studies. More fishery observers are needed on more boats. Additional open access to data that already exist in fisheries organizations and in national fisheries agencies is important. The limited data that we have already implicates fishing as a major, and today perhaps the major, source of mortality for leatherback turtles. When there are only 25 or 30 leatherbacks nesting in a season at Parque Nacional Las Baulas (PNMB) on the Pacific coast of Costa Rica, then every turtle killed by longlines and every turtle caught by an artisanal gill net in Peru is a critical loss to the population. If we can save 10 leatherbacks a year off Peru, we will see them nesting at PNMB and the population will start to increase.

Just because there are large numbers of leatherbacks nesting in the Atlantic Ocean we should not be fooled into thinking that they are not in danger as are leatherbacks in the Pacific. The Atlantic subpopulation is officially listed as vulnerable by the International Union for the Conservation of Nature (IUCN) (Wallace et al. 2013). But Wallace and coauthors cautioned that the Atlantic subpopulation can easily decline if conservation efforts are not vigorously maintained and if future threats from climate change and other human factors grow more severe. The Pacific populations were robust only 25 years ago and now they are "hanging on by a thread." The same can happen in the Atlantic if we become too complacent after learning of their large and increasing populations. Leatherback turtles are long-lived animals with a sizable generation time. The Atlantic populations will exhibit the same steep decline as the Pacific populations if eggs and beaches are threatened and if adults suffer from a high rate of bycatch in fisheries. Research and conservation should continue to focus on the Atlantic Ocean leatherback populations. Proactive efforts now will ensure that those populations remain robust, and will help avoid the more difficult task of recovering the populations after they suffer a large decline.

The take-home message from this book is that in the last 25 years we have acquired a tremendous amount of information on the biology and conservation of the leatherback turtle. We have learned what conservation actions work and which do not. Now we, and the leatherbacks, are faced with immense challenges as we move forward in a warming world filled with a rapidly increasing human population.

There is plenty of work to do for all of those interested in discovering more about leatherback biology and in conserving these magnificent turtles. Find a place where you can do your part as a scientist or conservationist, or both. Study and save your little part of the leatherback world and, if we all do our part, we will give these, the greatest of the turtles, a fighting chance to survive into the next millennium.

LITERATURE CITED

Alfaro-Shigueto, J., J. Mangel, F. Bernedo, P. H. Dutton, J. A. Seminoff, and B. J. Godley. 2011. Small-scale fisheries of Peru: a major sink for marine turtles in the Pacific. Journal of Applied Ecology 48: 1432–1440.

Eckert, S. A., and L. Sarti. 1997. Distant fisheries implicated in the loss of the world's largest leatherback nesting population. Marine Turtle Newsletter 78: 2–6.

Hirth, H. F., and L. H. Ogren. 1987. Some aspects of the ecology of the leatherback turtle, *Dermochelys coriacea*, at Laguna Jalova, Costa Rica. NOAA Technical Report NMFS 56. NOAA, Springfield, VA, USA.

Kaplan, I. C. 2005. A risk assessment for Pacific leatherback turtles (*Dermochelys coriacea*). Canadian Journal of Fisheries and Aquatic Science 62: 1710–1719.

Liew, H.-C. 2011. Tragedy of the Malaysian leatherback population: what went wrong. In: P. H. Dutton, D. Squires, and A. Mahfuzuddin (eds.), Conservation and sustainable management of sea turtles in the Pacific Ocean. University of Hawai'i Press, Oahu, HI, USA, pp. 97–107.

Pritchard, P.C.H. 1982. Nesting of the leatherback turtle, *Dermochelys coriacea* in Pacific Mexico, with a new estimate of the world population status. Copeia 1982: 741–747.

Saba, V. S., G. L. Shillinger, A. M. Swithenbank, B. A. Block, J. R. Spotila, J. A. Musick, and F. V. Paladino. 2008. An oceanographic context for the foraging ecology of eastern Pacific leatherback turtles: consequences of ENSO. Deep Sea Research Part I: Oceanographic Research Papers 55: 646–660.

Saba, V. S., C. A. Stock, J. R. Spotila, F. V. Paladino, and P. Santidrián Tomillo. 2012. Projected response of an endangered marine turtle population to climate change. Nature Climate Change 2: 814–820. doi: 10.1038/nclimate1582.

Santidrián Tomillo, P., V. S. Saba, R. Piedra, F. V. Paladino, and J. R. Spotila. 2008. Effects of illegal harvest of eggs on the population decline of leatherback turtles in Las Baulas Marine National Park, Costa Rica. Conservation Biology 22: 1216–1224.

Sarti Martínez, L., A. R. Barragán, D. Garcia-Muñoz, N. Garcia, P. Huerta, and F. Vargas. 2007. Conservation and biology of the leatherback turtle in the Mexican Pacific. Chelonian Conservation and Biology 6: 70–78.

Sinervo, B., F. Méndez-de-La-Cruz, D. B. Miles, B. Heulin, E. Bastiaans, M. Villagrán-Santa Cruz, R. Lara-Resendiz, et al. 2010. Erosion of lizard diversity by climate change and altered thermal niches. Science 328: 894–899.

Wallace, B. P., M. Tiwari, and M. Girondot. 2013. *Dermochelys coriacea*. In: IUCN 2013. IUCN Red List of Threatened Species. Version 2013.2. iucn-redlist.org. Accessed 27 February 2014.

Witt, M. J., B. Baert, A. C. Broderick, A. Formia, J. Fretey, A. Gibudi, C. Moussounda, et al. 2009. Aerial surveying of the world's largest leatherback turtle rookery: a more effective methodology for large-scale monitoring. Biological Conservation 142: 1719–1727.

Index

abducens nerve (cranial nerve VI), 46, 47, Plate 8
acetabulum, 36
acidification of oceans, 187
acromion process, 36, 39
adrenal artery, 45
adrenal glands, 41
Adriatic Sea, 97
Aegean Sea, 97
aerobic dive limit (ADL), 27
age at first reproduction (AFR), 15, 152–53, 154, 157
age at maturity, 153, 158, 173, 189
age estimation, 153
Agulhas Current System (ACS), 123, 126, 163, 175, 178, 179, 187, 210
albumin in eggs, 42, 51, 54, 57, 136–37, 145. *See also* shelled albumin gobs
alimentary canal, 38
Alligator mississippiensis, 56, 57, 139
anaerobic respiration, 27
anatomy, 3–4, 32–47, Plates 1–8; cardiovascular system, 42–46; coloration, 33–34; digestive system, 38–40; glands, 40–41; of hatchlings, 3, 4, 33; muscles, 38; nervous system, 46–47; respiratory system, 42; skeleton, 34–37; urogenital system, 41–43. *See also specific structures*
Andaman Islands, 123, 127–29
Angola, 98, 101
Angola Current, 166
anoxia tolerance, 27
aorta(e), 39, 45
appendicular bones, 34
aragonite crystals, 136, 138
Argentina, 14
arrested development, 77, 139–40
arterial system, 44–45. *See also specific arteries*
Ascension Island, 168, 169
assimilation efficiency, 152
Atlantic Northwest RMU, 98–101
Atlantic Ocean leatherbacks, 4, 7, 16, 97–104, 210, 211–12; abundance trends for, 7, 104, 110; barnacles on, 5; biometry of, 102–3; in Caribbean, 98; clutch frequency of, 69; clutch size of, 4, 138; distribution in water, 97–98; diving behavior of, 23–24; foraging areas of, 165–66, 175–79, 187; genetic and ecological distinctiveness of, 9, 10, 101–2; hatching success of, 75, 79; hatchling sex ratios for, 88; in Mediterranean Sea, 98; migration of, 154, 163–64; nesting behavior of, 64, 68, 180; nesting season of, 69, 101; nesting sites of, 98–102; nesting success of, 68; in North Atlantic, 98; in North Atlantic: bioenergetics of, 154–56; population genetics of, 12–14; Regional Management Units for, 98–99; reproduction of, 63, 68; sea surface temperature and movements of, 186–87; in South Atlantic, 98; temperature-dependent sex determination in, 85, 86; threats to, 103–4. *See also specific countries*
Atlantic Southeast RMU, 98, 101
Australia, 5, 59, 69, 116, 118, 125, 129, 187, 191, 204
Avens, L., 153
axillary artery, 40, 45
axillary vein, 46
azygos vein, 45

Bailey, H., 98
Baja California, Mexico, 6, 17, 111
Balasingam, E., 64
barnacles, 5, 63
beaches. *See* nesting beaches
Bell, B. A., 77
Benguela Current, 126, 164, 175
Benson, S., Plate 12, Plate 13
Bermuda, 21
bile, 40
Billes, A., 67
Binckley, C. A., 86, 88
bioenergetics, 149, 154–57
biogeochemical ocean general circulation models (BOGCMs), 175, 177, 178, 179
bio-loggers, 154; to study diving behavior, 21–22
Bird's Head Penninsula, Papua Barat, 114–15, 116–17
Block, B. A., 151
blubber layer, 34, 35, 36, 45
body mass, 22, 103, 150, 151, 152
body pit excavation, 66
body size, 3, 32; of Atlantic and Mediterranean leatherbacks, 102–3; large phytoplankton net primary productivity and, 178; of nesting females, 5. *See also* carapace length
body temperature, 25, 97, 150, 151, 162, 186; of juveniles, 22. *See also* thermoregulation
Bostrom, B. L., 151, 156
Botha, M., 126
bottleneck effect, 11, 12, 16, 210
Bowen, B. W., 9, 15, 17
Boyle's Law, 27, 141
brachial vein, 46
brachiocephalic artery, 41
Bradshaw, C.J.A., 27
brain, 34, 46, 47, Plate 3
braincase, 34, 35, 47, Plate 8
Brazil, 13, 14, 23, 69, 75, 98, 99, 101, 102, 104, 169, 189, 201
breathing, 21, 42; of hatchlings, 22; oxygen consumption rates, 26; oxygen stores, 25–26; suspension of, 25
breeding sex ratio, 15
bronchi, 42, 44, 45
bycatch. *See* fisheries impacts

calcitonin, 41
calcium blood level, 41, 51, 55–56, 60
calculated aerobic dive limit (cADL), 27
California Current, 165, 166, 175, 176, 181
Canada, 98, 103, 104, 156, 166
capture-mark-recapture (CMR) studies, 15–16, 100, 162
carapace, 3, 4, 5, 34, 36; ossicles of, 5, 28, 35, 46, 153; ridges on, 32, 33
carapace length, 5; age estimation and, 153–54; of Atlantic and Mediterranean leatherbacks, 102–3; follicle size and, 53; of hatchlings, 33; of juveniles, 22, 34, 35, 39, 153
carapace length, curved (CCL), 162; age estimation and, 153–54; diving patterns and, 22; of juveniles, 22, 102; of nesting female, 33, 125; of nesting female: follicle size and, 53
carapace length, straight (SCL): and age at maturity, 153; of hatchling, 33; of juveniles, 22, 34, 35, 39, 153

carbon dioxide (CO_2), 27, 140; in Earth's atmosphere, 185; in nest, 77, 136; in nest: hatching success and, 75, 139, 141, 144
cardiovascular system, 42–46
Caretta caretta (loggerhead turtle): cold stunning of, 4; diet of, 4; effect of ocean temperature changes on, 189; eggs of, 136, 137; hatching success of, 74; in Indian Ocean, 124; land walking of gravid females, 66; magnetic compass sense of, 163; multiple paternity in, 14; nesting of, 66; oxygen level in nests of, 140; phylogeny of, 10, 11; reducing bycatch of, 158; reproduction of, 51–52, 53, 54, 56, 58–59
Caribbean leatherbacks, 7, 98, 104; biometry of, 103; diving behavior of, 23–24; female fasting in nesting season, 53; hatching success of, 75, 77; nest density of, 68; nesting behavior of, 64, 65, 66, 79; population genetics of, 12, 14. *See also specific countries*
Caribbean Sea, 97; nesting sites in, 98, 100–102; Wider Caribbean Sea Turtle Conservation Network, 98
carotid arteries, 45, Plate 5
carpal bones, 36
carpal muscles, 38
Carr, A., 6, 21, 66, 168, 185
caruncle, 77
Caut, S., 69
Central American leatherbacks, 14, 80, 100, 112–13. *See also specific countries*
cephalic veins, 46
cerebellum, 46, Plate 3
cerebrum, Plate 3
cervical veins, transverse, 46
Chan, E. H., 86, 88
Charnier, M., 84
Chelonia mydas (green turtle): cold stunning of, 4; effects of ocean temperature changes on, 187; eggs of, 136, 137; foraging and behavioral plasticity of, 25; hatching success of, 74; magnetic compass sense of, 163; mating of, 64; multiple paternity in, 14; nest distribution of, 6; nesting of, 66; oxygen level in nests of, 140; phylogeny of, 9, 10, 11; reproduction of, 51, 52, 54–55, 56, 57, 58–59; sex ratios of, 189; thermoregulation in, 186
Chelonia mydas agassizii (black turtle), 6, 10
Cheloniidae, 3, 51
Chelydra serpentina, 136
Chevalier, J., 86
Chile, 3, 14, 113–14, 175, 180–81, 187
China, 3
chlorophyll concentration, 165, 175–78
chondrocranium, 34, 35, 47, Plate 8
Chrysaora hysoscella, 154
Chrysemys picta (painted turtle), 56, 136, 140
Chrysemys scripta, 26
circle hooks to reduce bycatch, 199, 201–2
climate change, 7, 15, 185–92, 210–11; conservation efforts and, 191–92; effects on eggs and hatchlings, 79–81, 188–91; effects on Pacific Ocean leatherbacks, 16–17, 80, 91, 118; El Niño Southern Oscillation and, 16, 79, 176–80, 187–89; Intergovernmental Panel on Climate Change, 80, 190; nesting ecology and, 187–89; population dynamics and, 79, 188, 190; rising sea levels, 118, 191, 192, 211; temperature-dependent sex determination and, 89–91, 189–91; thermoregulation and, 185–87; warming oceans, 187–89
clitoris, 41
cloaca, 41, 43, 67, 104
clutches: depth of burial of, 78, 87, 135; maternal investment in eggs and, 138; synchronous hatching of eggs in, 78
clutch frequency, 69, 135, 138
clutch relocation, 65, 100, 113; hatching success and, 74–75, 77, 79
clutch size, 4, 51, 57–58, 69, 135, 210; hatching success and, 75; hatchling emergence success and, 75; hatchling output and, 78; regional variations in, 5, 69, 138
Cnidaria, 175
coalescent theory, 12
coeliac artery, 45
cold tolerance, 4, 8, 11, 25, 32, 149, 186. *See also* thermoregulation
Colombia, 78, 88, 100, 102, 202
coloration, 33–34
Conchoderma, 63
Congdon, J. D., 57
conservation efforts, 211–12; climate change and, 191–92; clutch relocation, 65, 74–75, 77, 79, 100, 113; in Eastern Pacific, 113–14; hatcheries (*see* hatcheries); leatherback movements and, 169; to reduce fisheries bycatch, 203–4; in Western Pacific, 117–18
conservation genetics, 11
conservation physiology, 157–58
Convention on the Conservation of Migratory Species of Wild Animals (CMS), 203
corticosterone, 52, 59
Cosmochelys, 6
costal arteries, 45
Costa Rica, 5, 6, 7, 10, 13, 14, 17, 24, 52, 53, 56, 58, 59, 60, 64–65, 66, 69, 70, 75–80, 85–91, 100, 102, 103, 104, 110–13, 138–43, 145, 164–68, 175–78, 180–81, 187–88, 190, 191, 202, 209, 210, Plate 9, Plate 11, Plate 13
Costa Rica Coastal Current, 167
courtship behavior, 52, 63–64. *See also* mating behavior
cranial nerves, 46, 47, Plate 8
Crocodylus porosus, 68, 139
cryptophytes, 174
Ctenophora, 4, 175
Culebra, Puerto Rico, 68, 69, 100–101
Cyanea, 152; *C. capillata*, 154
cyanobacteria, 174

Davenport, J., 150
decompression sickness, 28
Delaware, United States, 191
Demographically Independent Populations (DIP), 8, 9
Deraniyagala, P.E.P., 6, 32, 66, 123, 129, 138
dermal bones, 34, 35
dermatocranium, 35
Dermochelyidae, 5, 8, 9, 51
Dermochelys coriacea (leatherback turtle), 3; anatomy of, 3–4, 32–47; in Atlantic Ocean, 97–104; diving behavior and physiology of, 21–28; effects of climate change on, 185–92; eggs, nests, and embryonic development of, 135–45; endangered status of, 5, 110, 118, Plate 11; fisheries bycatch of, 14, 98, 102, 103–4, 110, 113–14, 117–18, 126, 127, 158, 164, 168, 169, 178, 180, 196–204, 210, 212; fossils of, 5–6, 16; global distribution of, 6–7, 8; habitats for, 149; in Indian Ocean, 123–29; movement and behavior of, 162–69; in Pacific Ocean, 110–18; phylogeny and phylogeography of, 8–12; physiological ecology of, 149–58; population genetics of, 8, 9, 12–15; relation of marine primary productivity to biology and behavior of, 173–81; reproductive biology of, 51–60; sex determination and hatchling sex ratios of, 84–91; taxonomy of, 5–6; in today's world, 6–7; what we know about, 209–10
diatoms, 174, 175–76
diet, 4, 23, 149, 154, 173; gelatinous zooplankton, 23, 32, 150, 152, 154, 162, 165, 173–75, 176, 180; gelatinous zooplankton: biophysicals controls on, 174–75. *See also* foraging; jellyfish
digestive system, 38–40
digital arteries, 45
digits, 37
dinoflagellates, 174
diving behavior, 21–25, 173, 210; biologgers for studying, 21–22; dive depth and duration, 21, 23, 28; of hatchlings and juveniles, 22–23; in inter-nesting intervals, 22, 24–25; sexual maturity and, 23; while migrating or foraging, 23–24
diving physiology, 25–28; adaptations to hydrostatic pressures, 25, 27–28; aerobic dive limit and calculated aerobic dive limit, 27; nitrogen narcosis, 28; oxygen consumption rates, 26; oxygen stores, 25–26; regional blood flow, 26–27; withstanding hypoxia, 25, 27
Dominica, 65
Downie, J. R., 88
Doyle, T. K., 154
duodenum, 39–40, 45
Duron, M., 152
Dutton, P. H., 9, 10, 14, 15, 64, 88, 101

ear, 46–47
East Australian Current Extension, 166, 175
Eckert, K. L., 63, 69, 74, 77
Eckert, S. A., 3, 22, 63, 74, 77, 102, 113, 155, 163
Ecuador, 202
effective population size (N_e), 11–12
eggs, 4; biotic and abiotic factors affecting development of, 75, 140; composition of, 135–37; depth of burial of, 78, 87, 135; developmental stages of, 77; effects of climate change on, 189–91; "false" or "yolkless" (*see* shelled albumin gobs); fertilization of, 15, 57, 77, 84, 138, 139;

harvesting (poaching) of, 83, 103, 110, 113, 115, 117, 127, 187, 198, 212; hatching success of, 74–77, 138; incubation of (*see* incubation temperature); laying of, 4, 42, 53, 57–58, 63, 66, 67, 68, 77, 135, 137; maternal investment in, 138; movement-induced mortality of, 77, 138; oviductal development of, 53–54, 57; predation of, 4, 51, 65, 67, 68, 74, 79, 103, 110, 113, 115, 129, 189; SAGs interspersed among, 4, 54; size of, 53–54, 57, 135, 136–37; synchronous hatching of, 78; white spot on, 77, 138
eggshells, 57, 74, 77, 136, 137, 145
El Niño Southern Oscillation (ENSO), 16, 79, 187–89; foraging ecology and, 187; La Niña and El Niño phases of, 176, 178, 187, 189; large phytoplankton net primary productivity and, 176–80; Multivariate ENSO Index, 176–78; nesting ecology and, 187–88; reproductive output and, 188–89
El Salvador, 113, 202
embryonic development, 77, 135, 138–39; arrested, 77, 139–40; causes of early or late mortality, 77, 139; nest environment and, 140–41; stages of, 138–39
emergence of hatchlings: from nest, 78; rate of, 75; success of, 75, 139; timing of, 77
endocrinology, reproductive, 51–52, 60; estradiol, 51, 52, 53, 54–56, 58–59; follicle stimulating hormone, 51, 52; luteinizing hormone, 51, 52, 55, 57, 59; plasma calcium, 51, 55–56; progesterone, 51, 54–57, 58–59; testosterone, 51, 52, 53, 54–56, 58–59
energy budget, 154–57, 173
entoplastron, 4
Eocene, 6
Eosphargis, 6
epididymis, 41, 43, 52
epigastric arteries, 45
epiphysis, 40, 46
epiplastron, 35, 36
Erethmochelys imbricata (hawksbill turtle): diet of, 4; egg laying by, 4; eggs of, 136, 137; hatchling dispersal, 169; land walking of gravid females, 66; nest distribution of, 6; phylogeny of, 10, 11; reproduction of, 52, 59; sex determination in, 87, sex ratios of, 189
Escobilla, Mexico, 6
esophagus, 4, 35, 38, 39, 151, Plate 4
estradiol, 51, 52, 53, 54–56, 58–59, 60
euphotic zone, 174, 175, 176, 188
evolutionary history, 8–16
Evolutionary Significant Units (ESUs), 8, 9, 14
excavation: of body pit, 66; of nest chamber, 67
eye, 46, Plate 6

facial nerve (cranial nerve VII), 46, 47, Plate 8
fasting of nesting females, 53, 57
feeding, 152, Plate 16; energy costs and daily requirements for, 154–55; by juveniles, 152; metabolic rate and, 155, 156. *See also* diet; foraging
femur, 36–37
fertility, 77, 139
fertilization, 15, 57, 77, 84, 138, 139
fibula, 35, 37
Fick's Law, 141, 144
Fiji, 115
Fish, M. R., 191
fisheries impacts, 14, 98, 102, 103–4, 110, 113–14, 117–18, 126, 127, 158, 164, 168, 169, 178, 180, 196–204, 210–12; advances in bycatch reduction, 199–203; bycatch data on, 196–97; bycatch impact score, 198, 201; bycatch seascape, 204; current understanding of, 198–99; future of bycatch reduction, 203–4; information for evaluation of, 196–98; longline fisheries, 199–202; national / international policies for reduction of, 203; partnerships for reduction of, 204; passive nets and trawls, 202–3; small-scale fisheries, 203–4; telemetry data on, 197–98, 203
flatback turtle. See *Natator depressus*
flippers, 3, 32, 33, 34, 36; arteries of, 45; skeleton of, 36, 37; veins of, 46
Florida, United States, 4, 13, 17, 65, 69, 75, 98, 101, 102, 104, 163, 189, 191, 202, 209, 210
follicles, 42; atretic, 53, 57; number ovulated per nesting event, 58; ovulation of, 52, 55, 56–59, 136; preovulatory vitellogenic, 52–53, 58; size of, 53–54, 58; size of: relative to female size, 57
follicle stimulating hormone (FSH), 51, 52
Food and Agricultural Organization (FAO) International Plans of Action (IPOA), 203
foraging, 152; of hatchlings, 22; hotspots for, 174, 175–76, 187; during inter-nesting intervals, 25, 167; metabolic rate and, 155, 156; movements and behavior for, 23–24, 28, 165–67; of post-nesting females, 175–76. *See also* diet; feeding
Fossette, S., 152, 155, 156
fossil leatherbacks, 5–6, 16
founder effect, 11, 12, 13
France, 98, 104
Frazer, N. B., 52
French Guiana, 5, 6, 13, 24, 64, 65, 69, 75, 79, 85–86, 98–99, 101–2, 103–4, 162, 175, 209, 210
Fretey, J., 67
Fritz, U., 4

Gabon, 6, 13, 69, 88, 98, 101, 102, 103, 104, 164, 166, 167, 177, 180, 190
Galápagos Islands, 6, 164
Gamba Complex of Protected Areas, 101
gastric artery, 45
Gates, D., 185
gelatinous zooplankton, 23, 32, 150, 152, 154, 162, 165, 173–75, 176, 180; biophysical controls on, 174–75. *See also* jellyfish
genetic drift, 11–12
genetic sex determination (GSD), 84, 85
Georgia, United States, 191, 202
gigantotherms, 186
gill nets, 103, 113, 117, 118, 127, 188, 196–97, 198, 202–3, 204, 211
Girardin, N., 101
Girondot, M., 69
glands, 40–41; adrenal, 41; exocrine vs. endocrine, 40; liver, 40; orbital, 41; pancreas, 40; parathyroid, 41; thymus, 40–41; thyroid, 41; ultimobranchial bodies, 41
glenoid fossa, 36
Global Bycatch Assessment of Long-lived Species (Project GloBAL), 197
global distribution, 6–7, 8
Global Positioning System (GPS) tags, 65, 197
global warming. *See* climate change
glossopharyngeal nerve (cranial nerve IX), 46, Plate 8
Godfrey, M. H., 88
gonads, 41, 42
Gopherus agassizii, 55–57
Gopherus polyphemus, 57
Great Barrier Reef, 191
greenhouse gases, 185
green turtle. See *Chelonia mydas*
Grenada, 24, 75, 88
growth rates, 3, 4, 23, 152–54, 173; adult, 153–54; early, 152–53; von Bertalanffy growth function, 153–54
Guadeloupe, 65, 101, 102
Guatemala, 113, 202
Guinea Bissau, 168
Gulf of Mexico, 97, 98, 153, 163, 175
Gulf Stream, 168
Guyana, 5, 7, 99, 210

Handbook of Turtles (Carr), 6
haplotypes, 9–13, 87
Harderian glands, 40, 41
harvesting of eggs and adults, 83, 103, 110, 113, 115, 117, 127, 187, 198, 212
hatcheries, 91, 190; effects of tidal pumping on nest environments in, 141, 142; egg infections in, 79; "hatchery reflex," 211; hatching success in, 75, 138, 190, 211; hatchling sex ratios in, 91, 190, 211
hatching success, 74–77, 135, 138, 139, 210; calculation of, 74; clutch relocation and, 74–75, 77, 79; fertility and, 77; in hatcheries, 75, 138, 190, 211; intrinsic and extrinsic factors affecting, 75–76, 139; nest environment and, 140–41; predation and, 79; in Southwest Indian Ocean, 125; temperature and, 75, 76, 77
hatchlings, Plate 14; anatomy of, 3, 4, 33; breathing of, 22; capture-mark-recapture study of, 15–16; coloration of, 33; distribution in Southwest Indian Ocean, 125–26; diving behavior of, 22; effects of climate change on, 79–81, 189–91; emergence from eggs, 75, 77; emergence from nest and run to the ocean, 78, 139, Plate 10; gonads of, 42; heart of, 39; kidneys of, 41; movements and dispersal of, 168–69, 211; output of, 78–81; predation of, 78, 79, 110, 168, 189; sex ratios of, 15, 84, 87–89, 90, 139, 210; sex ratios of: climate change and, 189–91; sex ratios of: in hatcheries, 91, 190, 211; size of, 135; size of: survival related to, 57; synchronous hatching of, 78; thymus in, 40

Havas, P., 4
hawksbill turtle. See *Erethmochelys imbricata*
Hays, G. C., 27
head, 32, 33–34; arteries of, Plate 7; pink spot on, 34, 46, Plate 12
heart, 39, 42, 44
Heaslip, S. G., 152
heat exchange, 44, 151, 186. *See also* thermoregulation
hemoglobin oxygen stores, 26
Hendrickson, J. R., 64
hepatic arteries, 40
hepatic veins, 40, 45
hind limbs, 32, 33, 34, 35; arteries of, 45; metatarsals and digits of, 37; veins of, 46
Hirth, H. F., 57, 77
Hitipeuw, C., 114
Holding, J., Plate 9
homing to natal beaches, 8, 10, 11, 12, 17, 163
Houghton, J. D., 88
Hughes, G. R., 65, 126
humerus, 5, 36, 37
hurricanes, 80, 87
hydrostatic pressure adaptations, 25, 27–28
hyoid apparatus, 34, 35–36, Plate 4
hyoplastron, 35, 36
hypercapnia, 140
hypoglossal nerve (cranial nerve XII), 46
hypoxia, 25, 27, 139

ileum, 39–40
iliac arteries, 45
ilium, 34, 35, 36
incubation period, 75, 76
incubation temperature, 118; embryonic development and, 139, 140; hatching success and, 75, 76; hatchling sex ratio and, 85–86, 190, 191
India, 6, 69, 127–29
Indian Ocean leatherbacks, 6, 98, 123–29, 210; clutch frequency of, 69; clutch size of, 125, 138; diving behavior of, 23; foraging areas of, 187; hatching success of, 125; hatchling distribution of, 125–26; migration of, 164; monitoring of, 124–25; nesting behavior of, 180; nesting season of, 69, 124; nesting sites of, 7, 123–24, 127–29; Northeast Indian Ocean RMU, 123, 127–29; phylogeny of, 10, 12, 13; population genetics of, 13, 129; post- and inter-nesting movements of, 126–27; Southwest Indian Ocean RMU, 123–27; threats and prospects for, 127, 129. *See also specific countries*
Indonesia, 10, 14, 69, 75, 79, 110, 111, 114, 116, 117, 123, 127, 128, 129, 175, 210
infertility, 77, 139
ingestion: of fishing hooks, 201; of prey, 152, 155; of seawater, 25, 150, 152, 155
Inter-American Convention on the Protection and Conservation of Sea Turtles (IAC), 203
Inter-American Tropical Tuna Commission (IATTC), 201, 204
intercostal fontanelles, 4
Intergovernmental Panel on Climate Change (IPCC), 80, 190
International Union for the Conservation of Nature (IUCN), 110, 211
inter-nesting interval, 51, 53, 69–70; diving behavior in, 22, 24–25; foraging during, 167; movements during, 167–68; movements during: in Indian Ocean, 126–27
Ionian Sea, 97
ischium, 34, 36

James, M. C., 163
Jamursba-Medi, Papua Barat, 13, 17, 69, 114–15, 116
Janzen, F. J., 189, 191
Java, 123, 127, 129
jaw, 4, 32, 33, 34, 35, Plate 5
jejunum, 39–40
jellyfish, 4, 6, 149, 150, 152, 154, 173, Plate 16
Jones, T. T., 151, 152, 153, Plate 15
jugular veins, 45–46
Jurassic, 97
juveniles, 3, 211, Plate 15; coloration of, 33; diving behavior of, 22–23; feeding by, 152; gonads of, 42; growth of, 152–53; migration of, 163; plastron of, 35; sex ratios of, 15; thermoregulation in, 22

Kamel, S. J., 65
Kaplan, I. C., 117, 198
Karl, S. A., 15, 17
Katselidis, K. A., 191
Kemp's ridley turtle. See *Lepidochelys kempii*
kidneys, 41
Kinosternon, 4
Komodo dragon, 186
KwaZulu Natal, South Africa, 126, 127

lachrymal salt glands, 34, 35, 41, 47, 150–51, Plate 5, Plate 6
lactic acid, 25, 27
large intestine, 40
larynx, 27, 28, 38
Lazell, J. D., 63
leatherback turtle. See *Dermochelys coriacea*
Lepidochelys kempii (Kemp's ridley turtle): cold stunning of, 4; eggs of, 136, 137; mating of, 52; multiple paternity in, 14; nest distribution of, 6, 168; phylogeny of, 10, 11, 12; reproduction of, 51–52, 54–55, 56, 58
Lepidochelys olivacea (olive ridley turtle): eggs of, 136, 137; hatching success of, 74; hypercapnia and embryonic development of, 140; land walking of gravid females, 66; multiple paternity in, 14; nest distribution of, 6; nesting of, 13, 66; phylogeny of, 10, 11, 12; reproduction of, 51–52, 57, 58–59, 86
Lewison, R. L., 198
Liew, H. C., 86, 88
liver, 25, 27, 39, 40
Livingstone, S. R., 88
loggerhead turtle. See *Caretta caretta*
longline fisheries, 14, 103, 113–14, 117–18, 125, 127, 158, 166, 197, 198–202, 203, 204, 211
lost years, 3, 168–69
lungs, 42, 44, 150; collapse of, 25, 26, 27, 28; oxygen stores in, 25–26; pulmonary arteries, 45
luteinizing hormone (LH), 51, 52, 55, 57, 59, 60

Macrochelys temminckii, 57
Madagascar, 123, 125
magnetic compass sense, 163
Malaysia, 6, 10, 11, 12, 13, 16, 69, 79, 85, 86–87, 88, 110, 112, 114, 115, 117, 118, 145, 209, 210
Management Units (MU), 8, 9, 14
mandible, 33, 35, 38
Maputaland (Tongaland), South Africa, 69, 124, 126
marginocostal artery, 45
marine primary productivity, 173–74
Martinique, 65, 101, 102
Maryland, United States, 64, 191
maternal investment in eggs and clutches, 138
mating behavior, 8, 52, 60, 63–64; female avoidance behavior, 52, 64; genetic drift and, 11; hatching success and, 139; multiple paternity and, 8, 14–15, 64, 129, 139; ovulation and, 56; plasma calcium level and, 56; seasonal, 52; single paternity and, 14, 64, 139; testosterone level and, 54–55
Matura Beach, Trinidad, 68, 100
Mauritania, 98, 101, 175
maxilla, 35, 38
Mayumba National Park, Gabon, 101, 103, 166
McAllister, H. J., 124
Mediterranean leatherbacks, 98, 103, 104, 163, 189
Mediterranean Sea, 97; gelatinous zooplankton in, 174
medulla, 46
mesonephric duct, 41
metabolic rate (MR), 6, 10, 155, 156; diving and, 22, 25, 26, 27
metacarpal bones, 36
metacarpal muscles, 38
metatarsals, 37
Mexico, 5, 6, 7, 10, 14, 17, 69, 86, 97, 98, 110, 111–13, 153, 162, 163, 164, 175, 177, 178, 179–81, 191, 202, 210
Mexiquillo, Mexico, 112, 113
Michoacan, Mexico, 6
Mickelson, L. E., 88
Micronesia, 3
microsatellite markers, 9, 12, 13, 14, 16, 64, 101–2
migrations, 6, 8, 10, 28, 162–65, 173, 210; in Atlantic Ocean, 154, 163–64; for foraging, 23–24, 28, 165–67; in Indian Ocean, 164; of juveniles, 163; of males to breeding areas, 139, 163; metabolic rate and, 155, 156; navigation sense for, 163; in Pacific Ocean, 154, 164–65; remigration intervals, 70, 78, 128, 145, 154–57, 180, 188
Miller, J. D., 66, 77, 138
Miocene, 5, 97
Mitchell, N. J., 191
mitochondrial DNA (mtDNA), 9–14, 16, 101–2
molecular genetics, 9–10, 100, 210
Molony, B., 117
Morreale, S. J., 164
mouth, 35, 36, 38, 47, Plate 9
movements and behavior, 162–69; conservation implications of, 169; for foraging,

23–24, 28, 165–67; of hatchlings (lost years), 168–69, 211; during inter-nesting intervals, 167–68; migrations, 162–65; tracking to reduce fisheries bycatch, 197–98
Mozambique, 69, 123–27
Mrosovsky, N., 65, 87–88, 89
Müllerian duct, 41
multiple paternity, 8, 14–15, 64, 129, 139
Multivariate ENSO Index (MEI), 176–78
muscles, 38; deep jaw, Plate 5; dorsal neck, Plate 3; extrinsic eye, Plate 5; intermediate jaw, Plate 7; lateral neck and hyoid arch, Plate 2; lateral neck and throat, Plate 1
Myers, R. A., 163
myoglobin oxygen stores, 26

Naja naja, 56
Nambohri, N., 128
Namibia, 98
Naro-Maciel, E., 168
natal philopatry (homing), 8, 10, 11, 12, 17, 163
Natator depressus (flatback turtle): eggs of, 57, 136, 137; multiple paternity in, 14; nest distribution of, 6
Natemys peruvianus, 6
National Marine Fisheries Service, 102
navigation cues, 163
Neotethyan Ocean, 97
nervous system, 46–47
nest environment, 75, 77, 78, 79, 80, 137, 139–45, 189; effects of tidal pumping on, 75, 141–44; effects of tidal pumping on: at St. Croix, 144–45; embryonic development and, 140–41
nesting: density dependence of, 68–69; frequency of, 51, 52, 58; hatching success related to site of, 75–76; patterns of, 58–59; reproductive cycles and seasonality of, 69; success of, 68
nesting beaches, 4, 6–7, 64–65, 209; Atlantic Ocean and Caribbean, 98–102; climate change and, 210–11; erosion of, 65, 68, 78–79, 110, 115, 117, 118, 191; factors affecting hatchling output, 78–79; hatchling emergence from nest and run to the ocean, 78; homing to, 8, 10, 11, 12, 17, 163; Indian Ocean, 7, 123–24, 127–29; mating prior to arrival at, 63–64; nest density on, 68–69; Pacific Ocean, 111; proximity to favorable ocean currents, 168; rising sea levels and, 191; safeguarding of, 211; site selection for, 13, 64–65, 66, 87; tidal inundation of, 65, 74, 75–76, 78–79, 139, 191. *See also specific beaches*
nesting behavior, 64–68, 210; body pit excavation, 66; covering and concealing nest site, 67–68; depth of clutch burial, 78, 87, 135; emergence, movement up the beach, and nest site selection, 66; filling the nest, 67; nest chamber excavation, 67; nest size, 135; oviposition, 67; phases of, 65–68; returning to sea, 68; time period of, 66
nesting ecology, 51, 63–70, 187–88, 210; climate change and, 187–89
nesting season(s), 52, 53, 58, 64, 69; of Atlantic leatherbacks, 69, 101; corticosterone during, 59; El Niño Southern Oscillation and, 187–88; female fasting during, 53, 55; plasma testosterone, progesterone, and estradiol during, 54–56, 58–59; remigration intervals between, 70, 78, 128, 145, 154–57, 180, 188; reproductive success for clutches laid early vs. late in, 70
nest predation, 4, 51, 65, 67, 68, 74, 79, 103, 110, 113, 115, 129, 189
nest relocation. *See* clutch relocation
nest temperature, 70, 80, 84, 86, 87–89, 91, 140, 190
net primary productivity (NPP) of large phytoplankton, 173–81, 188–89; biogeochemical ocean general circulation models for estimation of, 175, 177, 178, 179; bottom-up forcing via, 174–75, 188; eastern Pacific leatherback nesting model and, 176–78; effects on interpopulation differences in body size, fecundity, and nesting trends, 178–81; El Niño Southern Oscillation and, 176–80; energy transfer efficiencies and, 174, 175, 180; euphotic zone for, 174, 175, 176, 188; foraging hotspots and, 174, 175–76, 187; resource availability ratio and, 178–79; satellite-derived ocean color models for estimation of, 175–76, 178; seasonal variability of, 179
neuromuscular system, 32, 47
New Guinea Costal Undercurrent (NGCUC), 188
New Jersey, United States, 191
Nicaragua, 6, 113, 168
Nicobar Islands, 7, 10, 11, 69, 123, 127–29
nitrogen narcosis, 28
Nordmoe, E. D., 65
North Atlantic Oscillation (NAO), 174
North Carolina, United States, 101, 190, 191, 202
Northeast Indian Ocean RMU, 123, 127–29
nuchal bones, 34, 36, 42

Oaxaca, Mexico, 6
ocean color models, 175–76, 178
oculomotor nerve (cranial nerve III), 46, 47, Plate 8
Ogren, L., 66, 77
olfactory bulbs, 46, Plate 3
olfactory nerve (cranial nerve I), 46, 47, Plate 8
Oligocene, 6
olive ridley turtle. See *Lepidochelys olivacea*
Oman, 6
operational sex ratio (OSR), 15
optic lobes, 46, Plate 3
optic nerve (cranial nerve II), 46, 47, Plate 8
orbital glands, 41
Orissa, India, 6, 128
osmoregulation, 149, 150–51, 152, 155
ossicles: dermal, 5, 28, 35; scleral, 46, 153
ossification, 33; of skull, 138
osteoderms, 33, 34, 35, 36
ovarian cycle, 52–53, 54
ovaries, 41, 42, 43, 52–53, 58
oviduct, 41, 42, 51, 77, 136, 137, 138, 139
oviductal egg development, 53–54, 57
oviposition, 4, 42, 53, 57–58, 63, 66, 67, 68, 77, 135, 137
ovulation, 52, 55, 56–59, 60, 136
Owens, D. W., 59
oxidative stress, 27
oxygenated blood, 40, 42, 44
oxygen consumption, 26, 78
oxygen in nest, 4, 135, 136, 139–41, 143–44
oxygen partial pressure (PO_2), 26
oxygen partial pressure difference between above-ground air and nest (DPO_2), 143, 144
oxygen partial pressure (PO_2) in nest, 139, 140–41, 143–44; embryonic development and, 77, 140–41; hatching success and, 75, 139
oxygen stores, 22, 25–26; aerobic dive limits and calculated aerobic dive limits, 27, 158; in hemoglobin, 26; in lung, 25–26, 150; in myoglobin, 26

Pacific Ocean leatherbacks, 4–5, 7, 110–18, 210, 211; barnacles on, 5; clutch frequency of, 69; clutch size of, 4, 69, 138; conservation efforts for, 113–14, 117–18; critically endangered status of, 110, 118, Plate 11; declining populations of, 12, 16, 69, 70, 110, 112, 113–15, 117–18, 157, 180, 181, 209, 211; distinct subpopulations of, 110; distribution of, 110; diving behavior of, 23–24; in Eastern Pacific, 110, 111–14; in Eastern Pacific: bioenergetics of, 154–56; eastern Pacific nesting model for, 176–78; effects of climate change on, 16–17, 80, 91, 118; foraging areas of, 165–66, 175–79, 187; hatching success of, 75, 77; hatchling output of, 78–79, 80; hatchling sex ratios in, 88–89; in Malaysia, 110, 114; migration of, 154, 164–65; nesting beaches of, 111; nesting behavior of, 64, 65, 68, 180; nesting behavior of: El Niño Southern Oscillation and, 188; nesting season of, 69; nesting success of, 68; oviposition and SAGs in, 58; phylogeny of, 9–11; population genetics of, 12–14; remigration intervals for, 70; reproduction of, 52, 60, 64, 189; sex determination in, 86–87, 91; threats to, 110, 117–18; in Western Pacific, 110, 114–18. *See also specific countries*
Paladino, F. V., 27
palate, 35, 38
Paleocene, 6
Panama, 11, 64, 75, 100, 102, 104, 113, 177, 180, 202
pancreas, 40
pancreaticoduodenal arteries, 45
Pangaea, 97
Panthera onca (jaguar), 68
papillae: oral and esophageal, 4, 38, 151; urogenital, 41
Papua Barat, 10, 13–14, 79, 110, 112, 114, 116–17, 175, 181
Papua New Guinea (PNG), 13, 14, 69, 79, 88–89, 90, 91, 111, 115–16
paramesonephric duct, 41, 42, 43
parathyroid glands, 41
parathyroid hormone, 41

Parham, J., 153
Parque Nacional Marino Las Baulas (PNMB), Costa Rica, 14, 52, 53, 56, 58, 59, 60, 65, 69, 86, 112, 113, 140, 141–45, 167, 187, 188, 190, 211
paternity: multiple, 8, 14–15, 64, 129, 139; single, 14, 64, 139
Patiño-Martínez, J., 77, 78, 88
pectodeltoid process, 36
pectoral girdles, 34, 35, 36
pectoral muscles, 37, 45, 186
pectoral veins, 45
Pehrson, T., 138
pelvis, 34, 35, 36
penis, 41, 43, 52, 64
pericardial sac, 42
Perrault, J. R., 75
Peru, 3, 14, 113–14, 175, 180–81, 187, 202, 211
phalanges, 35, 36, 37
photosynthesis, 173–74
phylogeny, 8–10
phylogeography, 8, 10–12
physiological ecology, 149–58; bioenergetics, 149, 154–57; conservation physiology, 157–58; feeding, ingestion, and assimilation efficiency, 152; growth and age at maturity, 152–54; osmoregulation, 149, 150–51, 152, 155; thermoregulation, 22, 25, 32, 149, 150, 151–52, 154–56, 209
phytoplankton, 173–81, 187. *See also* net primary productivity of large phytoplankton
pineal, 40, 46, 47, Plate 3
pink spot on head, 34, 46, Plate 12
"pipping," 77
pisiform, 35, 36
pituitary, 40, 46
pivotal temperature (PT), 85–87, 89
plastron, 3, 4, 34, 36; of juvenile, 35; of male leatherback, 52
Platform Transmitting Terminals (PTT), 197
Playa Grande, Costa Rica, 14, 24, 64, 65, 66, 67, 68, 69, 70, 75–77, 78, 79–80, 86, 87, 88, 89, 91, 112, 137, 139, 140, 141, 145, 164, 165, 167, 168, 177, 178, 188, 190, 191, Plate 9, Plate 11, Plate 13
Playa la Flor, Nicaragua, 6
Playa Langosta, Costa Rica, 112
Playa Nancite, Costa Rica, 6
Playa Ostional, Costa Rica, 6, 112
Pleistocene, 9, 11, 12, 16
Pliocene, 5, 185, 211
poaching (harvesting) of eggs, 83, 103, 110, 113, 115, 117, 127, 187, 198, 212
pollution of ocean, 103, 104, 110
Pongara National Park, Gabon, 101
population dynamics: climate change and, 79, 188, 190; physiological ecology and, 149–58
population genetics, 8, 9, 12–15, 210; of Atlantic Ocean leatherbacks, 101–2; of Indian Ocean leatherbacks, 13, 129
populations of leatherbacks, 12, 13, 209; in Atlantic Ocean, 97–104, 210; effective population size (N_e), 11–12; in Indian Ocean, 123–29, 210; in Pacific Ocean, 110–18, 210
population vital rates, 15–16
precipitation, 185, 210–11; hatchling output and, 79–80; hatchling sex ratio and, 87–91, 140, 189–90; reproductive output and, 70
predation: egg/nest, 4, 51, 65, 67, 68, 74, 79, 103, 110, 113, 115, 129, 189; hatchling, 78, 79, 110, 168, 189
Price, E. R., 153, 155
Pritchard, P.C.H., 66, 111, 162, 163
procoracoids, 34, 36, 39
progesterone, 51, 54–57, 58–59, 60
Project GloBAL, 197
Proyecto Láud in Mexico, 113
pubis, 34, 35, 36
Puerto Rico, 68, 69, 100–101
pulmonary arteries, 25, 42, 45
pulmonary trunk, 45
Putnam, N. F., 168
Pyrosomas atlantica, 154

radius, 36
Rafferty, A. R., 70, 77
rainfall. *See* precipitation
recolonizations, 12, 16
regional fishery management bodies (RFMOs), 203
Regional Management Units (RMUs), 8, 14; Atlantic Northwest, 98–101; Atlantic Southeast, 98, 101; fishery bycatch data for, 198–201; Indian Southwest, 98; Northeast Indian Ocean, 123, 127–29; Southwest Indian Ocean, 123–27
Reina, R. D., 64, 188
remigration interval (RI), 70, 78, 128, 145, 180, 188; bioenergetics and, 154–57
renal arteries, 45
renal portal vein, 39
Renous, S., 138
reproductive behavior, 8, 14–15; courtship and mating, 52, 63–64; factors influencing reproductive output, 70; factors influencing reproductive output: climate change, 188–89; multiple paternity, 8, 14–15, 64, 129, 139; nesting behavior, 64–68; remigration intervals between nesting seasons, 70, 78, 128, 145, 154–57, 180, 188; single paternity, 14, 64, 139
reproductive biology, 51–60, 210; age at first reproduction, 15, 152–53, 154, 157; breeding sex ratio, 15; female reproductive system, 52; generalized model and future research needs, 59–60; male reproductive system, 52; nesting patterns, 58–59; ovarian cycle, 52–53; oviposition, 4, 42, 53, 57–58, 63, 66, 67, 68, 77, 135, 137; ovulation, 55, 56–59, 136; plasma testosterone, progesterone, and estradiol, 54–56, 60; vitellogenesis, 53, 55–56, 57, 145, 173, 178. *See also* endocrinology, reproductive
reproductive energy (RE) costs, 154–56
reproductive investment, 69–70
Republic of Congo, 101, 103, 104, 166
resource availability ratio (RAR), 178–79
respiratory system, 42; diving and, 25–28, 42
respiratory water loss (RWL), 150
rete (retia), 38, 39, 44, 45
Rhizostoma octupus, 154
Rhodin, A. G., 153
ribs, 4, 5, 33–36
Rimblot, F., 85
Rimblot-Baly, F., 85
rising sea levels, 118, 191, 192, 211
Roden, S. E., 14
Rostal, D. C., 52–53
Ryther, J. H., 174, 176

Saba, V. S., 80, 118, 156, 176–77, 178, 188, 190
salinity of seawater, 149, 150, 191
Salm, R. V., 114
salt glands, 34, 35, 41, 47, 150–51, Plate 5, Plate 6
"sand bath," 66
Santidrián Tomillo, P., 70, 80, Plate 9
Sarti, L., 113
satellite-derived ocean color models, 175–76, 178
satellite tracking, 14, 22; in Atlantic Ocean, 104, 163–64, 186; to define migratory pathways, 162, 163–64, 168; in Indian Ocean, 126, 129, 164; to locate foraging hotspots, 175–76, 187; to monitor fishery bycatch, 197, 203; in Pacific Ocean, 164; of phytoplankton, 178. *See also* tagging data
scapular artery, anterior, 45
scapular musculature, 45, 46
scapular vein, transverse, 46
Schumacher, G.-H., 32, 38, 46, Plates 1–8
Schwanz, L. E., 189
sea surface temperature (SST), 25, 158, 175, 176, 185–89, 199, 204
sense organs, 46–47
sex determination: climate change and, 89–91; genetic, 84, 85; temperature-dependent, 4, 11, 84–87, 89, 189, 192, 210
sex ratios of hatchlings, 15, 84, 87–89, 90, 139, 210; climate change and, 189–91; in hatcheries, 91, 190, 211
sexual maturity, 23, 153; age at, 153, 158, 189; diving behavior and, 23
Shanker, K., 14, 128
shelled albumin gobs (SAGs), 4, 51, 57–58, 67, 69, 74, 135, 137–38, 145; functions of, 51, 58, 137; mass of, 137; in oviduct, 53–54; quantity per clutch, 78
Shillinger, G. L., 65, 168, 188
shrimp trawling–related bycatch, 202
Sieg, A. E., 88
Sinervo, B., 190
single paternity, 14, 64, 139
skeletochronology, 153–54
skeleton, 3, 4, 6, 34–37; axial and integumentary, 36; of juveniles, 33, 34, 35; limbs and limb girdles, 36–37
skull, 5, 34–35, 46, Plate 3; embryonic, 139; ossification of, 138
small intestine, 39–40
Smelker, K., 53
sodium balance, 150
Solomon Islands, 10, 13–14, 86, 87, 115–17, 152, Plate 13
South Africa, 7, 10, 12, 13, 64, 69, 98, 102, 123–27, 129, 164, 166, 175, 177, 178, 180, 181, 210

South Carolina, United States, 6, 191, 202
South China Sea, 3, 166, 175, 181, 187
South Equatorial Current, 166
Southwest Indian Ocean RMU, 123–27
spermatic sulcus, 41, 43
spermatogenesis, 52, 59
sperm storage, 14, 15, 52, 64
Sphenodon punctatus (tuatara), 56
spinal accessory nerve (cranial nerve XI), 46
spinal cord, 46, Plate 3
splanchnocranium, 35
sponges, 4
Spotila, J. R., 178, 185
Sri Lanka (Ceylon), 6, 69, 123, 127, 129
Standora, E. A., 185
State of the World's Sea Turtles (SWOT), 98
statoaccoustic nerve (cranial nerve VIII), 46, 47, Plate 8
St. Croix, US Virgin Islands, 12, 13, 14, 15, 18, 24, 33, 66, 68, 69, 75, 77, 79, 86, 100, 102, 103, 113, 139, 144–45, 175, 177, 180, 181, 210
Steckenreuter, N. P., 88
Sternotherus odoratus, 55
Stewart, K. R., 15
stomach, 39
stranding events: on Indian coasts, 127–28; in Mediterranean, 163; related to shrimp trawls, 202; on South African coast, 126; in Wales, 162
stress response during nesting season, 59
subclavian arteries, 40, 45
subclavian veins, 45, 46
Sumatra, 123, 127, 129
superior mesenteric artery, 45
Suriname, 13, 24, 64, 69, 75, 79, 87–88, 89–90, 98–99, 102, 103–4, 177, 180, 189, 209, 210
swallowing, 38
Swaminathan, A., 128
swimming speeds, 126, 155, 156, 173
swordfish, 114, 118, 127, 180, 201
synchronous hatching, 78

tagging data, 9, 190; in Atlantic Ocean and Caribbean, 100, 163–64, 191; on climate effects on reproductive ecology, 188; for hatchlings, 15; in Indian Ocean, 125, 126, 128; on migration, 162; on migration: El Niño Southern Oscillation and, 178; on migration: males, 126, 163; on migration: post- and inter-nesting movements, 126, 163; for nesting females, 65, 67; in Pacific Ocean, 117, 165. *See also* satellite tracking
tail, 36
Tamaulipas, Mexico, 6
Tapilatu, R., 115
tarsals, 35, 37
taxonomy, 5–6
temperature: air, 185; effect on gelatinous zooplankton growth, 175; hatchling emergence and, 78; hatchling output and, 79–81; hatchling sex ratios and, 87, 189–90; transitional range of, 85–87, 89; water, 25, 149, 151, 154, 155, 156, 157, 167, 186; water: sea surface temperature, 25, 158, 175, 176, 185–89, 199, 204. *See also* body temperature; climate change
temperature-dependent sex determination (TSD), 4, 11, 84–87, 89, 189, 192, 210; climate change and, 89–91, 189–91; patterns of, 84–87; pivotal temperature, 85–87, 89; transitional range of temperature, 85–87, 89
temperature in nest, 70, 80, 84, 86, 87–89, 91, 139–40; embryonic development and, 140; hatching success and, 75, 76, 77
Terengganu, Malaysia, 6, 7, 13, 69, 79, 114, 127
testes, 41, 42, 43, 52
testosterone, 51, 52, 53, 54–56, 58–59, 60
Tethys Sea, 97
The Tetrapod Reptiles of Ceylon (Deraniyagala), 6
Thailand, 114, 123, 127, 129
thermoregulation, 25, 32, 149, 150, 151–52, 209; bioenergetics and, 154–56; climate change and, 185–87; in juveniles, 22
thoracic squeeze, 27–28
threats to leatherbacks, 210–11; in Atlantic Ocean, 103–4; beach erosion and inundation, 65, 68, 74, 75–76, 78–79, 110, 115, 117, 118, 139, 191; climate change, 185–92, 210–11; fisheries, 14, 103, 110, 113–14, 117–18, 126, 158, 164, 168, 178, 180, 196–204, 210–12; harvesting of eggs and adults, 103, 113, 115, 117, 127, 187, 198; in Indian Ocean, 127, 129; ocean pollution, 103, 104, 110; in Pacific Ocean, 110, 117–18, 181; predation, 4, 51, 65, 67, 68, 74, 78, 79, 103, 110, 113, 115, 129, 168, 189; rising sea levels, 118, 191, 192, 211
thymus, 40–41
thyroid, 39, 41
thyroid arteries, 45
thyroscapular vein, 46
tibia, 35, 37
tidal inundation of nesting beaches, 65, 74, 75–76, 78–79, 139, 191
tidal pumping effects on nest environments, 75, 141–44; at St. Croix, 144–45
time-depth recorders (TDRs), 197
Tobago, 12, 88, 98, 99, 100, 102
tongue, 28, 35, 38, 46
total body water (TBW), 150–51
Tracers of Phytoplankton with Allometric Zooplankton (TOPAZ) model, 175, 177, 178, 179
trachea, 42, 44, 150, 151, Plate 4; collapse of, 7, 22, 26, 28, 42
Trachemys scripta scripta, 57
transitional range of temperature (TRT), 85–87, 89
Trebbau, P., 66
trigeminal nerve (cranial nerve V), 46, 47, Plate 5
Trinidad, 6, 7, 12, 13, 65, 68, 86, 98, 99, 100, 102–4, 163, 175, 177, 180, 210
trochlear nerve (cranial nerve IV), 46, 47, Plate 8
tsunami of 2004, 7, 10, 127, 128, 129
Tucker, A. D., 52
turtle excluder devices (TEDs), 202
Turtle Watch, 204
Tyrrhenian Sea, 97

ulna, 36
ulnar artery, 45
ultimobranchial bodies, 41
Units to Conserve (UTC), 8
ureters, 41, 43
urinary bladder, 39, 41–42
Urochordata, 4
urogenital system, 32, 41–42, 43
Uruguay, 98, 166
US Endangered Species Act, 9
uterus, 41

vagus nerve (cranial nerve X), 46, Plate 8
Vandelli, D., 98
Vanuatu, 79, 115
vas deferens, 42, 43
vena cava, superior and inferior, 45
Venezuela, 69, 75, 99, 100
venous system, 45–46. *See also specific veins*
vertebrae, 34–36, 39; caudal, 36; cervical, 36, 46; sacral, 36
vertebral artery, 45
vital rates, 15
vitellogenesis, 53, 55–56, 57, 60, 145, 173, 178
vitellogenin, 40, 53
von Bertalanffy growth function (VBGF), 153–54

Walker, 38
Wallace, B. P., 27, 87, 140, 150, 151, 154, 155, 156, 198, 211
water balance, 150–51
water salinity, 149, 150, 191
water temperature, 25, 149, 151, 154, 155, 156, 157, 167, 186; El Niño Southern Oscillation and, 176; foraging success and, 179–80; sea surface temperature, 25, 158, 175, 176, 185–89, 199, 204; warming oceans, 187–89
Wermon Beach, Papua Barat, 13, 17, 114–15, 116
Western and Central Pacific Fisheries Commission, 202
"whale turtle," 3
Whittier, J. M., 59
Wider Caribbean Sea Turtle Conservation Network (WIDECAST), 98
The Windward Road (Carr), 185
Wolffian duct, 41
Wood, Roger, 5, 7
World Wildlife Fund (WWF), 191, 201, 204
Wyneken, J., 32

xiphiplastron, 35, 36

Ya:lima:po-Awa:la, French Guiana, 65, 69, 79, 99, 102
Yap, Micronesia, 3

Zanclean flood, 97
zooplankton, gelatinous, 23, 32, 150, 152, 154, 162, 165, 173–75, 176, 180; biophysical controls on, 174–75. *See also* jellyfish
Zug, G., 153